Encyclopedia of Salmonella: Researches and Case Studies

Volume IV

Encyclopedia of Salmonella: Researches and Case Studies Volume IV

Edited by **Alan Klein**

New York

Published by Callisto Reference,
106 Park Avenue, Suite 200,
New York, NY 10016, USA
www.callistoreference.com

Encyclopedia of Salmonella: Researches and Case Studies
Volume IV
Edited by Alan Klein

International Standard Book Number: 978-1-63239-293-0 (Hardback)

Printed in the United States of America.

Contents

Preface

This book was inspired by the evolution of our times; to answer the curiosity of inquisitive minds. Many developments have occurred across the globe in the recent past which has transformed the progress in the field.

Salmonella comprises of two species (enterica and bongori) and is said to be an intensely variegated genus which infects a variety of hosts. This group is made up of 2579 serovars which fascinates researchers and draws their attention towards its different characteristics. Salmonella causes problems of zoonoses and also leads to food borne illness. Diseases caused by salmonella are becoming a matter of concern for developed and developing countries because of their impact on economy and other important sectors. Antimicrobial resistance in salmonella makes it difficult to reveal different mechanisms involved and this problem seems to increase further. This book presents new developments and researches in the field of salmonella from across the globe. It includes important topics such as immunological responses to salmonella, mechanisms of pathogenesis and diagnosis of salmonella infections. Internationally acclaimed researchers and practitioners in this field have made their valuable contributions to this book.

This book was developed from a mere concept to drafts to chapters and finally compiled together as a complete text to benefit the readers across all nations. To ensure the quality of the content we instilled two significant steps in our procedure. The first was to appoint an editorial team that would verify the data and statistics provided in the book and also select the most appropriate and valuable contributions from the plentiful contributions we received from authors worldwide. The next step was to appoint an expert of the topic as the Editor-in-Chief, who would head the project and finally make the necessary amendments and modifications to make the text reader-friendly. I was then commissioned to examine all the material to present the topics in the most comprehensible and productive format.

I would like to take this opportunity to thank all the contributing authors who were supportive enough to contribute their time and knowledge to this project. I also wish to convey my regards to my family who have been extremely supportive during the entire project.

Editor

Part 1

Immunology

Salmonella as Live Carrier of Antigens in Vaccine Development

César Gonzalez Bonilla, Ericka Pompa Mera,
Alberto Diaz Quiñonez and Sara Huerta Yepez
División de Laboratorios de Vigilancia e Investigación Epidemiológica
Coordinación de Vigilancia Epidemiológica y Apoyo en Contingencias
Instituto Mexicano del Seguro Social
México

1. Introduction

Attenuated *Salmonella* strains have been extensively used as live carriers of heterologous antigens. In animal models they elicit strong mucosal and systemic immune responses to passenger antigens of a broad variety of pathogens. *Salmonella* has several theoretical advantages over other vaccine vector systems. Among them, especially attractive is the bacterial ability to interact with mucosal and systemic compartments of the immune system and deliver passenger antigens directly to antigen presenting cells (APC) when administered by the oral route. However, despite the promising results in animal models, clinical trails have been disappointing and more research is needed in order to understand the protective immunity mechanisms and solve the main drawbacks of vaccine design. Herein, we summarize the accumulated experience using *Salmonella* as live carrier emphasizing the role of passenger antigen localization into different bacterial compartments. This is an important factor determining the type and quality of the host immune response. The evidence suggests that antigens located on the bacterial surface induce higher antibody responses whereas those located in the cytoplasm elicit better cellular immunity. In order to display recombinant antigens on the bacterial surface we have used outer membrane (OMPs) or autotransporter proteins. Mice immunized with *Salmonella* strains expressing the main B cell epitope from the *Plasmodium falciparum* circumspozoite protein, presented better antibodies response when placed on the bacterial surface as a fusion proteins with OMPs or autotransporters, than the whole recombinant protein located in the bacterial cytoplasm.

2. *Salmonella* as live carrier of antigens

Vaccination is the public health intervention intended to control infectious diseases with the best cost-effective ratio. Among the number of different approaches for vaccine development, live vectors stand as one of the most promising options. They may be defined as attenuated microorganisms which produce heterologous recombinant proteins (or bear plasmids DNA with eukaryote transcription machinery) and elicit immune response to the passenger antigen. Extensive research in the last decade in the fields of molecular biology, bacterial genetics and immunology has significantly accelerated the use of a number of microorganisms as vaccine vectors. Several pathogenic bacteria can be engineered for attenuation and expression of foreign proteins originally encoded by other microorganisms. Although several attenuated or commensal nonpathogenic bacteria have been employed to express passenger antigens (Kottonet al. 2004; Medinaet al. 2001), enteric Gram negative bacteria, especially *Salmonella*, have been preferred because they are easy to manipulate. However, these vaccine candidates raise safety concerns associated with the release of genetically modified microorganisms into the environment. (Abd Elet al. 2007; Kottonet al. 2004).

An ideal vaccine has to fulfill not only immunogenicity, efficacy and safety requirements, but also has to be inexpensive, stable at room temperature and easy to administer (Polandet al. 2010). Bacterial carriers specifically *Salmonella* strains, attenuated by genetic engineering, have several advantages over other vaccine types because they can be administered by the oral route and are more stable at room temperature than other vectors, such as viruses. Recent research have developed more stable strains, one example is a live attenuated *S.* Typhi Ty21a vaccine treated by foam drying, a modified freeze drying process, formulated with trehalose, methionine, and gelatin, is stable for approximately 12 weeks at 37°C and its derivatives expressing foreign antigens, such as anthrax, are immunogenic (Ohtakeet al. 2011).

Persistent and effective immune responses require initial activation of innate immunity. Bacterial pathogen associated molecular patterns (PAMPS) recognized by pattern recognition receptors (PRRs) in macrophages, dendritic cells (DC) or epithelial cells (ECs) determine the nature and extent of the adaptive response through membrane associated and soluble cytokine signaling. For example, Toll-like receptors (TLRs) TLR4 and TLR2 deficient mice immunized with *S* Typhi porins, which have been found to elicit maturation of CD11c[+] conventional DC, showed impaired B-cell response, characterized by reduction of IgG antibody titers to porins, specially of the IgG3 isotype (Cervantes-Barraganet al. 2009). Other authors have investigated the role of TLRs responses to an attenuated *S.* Typhimurium BRD 509 expressing the saliva-binding region (SBR) of *Streptococcus mutans*. Using TLR2, TLR4 and MyD88 deficient mice, they demonstrated that the induction of a serum IgG2a (type 1 response) to the passenger antigen involved TLR2 signaling, whereas the response to *Salmonella* involved signaling through TLR4 (Salamet al. 2010). Thus, antigen specific T and B cells are activated in a coordinated manner to achieve optimal primary and secondary immune responses. The induction of memory is a key characteristic of an effective vaccine (Chenet al. 2010; Medinaet al. 2001). It has been demonstrated that primary effector T cell activation to Salmonella depends on the innate function of B cells (though TLR signaling mediated by MyD88) whereas the induction of T cell memory mediated by antigen specific presentation by the BCR (Barret al. 2010). Moreover, the cell wall of Gram negative bacteria promotes Th1 responses. For instance, an *Escherichia coli* strain expressing an ovalbumin (OVA) allergenic peptide on the bacterial surface, as a fusion protein with the *S.* Typhi OmpC porin, reduced the lung inflammatory response in mice allergic to OVA, with a significant decrease of IL-5 mRNA and induction of IFN-γ mRNA in cells from bronchio alveolar lavages and specific anti-OVA IgE reduction (Yepezet al. 2003).

Several routes of immunization have been assessed with *Salmonella* live carriers. The oral route elicits protective immunity. For instance, mice immunized orally with *S.* Typhiumurium or *S.* Typhi expressing the fullength *B. anthracis* protective antigen (PA) were protected against a lethal challenge with aerosolized *B. anthracis* spores (Stokeset al. 2007). However, the intranasal route is more immunogenic. Indeed, mucosal immunity is most effectively induced when antigens are delivery directly in mucosa, i.e. by oral, intranasal, intrarectal, or intravaginal routes. (Galenet al. 1997). The eye conjunctiva has also demonstrated to be a feasible administration route. Attenuated *Salmonella* vaccine strains administered by eyedrops induced LPS-specific antibodies and protection to the oral challenge with virulent *Salmonella* in mice. Eyedrop vaccinations do not deliver antigens into the CNS as noted with the intranasal route.(Seoet al. 2010)

A needleless delivery system has been developed recently, consisting in a micro wave controlled explosion which disrupts the skin barrier. This system was used to immunize mice with a *S.* Typhimurium vaccine strain pmrG-HM-D (DV-STM-07) with the idea to place the bacteria in the epidermis where resident Langerhans cells may uptake them more efficiently and present the bacterial antigens to lymphocytes (Jagadeeshet al. 2011)

There are a number of attenuated *S.* Typhi strains with defined mutations constructed by genetic engineering. Some have been tested in humans demonstrating immunogenicity and acceptable safety profile, such as *S.* Typhi Ty800, which is mutated in *phoP/phoQ*.(*Hohmannet al. 1996)* or M01ZH09, a *S.* Typhi (Ty2 aroC-ssaV-) ZH9 (Tranet al. 2010). *Other attenuated Salmonella strains* have disrupted the *aroC* and *aroD* genes. The interruption of the biosynthetic pathway of aromatic metabolites results in a bacterial nutritional dependence on *p*-aminobenzoic acid and 2,3-dihydroxybenzoate, substrates not available to bacteria in mammalian tissues (Hoisethet al. 1981). As result, the *aro*-deleted bacteria are not able to proliferate within mammalian cells. However, the organisms survive intracellularly long enough to stimulate immune responses. Inactivation of either *aroC* or *aroD* independently results in attenuation, but deletions in both genes reduce the possibility of virulence restoration by recombination. Two vaccine strains harboring deletion mutations in *aroC* and *aroD* have been evaluated as candidate live oral vaccines in adult volunteers (Bumannet al. 2010; Gonzalezet al. 1994; Tacketet al. 2000; Tacketet al. 2007).

Attenuated *S.* Typhi vaccines have been engineered to express and deliver passenger antigens (proteins and DNA encoded) of a number of pathogens, as the measles virus hemagglutinin, the *Bacillus anthracis* protective antigen (PA), the *Plasmodium falciparum* circumsporozoite surface protein (tCSP), the nucleocapsid (N) protein of severe acute respiratory syndrome-associated coronavirus (SARS-CoV), or the HPV16 L1 protein (L1S).(Chinchillaet al. 2007; Frailleryet al. 2007; Galenet al. 2004; Luoet al. 2007; Pasettiet al. 2003)

3. Influence of passenger antigen location on the immune response

There is increasing evidence that expression level and antigen location determine vaccines efficacy. It is important to achieve passenger antigen expression in the desired bacterial compartment under constitutive or inducible conditions, in order to regulate antigen production. (Bumann 2001; Galenet al. 2001; Kurlandet al. 1996; Pathangeyet al. 2009). Insufficient expression interferes with the immune response to passenger antigens and there is a general notion that high antigen production by live vectors may result in better immune response. Therefore, the production of antigens from high copy number plasmids is apparently the best designing approach. However, excessive expression drives to increased metabolic load, plasmid loss and toxicity (Galenet al. 2001; Pathangeyet al. 2009) Antibody responses to antigens delivered by S. Typhi live vectors are inversely related to the metabolic burden imposed by antigen production, and may be improved when antigens are expressed from low-copy-number plasmids and exported out of the cytoplasm (Galenet al. 2010). Three solutions are proposed to solve this problem: 1) chromosomal integration of heterologous genes, 2) On/off recombinant protein production by using *in vivo*-inducible promoters, and 3) Plasmid stabilization systems. Although both strategies are intended to limit heterologous gene expression, the second strategy has the additional advantage that protein is preferentially produced at the appropriate host environment, such as acidic vacuoles in macrophages. In addition, when plasmid stability is maintained in the absence of antibiotics, there is flexibility for the introduction of a variety of passenger genes without the need to use chromosomal integration systems.

Thus, S Typhi ZH9 (Ty2 Delta aroC Delta ssaV) producing the B subunit of *Escherichia coli* heat-labile toxin or hepatitis B virus core antigen from the bacterial chromosome using the *in vivo* inducible ssaG promoter, stimulated potent antigen-specific serum IgG antibodies to the heterologous antigens (Stratfordet al. 2005).

S. Typhimurium aroA (STM-1) expressing *Mycoplasma hyopneumoniae* antigens form plasmid or chromosomal systems were administered to mice. Whereas no significant immune response was detected with the plasmid based expression, systemic IgM and IgG responses were detected with the chromosomal integration system which used strong promoters (Maticet al. 2009).

A plasmid maintenance system has been tested in S. Typhi CVD 908-htrA consisting in the deletion of genes encoding catalytic enzymes and addition of random segregation function of multicopy plasmid (Galenet al. 2010). Other option to achieve plasmid stability relies on the development of plasmid trans- complementation of lethal deletions in the live vector. Thus, plasmids encoding the single-stranded binding protein (SSB), an protein involved in DNA replication were used to transform S. Typhi CVD 908-htrA and CVD 908, and used to deliver anthrax toxin from *Bacillus anthracis* as a foreign antigen in mice (Galenet al. 2010).

A dual system to achieve increased antigen expression was developed by chromosomal integration of the T7 RNA polymerase gene (T7pol) in S. Typhi CVD908. The T7pol gene was amplified from *Escherichia coli* BL21(DE3) and inserted by homologous recombination in the bacterial chromosome under the control of the inducible nirB promoter. The resulting strain, S. typhi CVD908-T7pol, was able to trans-complement two plasmids bearing the luc or the lacZ reporter genes controlled by the T7 promoter under anaerobic culture conditions (Santiago-Machucaet al. 2002)

Other factor influencing protein expression efficiency include differences of codon usage between the native passenger gene and that host chromosome. Although so far exclusively applied to *Escherichia coli*, codon harmonization may provide a general strategy for improving the expression of soluble, functional proteins during heterologous host expression (Angovet al. 2011)

It has been suggested that passenger antigens delivered by attenuated *Salmonella* strains induce better systemic and mucosal immune responses when displayed on the bacterial surface (Chenet al. 2000; Leeet al. 2000; Ruiz-Perezet al. 2002). A variety of surface display systems have been described (Samuelsonet al. 2002). The most widely used have been fimbria and outer membrane proteins (OMPs), including porins and autotransporters. (Kjaergaardet al. 2002; Klemmet al. 2000; Krameret al. 2003; Rizoset al. 2003)

Passenger fusion proteins (peptides-flagellin) have demonstrated to enhance the immunogenicity of vaccine peptides (Newtonet al. 1989; Newtonet al. 1991a; Newtonet al. 1991b; Stocker 1990; Stockeret al. 1994). In these models the heterologous peptide is fused in-frame to the central hypervariable domain of *Salmonella* FliCd flagellin, which is derived from *S.* Müenchen and expressed by an attenuated *S.* Dublin strain. The chimeric flagellins are exported to the bacterial surface where the subunits assemble into the flagellar shaft without a significant impact on bacterial motility and host tissue colonization (Newtonet al. 1989; Stockeret al. 1994). Nonetheless, previous results showed that the genetic fusion may not enhance antigen-specific antibody responses in mice immunjized by the oral route with recombinant *S.* Dublin (De Almeidaet al. 1999; Sbrogio-Almeidaet al. 2001). Interestingly, the genetic background of both the mice and the *Salmonella* strains affected the immunogenicity of flagellins (Sbrogio-Almeidaet al. 2004). Indeed, recent evidence indicates that *Salmonella* flagellin administered by the oral route may trigger immunological tolerance in healthy mice, although the precise mechanism underlying this response remains unknown (Sanderset al. 2006)

However, in some cases exported, secreted proteins may induce stronger antibody response. (Galenet al. 2001).

Table 1 shows some examples of the influence that heterologous protein location in *Salmonella* has on the type of immune response, which are described as follows.

Antigen displayed	Location in bacterial carrier	Immune response	Reference
TGEV C and A epitopes	Fimbria and outer membrane (MisL)	Humoral	(Chenet al. 2007)
NS3 **Dengue Virus**	Outer membrane (MisL)	CTL	(Luria-Perezet al. 2007)
Ea1A y EaSC2 *Eimeria stiedae*	Cytoplasm	CTL	(Vermeulen 1998)
LTB *E. coli*	Periplasm	Humoral and cellular	(Takahashiet al. 1996)
SERP and HRPII *P. falciparum* antigens	Outer membrane (OmpA)	Humoral	(Schorret al. 1991)
p60 *L. monocytogenes*	Cytoplasm	CTL	(Gentschevet al. 1995)

Table 1. Influence of heterologous protein location in *Salmonella* carrier, on the immune response.

Attenuated *S.* Typhimurium CS4552 (crp cya asd pgtE) was constructed expressing transmissible gastroenteritis virus (TGEV) C and A epitopes fused to the passenger domain of the MisL autotransporter or to the 987P FasA fimbriae subunit under the control of in vivo-induced promoters. The antibody response between both expression systems was compared. Mice vaccinated with the recombinant bacteria displaying the antigens in fimbriae presented the highest level of anti-TGEV antibodies with the epitopes expressed in fimbriae. This result suggests that polymeric display could induce better immune responses towards specific epitopes (Chenet al. 2007).

The second example is a *S.* Typhimurium SL3261 producing a fusion protein designed to destabilize the phagosome membrane and allow a dengue epitope to reach the cytosol. The fusion protein was displayed on the bacterial surface though MisL and the passenger alpha domain contained a fusogenic sequence, a NS3 protein CTL epitope from the dengue virus type 2 and a recognition site for the protease OmpT. The passenger antigen was released to the milieu, processed through the MHC class I-dependent pathway and simulated cytotoxic T lymphocytes (CTLs).(Luria-Perezet al. 2007)

Eimeria stiedae antigens Ea1A and EaSC2, a parasite refractile body transhydrogenase and a lactate dehydrogenase, respectively were expressed in *S.* Typhimurium, and used to immunize chickens. The challenge with the parasite demonstrated oocyst output reduction related with CD4[+] and CD8[+] T cells activation (Vermeulen 1998)

The fourth example in **Table 1** is a comparison between the immune response elicited in mice immunized with native *Escherichia coli* enterotoxin (LT) or with *S.* enteritidis expressing the heat-labile toxin B subunit (LT-B) in the bacterial periplasm. Both antigens elicited mucosal IgA antibodies directed to different LT-B epitopes, and serum IgG antibodies to the same immunodominant LT-B epitopes. The same single T-cell epitope was recognized by immune lymphocytes purified from mice immunized with either antigen (Takahashiet al. 1996)

Immunogenic epitopes of the *Plasmodium falciparum* blood stage antigens SERP and HRPII were expressed on the surface of the attenuated *S.* Typhimurium SR-11 strain as fusion proteins with OmpA from *Escherichia coli*. Mice immunized orally with the bacterial recombinants produced anti-SERP and anti-HRPII IgG and IgM antibodies. (Schorret al. 1991)

Finally, an attenuated *S.* Dublin aroA strain which secretes an active listeriolysin from *Listeria monocytogenes* is partially released from the phagosome into the cytoplasm after uptake by J774 macrophage cells. This is an attractive approach to evoke CTLs responses to passenger antigens through the MHC class I-dependent antigen processing pathway (Gentschevet al. 1995)

4. Display of antigens on the bacterial surface

Following, some of our experience with cytosolic versus surface display of antigens will be described with more detail. The same antigen and live vector were used with the only difference in the bacterial compartment where the antigen was expressed.

It has been demonstrated that *Salmonella typhi* OMPs, porins, and particularly OmpC induce protective immune response in a murine model of infection (Gonzalezet al. 1993; Gonzalezet al. 1995; Isibasiet al. 1988; Isibasiet al. 1992; Isibasiet al. 1994). Porins are OMPs which conform diffusion channels for low molecular weight molecules into the bacterial cell. The tertiary structure is a barrel conformed by 16 anti parallel β sheets with external and internal loops. These external loops have permissive regions were heterologous peptides can be inserted (Vegaet al. 2003; Yepezet al. 2003; Zenteno-Cuevaset al. 2007)

Considering that the major B cell epitope from the *Plasmodium falciparum* circumpsporozoite protein (CSP), the Asp-Ala-Asp-Pro (NANP) repeating sequence, has been inserted in permissive sites of *Pseudomonas aeruginosa* OMP OprF (Wonget al. 1995) we decided to introduce the NANP encoding sequence in the of *S.* Typhi OmpC porin. The NANP3 *P.*

falciparum CSP was inserted in the predicted external loop 5. A site directed mutagenesis was achieved by an overlapping PCR in two amplification rounds. In the first amplification two products were generated separately from plasmid pST13 (bears the complete *S.* Typhi OmpC porin and was kindly donatd by Dr. Felipe Cabello Felipe C. Cabello, New York Medical College) a 5` *omp*C moiety bearing the NANP3 sequence in the 3`end, and a 5´moiety with the NANP3 sequence in the 5´end. In the second amplification both moieties were used to generate a fusion product which was digested and religated to pST13, resulting plasmid pST13-NANP. Functionality of the hybrid *omp*C-NANP gene was assessed by Northern blot using RNA obtained from *Escherichia coli* UH312 transformed with pST13 or pST13-NANP. The autoradiography revealed a more intense band in *Escherichia coli*-pST13 as compared with the same strain transformed with pST13-NANP. Thus, suggesting that the *omp*C-NANP hybrid gene is transcribed less efficiently than the *omp*C native gene.

Protein extracts were obtained from *Escherichia coli* UH302 and *Salmonella typhi* CVD908 transformed with pST13 or pST13-NANP, and the OmpC-NANP fusion protein expression was estimated by SDS-PAGE. When the porinless *Escherichia coli* UH302 strain was transformed with pST13, produced large amounts of the 36 kDa protein. Nevertheless, when transformed with pST13-NANP a faint 36 kDa band was observed. The lower protein production found in the strains transformed with pST13-NANP is consistent with the Northern blot analysis. No differences in OmpC expression was observed between the *Salmonella typhi* CVD908 strains transformed with pST13 or pST13-NANP

Groups of five BALB/c mice were immunized with *Escherichia coli* UH302 transformed with pST13 or with pST13-NANP and *S.* Typhi CVD908, CVD908-pST13-NANP, CVD908ΩCSP (bears the whole CSP integrated in the bacterial chromosome)(Gonzalezet al. 1994), CVD908ΩCSP-pST13-NANP. Seven days after the last immunization, antibodies against *Plasmodium falciparum* sporozoites were assessed by IFA. Neither preimmune sera nor sera from mice immunized with *E. coli* UH302-pST13 or *Salmonella typhi* CVD908 recognized sporozoites, and some structures suggesting sporozoites were observed with sera from mice immunized with *E. coli* UH302-pST13-NANP (**Figure 1**). Interestingly, *S. typhi* CVD908ΩCSP (cytosolic CSP expression) was unable to induce antibodies against the parasite and similarly to *E. coli* UH302-pSt13-NANP, sera from mice immunized with *S. typhi* CVD908-pST13-NANP (epitope surface expression) showed some structures resembling parasites. Nevertheless, only mice immunized with *Salmonella typhi* CVD908ΩCSP-pST13-NANP (both surface and cytosolic expression) clearly depicted sporozoites comparable to the positive controls reveled with 2A10 antibody (**Figure 1**). These data were consistent with the measurement of antibodies by ELISA. Serum of BALB/c mice immunized with *S.* Typhi CVD908, CVD908ΩCSP, CVD908-pST13-NANP, and CVD908ΩCSP-pST13-NANP were collected, and antibodies to (B)4MAPs (a branched peptide containing the NANP) sequence were determined. *S.* Typhi CVD908ΩCSP, which produce a cytosolic CSP from a chromosomal integrated gene, did not elicit measurable antibodies under the experimental conditions for this experiment (1:25 to 1:250). *S.* Typhi CVD908-pST13-NANP, which displayed the NANP epitope on the bacterial surface, elicited mild antibody response (titer 1:25), whereas *S.* Typhi CVD908ΩCSP-pST13-NANP, which produced the epitope on the bacterial surface and the whole CSP in the cytosol, raised the highest antibody response (titer 1:200) after both o.g or i.p immunization (**Figure 2**). Taken together these data suggest, as stated earlier, that the epitope expressed on the bacterial surface may exhibit antigenic identity with the native CSP in *P. falciparum* sporozoites and that both surface and cyotosolic expression elicited better antibody response.

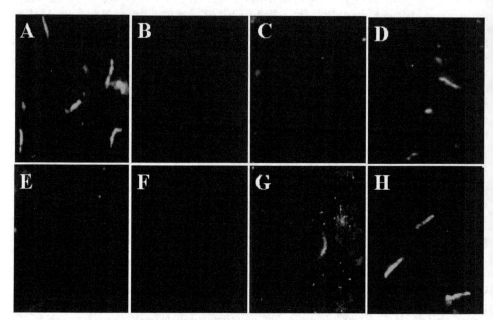

A) Positive control, sporozoites revealed with the 2A10 monoclonal antibody (which recognizes the (NANP)n repeat of Plasmodium falciparum circumsporozoite surface protein); **B)** Negative control, preimmune sera; **C)** *Escherichia coli* UH302,-pST13; **D)** *Escherichia coli* UH302-pST13-NANP; **E)** *Salmonella typhi* CVD908; **F)** *Salmonella typhi* CVD908ΩCSP; **G)** *Salmonella typhi* CVD908-pST13-NANP; **H)** *Salmonella typhi* CVD908ΩCSP-pST13-NANP.

Fig. 1. Antibodies against *Plasmodium falciparum* sporozoites elicited by the immunization of BABL/c mice with bacterial strains expressing the NANP epitope on the bacterial surface were assessed by immunofluorescence assay. The antibody response was compared between strains with cytosolic expression of CSP, surface expression of (NANP)$_3$, or both surface and cytosolic expression.

Western blot was performed revealing OmpC with a rabbit hyperimmune anti-OmpC serum or a cocktail of monoclonal antibodies against OmpC. The rabbit anti-OmpC serum showed two clear bands in the *Salmonella typhi* strains due to cross reactivity with other porin (probably OmpF, which is 35KDa). *Escherichia coli* UH302-pST13 and pST13-NANP revealed a single band. The cocktail of monoclonal antibodies against OmpC revealed a single band and demonstrated that native OmpC and the OmpC-NANP fusion protein display similar molecular weight. It is important to notice that the western blot from these protein extracts using monoclonal antibody 2A10, which recognizes the *Plasmodium falciparum* CSP, failed to recognize the (NANP)$_3$ epitope in the fusion OmpC-NANP protein. Nevertheless, the flow cytometry performed with *Escherichia coli* UH302 transformed with pST13 or pST13-NANP using the same 2A10 monoclonal antibody revealed that *Escherichia coli* UH302-pST13-NANP displays the chimeric OmpC with the *Plasmodium falciparum*

NANP epitope on the bacterial surface. These data could be explained by the low expression levels of the ompC-NANP fusion protein, but may be related to conformational changes in the SDS-PAGE.

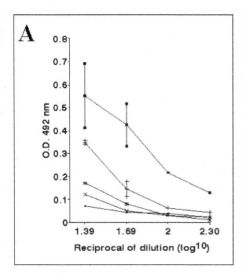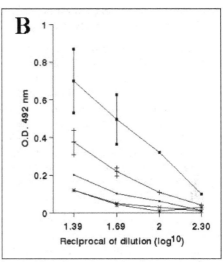

A). Antibodies against (B)₄MAPs in mice immunized orogastrically with *Salmonella typhi* strains as described elsewhere (Gonzalezet al. 1998) assessed by ELISA; **B).** Mice immunized intraperitoneally with (•) CVD908; (+) CVD908-pST13-NANP; (∗) CVD908ΩCSP; (■) CVD908ΩCSP-pST13-NANP; (×) Preimmune sera.

Fig. 2. Comparison between the antibody response against (B)₄MAPs, a tetramer branched synthetic peptide containing (NANP)₃ in each of the four branches (kindly donated by Dr. Elizabeth Nardin, Department of Medical and Molecular Parasitology. New York University School of Medicine, New York, NY), elicited in mice immunized with *Salmonella typhi* CVD908 expressing the CSP in the cytosol, the (NANP)₃ epitope on the bacterial surface or from both bacterial compartments.

Finally, we will describe some experience with autotransporters for autodisplay of antigens. Autotransporters belong to a family of OMPs, which lack the requirement of specific accessory molecules for secretion through the outer membrane. These proteins bear all necessary signals encoded within the polypeptide itself. They contain a C-terminal domain, (β-domain or translocator domain) which allows the N-terminal α passenger domain to cross from the inner membrane to the periplasmic space. The α-passenger domain is flanked by an N-terminal signal sequence responsible for initial export into the bacterial perplasmic space by a *sec* dependent mechanism. Once in the periplasmic space the C-terminal translocator β-domain forms a barrel and inserts in the outer membrane, and the N-terminal passenger α passenger domain travels through the central pore to the external milieu where exerts its biological function. Once on the surface, the final fate of the N-terminal passenger α passenger domain is determined by the presence of autoproteolytic mechanisms or surface proteases, which cleavage and release the α passenger domain to the external environment. (Finket al. 2001). More than 40 proteins with autotransporting properties have been characterized (Desvauxet al. 2004; Hendersonet al. 2001). Due the relative simplicity of their transporting mechanism, the β-domain from several autotransporters has been employed translocate and display recombinant passenger proteins on the surface of enterobacteria. We already reported the use of MisL (another member of the AIDA-subfamily) to express foreign immunogenic epitopes on the surface of gramnegative bacteria (Luria-Perezet al. 2007; Ruiz-Olveraet al. 2003; Ruiz-Perezet al. 2002).

ShdA is other large autotransporter, (Desvauxet al. 2004) identified in *S. enterica* subespecies (Kingsleyet al. 2000), with similar structure to AIDA-I, TibA, and MisL, therefore it has been included also in the AIDA-subfamily. The α-domain is an adhesin (Kingsleyet al. 2000) that mediates bacterial colonization in the host cecum, the main reservoir for *S.* Typhimurium during infection in mice (Kingsleyet al. 2002) In fact, the inactivation of *shdA* produces bacterial number and bacterial permanence in the intestinal mucosa (shedding reduction) (Kingsleyet al. 2000; Kingsleyet al. 2002). The extracellular matrix protein fibronectin is a receptor for the ShdA passenger domain. This was demonstrated by a ShdA–GST (glutathione *S*-transferase) fusion protein which bound fibronectin *in vitro* in a dose dependent manner and was partially inhibited by anti-fibronectin antibodies, suggesting that other receptors may also play a role in ShdA-mediated adherence to the intestinal mucosa (Kingsleyet al. 2004).

Several autotransporters (Maureret al. 1999) require a link region between the α and β domains for autodisplay. This minimal translocation unit (TU) is necessary to allow folding of the passenger α -domain (Oliveret al. 2003). The role of TU in ShdA still remains to be show. Since autotransporters are able to display heeterologous peptide substituting the α - domain they have been used for the construction of bacterial whole-cell absorbents, study receptor-ligand interactions surface display of random peptide libraries and vaccine development (Lattemannet al. 2000).

We describe here an example of the latter application exposing the NANP immunodominat epitope from *Plamodium falciparum* CSP on the surface of *Salmonella* using an autotransporter. We generated a series of NANP-ShdA fusion proteins containing the β-domain and different truncated α-domains forms under the control of *nirB* promoter (Chatfieldet al. 1992), using the technical approach described elsewhere (Ruiz-Perezet al. 2002).

The flow cytometry **in Figure 3** presents the summary of several assays performed to identify the minimal α-domain amino acid strand necessary for translocation through the ShdA β-domain. *S.* Typhimurium SL3261 was transformed with plasmids bearing different truncated α-domain forms fused to three repeats of NANP [(NANP)3] or the complete CSP. NANP expression on to the surface of the bacteria was determined with a monoclonal antibody. We identified that the minimum translocation unit necessary to translocate the epitope is conformed 16 residues in the α-domain. Interestingly only around 45% of the bacterial strains expressed the antigen on their surface.

BALB/c mice were immunized with different S. Typhimurium SL3261 expressing the full length CSP or the (NANP)3 epitope on the surface and compared with a strain producing the antigen in the bacterial cytosol **(Figure 3 A-E)**. As expected, the strain expressing only ShdA did not elicit antibodies. The strains expressing the NANP or the CSP elicited good antibody response **(Figure 3 B-C)**, whereas the strain producing the CSP in the cytosol was unable to elicit antibodies **(Figure 3 E)**. An additional control, autotransporter MisL expressing the NANP epitope, was able as well to elicit antibodies **(Figure 3 D)**.

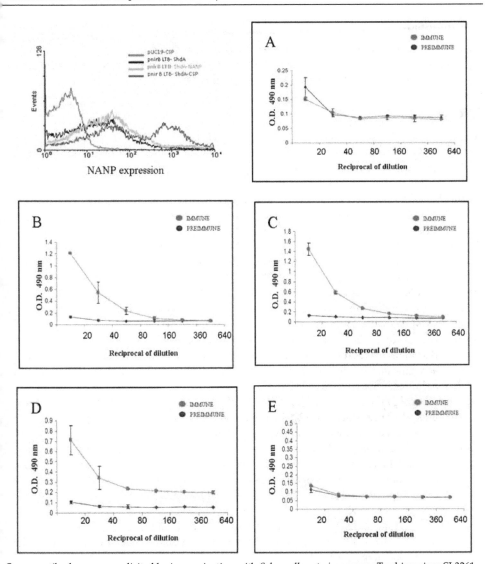

Serum antibody response elicited by immunization with *Salmonella enterica* serovar Typhimurium SL3261 transformed with differents plasmids. BALB/c mice were immunized o.g. as described elsewhere (Gonzalezet al. 1998) (A) pnirB LTB- ShdA (negative control) (B) pnirB-LTB NANP ShdA; (C) pnirB-LTB CSP ShdA.; (D) pnirB-LTB NANP MisL; (E) pUC19 CSP. Groups of 5 BALB/c mice were immunized orally with two doses of 1x10^10 C.F.U. (15-day interval) of the *Salmonella SL3261* strain transformed with different plasmids. IgG levels were determined one week after last immunization by ELISA as previously described (González et al., 1998). Each graphic represents the serum IgG from one mouse.

Fig. 3. Flow cytometry analysis of strains of *Salmonella enterica* serovar Typhimurium SL3261 transformed with differents plasmids. Plasmid pUC19 CSP corresponding to cytosolic form of antigen, whereas pnir B LTB ShdA-CSP and pnir B LTB ShdA-NANP corresponding to antigen display on the bacterial surface.

5. Conclusion

In summary, there is increasing evidence that antigen location in live bacterial carrier vaccines, in this case attenuated *Salmonella* strains is an important factor determining the type of immune response elicited to the passenger antigen.

6. References

Abd El, Ghany M., Jansen, A., Clare, S., Hall, L., Pickard, D., Kingsley, R. A., and Dougan, G.(2007), Candidate live, attenuated Salmonella enterica serotype Typhimurium vaccines with reduced fecal shedding are immunogenic and effective oral vaccines. *Infect.Immun.*75,4,1835-1842.

Angov, E., Legler, P. M., and Mease, R. M.(2011), Adjustment of codon usage frequencies by codon harmonization improves protein expression and folding. *Methods Mol.Biol.*705,1-13.

Barr, T. A., Brown, S., Mastroeni, P., and Gray, D.(2010), TLR and B cell receptor signals to B cells differentially program primary and memory Th1 responses to Salmonella enterica. *J.Immunol.*185,5,2783-2789.

Bumann, D.(2001), Regulated antigen expression in live recombinant Salmonella enterica serovar Typhimurium strongly affects colonization capabilities and specific CD4(+)-T-cell responses.*Infect.Immun.*69,12,7493-7500.

Bumann, D., Behre, C., Behre, K., Herz, S., Gewecke, B., Gessner, J. E., von Specht, B. U., and Baumann, U.(2010), Systemic, nasal and oral live vaccines against Pseudomonas aeruginosa: a clinical trial of immunogenicity in lower airways of human volunteers. *Vaccine.*28,3,707-713.

Cervantes-Barragan, L., Gil-Cruz, C., Pastelin-Palacios, R., Lang, K. S., Isibasi, A., Ludewig, B., and Lopez-Macias, C.(2009), TLR2 and TLR4 signaling shapes specific antibody responses to Salmonella typhi antigens. *Eur.J.Immunol.*39,1,126-135.

Chatfield, S. N., Charles, I. G., Makoff, A. J., Oxer, M. D., Dougan, G., Pickard, D., Slater, D., and Fairweather, N. F.(1992), Use of the nirB promoter to direct the stable expression of heterologous antigens in Salmonella oral vaccine strains: development of a single-dose oral tetanus vaccine. *Biotechnology (N.Y.).*10,8,888-892.

Chen, H. and Schifferli, D. M.(2000), Mucosal and systemic immune responses to chimeric fimbriae expressed by Salmonella enterica serovar typhimurium vaccine strains. *Infect.Immun.*68,6,3129-3139.

Chen, H. and Schifferli, D. M.(2007), Comparison of a fimbrial versus an autotransporter display system for viral epitopes on an attenuated Salmonella vaccine vector.*Vaccine.*25,9,1626-1633.

Chen, K. and Cerutti, A.(2010), Vaccination strategies to promote mucosal antibody responses. *Immunity.*33,4,479-491.

Chinchilla, M., Pasetti, M. F., Medina-Moreno, S., Wang, J. Y., Gomez-Duarte, O. G., Stout, R., Levine, M. M., and Galen, J. E.(2007), Enhanced immunity to Plasmodium falciparum circumsporozoite protein (PfCSP) by using Salmonella enterica serovar

Typhi expressing PfCSP and a PfCSP-encoding DNA vaccine in a heterologous prime-boost strategy.*Infect.Immun.*75,8,3769-3779.

De Almeida, M. E., Newton, S. M., and Ferreira, L. C.(1999), Antibody responses against flagellin in mice orally immunized with attenuated Salmonella vaccine strains.*Arch.Microbiol.*172,2,102-108.

Desvaux, M., Parham, N. J., and Henderson, I. R.(2004), Type V protein secretion: simplicity gone awry?*Curr.Issues Mol.Biol.*6,2,111-124.

Fink, D. L., Cope, L. D., Hansen, E. J., and Geme, J. W., III.(2001), The Hemophilus influenzae Hap autotransporter is a chymotrypsin clan serine protease and undergoes autoproteolysis via an intermolecular mechanism. *J. Biol. Chem.* 276,42,39492-39500.

Fraillery, D., Baud, D., Pang, S. Y., Schiller, J., Bobst, M., Zosso, N., Ponci, F., and Nardelli-Haefliger, D.(2007), Salmonella enterica serovar Typhi Ty21a expressing human papillomavirus type 16 L1 as a potential live vaccine against cervical cancer and typhoid fever.*Clin.Vaccine Immunol.*14,10,1285-1295.

Galen, J. E., Gomez-Duarte, O. G., Losonsky, G. A., Halpern, J. L., Lauderbaugh, C. S., Kaintuck, S., Reymann, M. K., and Levine, M. M.(1997), A murine model of intranasal immunization to assess the immunogenicity of attenuated Salmonella typhi live vector vaccines in stimulating serum antibody responses to expressed foreign antigens.*Vaccine.*15,6-7,700-708.

Galen, J. E. and Levine, M. M.(2001), Can a 'flawless' live vector vaccine strain be engineered? *Trends Microbiol.*9,8,372-376.

Galen, J. E., Wang, J. Y., Chinchilla, M., Vindurampulle, C., Vogel, J. E., Levy, H., Blackwelder, W. C., Pasetti, M. F., and Levine, M. M.(2010), A new generation of stable, nonantibiotic, low-copy-number plasmids improves immune responses to foreign antigens in Salmonella enterica serovar Typhi live vectors. *Infect. Immun.* 78,1,337-347.

Galen, J. E., Zhao, L., Chinchilla, M., Wang, J. Y., Pasetti, M. F., Green, J., and Levine, M. M.(2004), Adaptation of the endogenous Salmonella enterica serovar Typhi clyA-encoded hemolysin for antigen export enhances the immunogenicity of anthrax protective antigen domain 4 expressed by the attenuated live-vector vaccine strain CVD 908-htrA.*Infect.Immun.*72,12,7096-7106.

Gentschev, I., Sokolovic, Z., Mollenkopf, H. J., Hess, J., Kaufmann, S. H., Kuhn, M., Krohne, G. F., and Goebel, W.(1995), Salmonella strain secreting active listeriolysin changes its intracellular localization.*Infect.Immun.*63,10,4202-4205.

Gonzalez, C., Hone, D., Noriega, F. R., Tacket, C. O., Davis, J. R., Losonsky, G., Nataro, J. P., Hoffman, S., Malik, A., Nardin, E., Sztein, M. B., Heppner, D. G., Fouts, T. R., Isibasi, A., Levine, M. M., and.(1994), Salmonella typhi vaccine strain CVD 908 expressing the circumsporozoite protein of Plasmodium falciparum: strain construction and safety and immunogenicity in humans.*J.Infect.Dis.*169,4,927-931.

Gonzalez, C. R., Isibasi, A., Ortiz-Navarrete, V., Paniagua, J., Garcia, J. A., Blanco, F., and Kumate, J.(1993), Lymphocytic proliferative response to outer-membrane proteins isolated from Salmonella. *Microbiol.Immunol.*37,10,793-799.

Gonzalez, C. R., Mejia, M. V., Paniagua, J., Ortiz-Navarrete, V., Ramirez, G., and Isibasi, A.(1995), Immune response to porins isolated from Salmonella typhi in different mouse strains.*Arch.Med.Res.*26 Spec No,S99-103.

Gonzalez, C. R., Noriega, F. R., Huerta, S., Santiago, A., Vega, M., Paniagua, J., Ortiz-Navarrete, V., Isibasi, A., and Levine, M. M.(1998), Immunogenicity of a Salmonella typhi CVD 908 candidate vaccine strain expressing the major surface protein gp63 of Leishmania mexicana mexicana.*Vaccine.*16,9-10,1043-1052.

Henderson, I. R. and Nataro, J. P.(2001), Virulence functions of autotransporter proteins.*Infect.Immun.*69,3,1231-1243.

Hohmann, E. L., Oletta, C. A., Killeen, K. P., and Miller, S. I.(1996), phoP/phoQ-deleted Salmonella typhi (Ty800) is a safe and immunogenic single-dose typhoid fever vaccine in volunteers.*J.Infect.Dis.*173,6,1408-1414.

Hoiseth, S. K. and Stocker, B. A.(1981), Aromatic-dependent Salmonella typhimurium are non-virulent and effective as live vaccines.*Nature.*291,5812,238-239.

Isibasi, A., Ortiz, V., Vargas, M., Paniagua, J., Gonzalez, C., Moreno, J., and Kumate, J.(1988), Protection against Salmonella typhi infection in mice after immunization with outer membrane proteins isolated from Salmonella typhi 9,12,d, Vi. *Infect. Immun.* 56,11,2953-2959.

Isibasi, A., Ortiz-Navarrete, V., Paniagua, J., Pelayo, R., Gonzalez, C. R., Garcia, J. A., and Kumate, J.(1992), Active protection of mice against Salmonella typhi by immunization with strain-specific porins.*Vaccine.*10,12,811-813.

Isibasi, A., Paniagua, J., Rojo, M. P., Martin, N., Ramirez, G., Gonzalez, C. R., Lopez-Macias, C., Sanchez, J., Kumate, J., and Ortiz-Navarrete, V.(1994), Role of porins from Salmonella typhi in the induction of protective immunity.*Ann.N.Y.Acad.Sci.*730,350-352.

Jagadeesh, G., Prakash, G. D., Rakesh, S. G., Allam, U. S., Krishna, M. G., Eswarappa, S. M., and Chakravortty, D.(2011), Needleless vaccine delivery using micro-shock waves. *Clin.Vaccine Immunol.*18,4,539-545.

Kingsley, R. A., Abi, Ghanem D., Puebla-Osorio, N., Keestra, A. M., Berghman, L., and Baumler, A. J.(2004), Fibronectin binding to the Salmonella enterica serotype Typhimurium ShdA autotransporter protein is inhibited by a monoclonal antibody recognizing the A3 repeat.*J.Bacteriol.*186,15,4931-4939.

Kingsley, R. A., Santos, R. L., Keestra, A. M., Adams, L. G., and Baumler, A. J.(2002), Salmonella enterica serotype Typhimurium ShdA is an outer membrane fibronectin-binding protein that is expressed in the intestine. *Mol. Microbiol.* 43,4,895-905.

Kingsley, R. A., van, Amsterdam K., Kramer, N., and Baumler, A. J.(2000), The shdA gene is restricted to serotypes of Salmonella enterica subspecies I and contributes to efficient and prolonged fecal shedding.*Infect.Immun.*68,5,2720-2727.

Kjaergaard, K., Hasman, H., Schembri, M. A., and Klemm, P.(2002), Antigen 43-mediated autotransporter display, a versatile bacterial cell surface presentation system.*J.Bacteriol.*184,15,4197-4204.

Klemm, P. and Schembri, M. A.(2000), Fimbrial surface display systems in bacteria: from vaccines to random libraries.*Microbiology.*146 Pt 12,3025-3032.

Kotton, C. N. and Hohmann, E. L.(2004), Enteric pathogens as vaccine vectors for foreign antigen delivery.*Infect.Immun.*72,10,5535-5547.

Kramer, U., Rizos, K., Apfel, H., Autenrieth, I. B., and Lattemann, C. T.(2003), Autodisplay: development of an efficacious system for surface display of antigenic determinants in Salmonella vaccine strains.*Infect.Immun.*71,4,1944-1952.

Kurland, C. G. and Dong, H.(1996), Bacterial growth inhibition by overproduction of protein. *Mol.Microbiol.*21,1,1-4.

Lattemann, C. T., Maurer, J., Gerland, E., and Meyer, T. F.(2000), Autodisplay: functional display of active beta-lactamase on the surface of Escherichia coli by the AIDA-I autotransporter.*J.Bacteriol.*182,13,3726-3733.

Lee, J. S., Shin, K. S., Pan, J. G., and Kim, C. J.(2000), Surface-displayed viral antigens on Salmonella carrier vaccine.*Nat.Biotechnol.*18,6,645-648.

Luo, F., Feng, Y., Liu, M., Li, P., Pan, Q., Jeza, V. T., Xia, B., Wu, J., and Zhang, X. L.(2007), Type IVB pilus operon promoter controlling expression of the severe acute respiratory syndrome-associated coronavirus nucleocapsid gene in Salmonella enterica Serovar Typhi elicits full immune response by intranasal vaccination. *Clin.Vaccine Immunol.*14,8,990-997.

Luria-Perez, R., Cedillo-Barron, L., Santos-Argumedo, L., Ortiz-Navarrete, V. F., Ocana-Mondragon, A., and Gonzalez-Bonilla, C. R.(2007), A fusogenic peptide expressed on the surface of Salmonella enterica elicits CTL responses to a dengue virus epitope.*Vaccine.*25,27,5071-5085.

Matic, J. N., Terry, T. D., Van, Bockel D., Maddocks, T., Tinworth, D., Jennings, M. P., Djordjevic, S. P., and Walker, M. J.(2009), Development of non-antibiotic-resistant, chromosomally based, constitutive and inducible expression systems for aroA-attenuated Salmonella enterica Serovar Typhimurium. *Infect. Immun.* 77,5,1817-1826.

Maurer, J., Jose, J., and Meyer, T. F.(1999), Characterization of the essential transport function of the AIDA-I autotransporter and evidence supporting structural predictions.*J.Bacteriol.*181,22,7014-7020.

Medina, E. and Guzman, C. A.(2001), Use of live bacterial vaccine vectors for antigen delivery: potential and limitations.*Vaccine.*19,13-14,1573-1580.

Newton, S. M., Jacob, C. O., and Stocker, B. A.(1989), Immune response to cholera toxin epitope inserted in Salmonella flagellin.*Science.*244,4900,70-72.

Newton, S. M., Kotb, M., Poirier, T. P., Stocker, B. A., and Beachey, E. H.(1991a), Expression and immunogenicity of a streptococcal M protein epitope inserted in Salmonella flagellin.*Infect.Immun.*59,6,2158-2165.

Newton, S. M., Wasley, R. D., Wilson, A., Rosenberg, L. T., Miller, J. F., and Stocker, B. A.(1991b), Segment IV of a Salmonella flagellin gene specifies flagellar antigen epitopes.*Mol.Microbiol.*5,2,419-425.

Ohtake, S., Martin, R., Saxena, A., Pham, B., Chiueh, G., Osorio, M., Kopecko, D., Xu, D., Lechuga-Ballesteros, D., and Truong-Le, V.(2011), Room temperature stabilization of oral, live attenuated Salmonella enterica serovar Typhi-vectored vaccines. *Vaccine.* 29,15,2761-2771.

Oliver, D. C., Huang, G., and Fernandez, R. C.(2003), Identification of secretion determinants of the Bordetella pertussis BrkA autotransporter.*J.Bacteriol.*185,2,489-495.

Pasetti, M. F., Barry, E. M., Losonsky, G., Singh, M., Medina-Moreno, S. M., Polo, J. M., Ulmer, J., Robinson, H., Sztein, M. B., and Levine, M. M.(2003), Attenuated

Salmonella enterica serovar Typhi and Shigella flexneri 2a strains mucosally deliver DNA vaccines encoding measles virus hemagglutinin, inducing specific immune responses and protection in cotton rats.*J. Virol.*77,9,5209-5217.

Pathangey, L., Kohler, J. J., Isoda, R., and Brown, T. A.(2009), Effect of expression level on immune responses to recombinant oral Salmonella enterica serovar Typhimurium vaccines.*Vaccine.*27,20,2707-2711.

Poland, G. A., Levine, M. M., and Clemens, J. D.(2010), Developing the next generation of vaccinologists.*Vaccine.*28,52,8227-8228.

Rizos, K., Lattemann, C. T., Bumann, D., Meyer, T. F., and Aebischer, T.(2003), Autodisplay: efficacious surface exposure of antigenic UreA fragments from Helicobacter pylori in Salmonella vaccine strains.*Infect.Immun* 71,11,6320-6328.

Ruiz-Olvera, P., Ruiz-Perez, F., Sepulveda, N. V., Santiago-Machuca, A., Maldonado-Rodriguez, R., Garcia-Elorriaga, G., and Gonzalez-Bonilla, C.(2003), Display and release of the Plasmodium falciparum circumsporozoite protein using the autotransporter MisL of Salmonella enterica.*Plasmid.*50,1,12-27.

Ruiz-Perez, F., Leon-Kempis, R., Santiago-Machuca, A., Ortega-Pierres, G., Barry, E., Levine, M., and Gonzalez-Bonilla, C.(2002), Expression of the Plasmodium falciparum immunodominant epitope (NANP)(4) on the surface of Salmonella enterica using the autotransporter MisL.*Infect.Immun.*70,7,3611-3620.

Salam, M. A., Katz, J., and Michalek, S. M.(2010), Role of Toll-like receptors in host responses to a virulence antigen of Streptococcus mutans expressed by a recombinant, attenuated Salmonella vector vaccine.*Vaccine.*28,31,4928-4936.

Samuelson, P., Gunneriusson, E., Nygren, P. A., and Stahl, S.(2002), Display of proteins on bacteria.*J.Biotechnol.*96,2,129-154.

Sanders, C. J., Yu, Y., Moore, D. A., III, Williams, I. R., and Gewirtz, A. T.(2006), Humoral immune response to flagellin requires T cells and activation of innate immunity.*J.Immunol.*177,5,2810-2818.

Santiago-Machuca, A. E., Ruiz-Perez, F., Delgado-Dominguez, J. S., Becker, I., Isibasi, A., and Gonzalez-Bonilla, C. R.(2002), Attenuated Salmonella enterica serovar typhi live vector with inducible chromosomal expression of the T7 RNA polymerase and its evaluation with reporter genes.*Plasmid.*47,2,108-119.

Sbrogio-Almeida, M. E. and Ferreira, L. C.(2001), Flagellin expressed by live Salmonella vaccine strains induces distinct antibody responses following delivery via systemic or mucosal immunization routes.*FEMS Immunol.Med.Microbiol.*30,3,203-208.

Sbrogio-Almeida, M. E., Mosca, T., Massis, L. M., Abrahamsohn, I. A., and Ferreira, L. C.(2004), Host and bacterial factors affecting induction of immune responses to flagellin expressed by attenuated Salmonella vaccine strains.*Infect.Immun.*72,5,2546-2555.

Schorr, J., Knapp, B., Hundt, E., Kupper, H. A., and Amann, E.(1991), Surface expression of malarial antigens in Salmonella typhimurium: induction of serum antibody response upon oral vaccination of mice.*Vaccine.*9,9,675-681.

Seo, K. Y., Han, S. J., Cha, H. R., Seo, S. U., Song, J. H., Chung, S. H., and Kweon, M. N.(2010), Eye mucosa: an efficient vaccine delivery route for inducing protective immunity.*J.Immunol.*185,6,3610-3619.

Stocker, B. A.(1990), Aromatic-dependent Salmonella as live vaccine presenters of foreign epitopes as inserts in flagellin.*Res.Microbiol.*141,7-8,787-796.

Stocker, B. A. and Newton, S. M.(1994), Immune responses to epitopes inserted in Salmonella flagellin.*Int.Rev.Immunol.*11,2,167-178.

Stokes, M. G., Titball, R. W., Neeson, B. N., Galen, J. E., Walker, N. J., Stagg, A. J., Jenner, D. C., Thwaite, J. E., Nataro, J. P., Baillie, L. W., and Atkins, H. S.(2007), Oral administration of a Salmonella enterica-based vaccine expressing Bacillus anthracis protective antigen confers protection against aerosolized B. anthracis. *Infect. Immun.* 75,4,1827-1834.

Stratford, R., McKelvie, N. D., Hughes, N. J., Aldred, E., Wiseman, C., Curtis, J., Bellaby, T., Bentley, M., Hindle, Z., Brennan, F. R., Chatfield, S. N., Dougan, G., and Khan, S. A.(2005), Optimization of Salmonella enterica serovar typhi DeltaaroC DeltassaV derivatives as vehicles for delivering heterologous antigens by chromosomal integration and in vivo inducible promoters.*Infect.Immun.*73,1,362-368.

Tacket, C. O. and Levine, M. M.(2007), CVD 908, CVD 908-htrA, and CVD 909 live oral typhoid vaccines: a logical progression.*Clin.Infect.Dis.*45 Suppl 1,S20-S23.

Tacket, C. O., Sztein, M. B., Wasserman, S. S., Losonsky, G., Kotloff, K. L., Wyant, T. L., Nataro, J. P., Edelman, R., Perry, J., Bedford, P., Brown, D., Chatfield, S., Dougan, G., and Levine, M. M.(2000), Phase 2 clinical trial of attenuated Salmonella enterica serovar typhi oral live vector vaccine CVD 908-htrA in U.S. volunteers. *Infect. Immun.*68,3,1196-1201.

Takahashi, I., Kiyono, H., Jackson, R. J., Fujihashi, K., Staats, H. F., Hamada, S., Clements, J. D., Bost, K. L., and McGhee, J. R.(1996), Epitope maps of the Escherichia coli heat-labile toxin B subunit for development of a synthetic oral vaccine. *Infect. Immun.* 64,4,1290-1298.

Tran, T. H., Nguyen, T. D., Nguyen, T. T., Ninh, T. T., Tran, N. B., Nguyen, V. M., Tran, T. T., Cao, T. T., Pham, V. M., Nguyen, T. C., Tran, T. D., Pham, V. T., To, S. D., Campbell, J. I., Stockwell, E., Schultsz, C., Simmons, C. P., Glover, C., Lam, W., Marques, F., May, J. P., Upton, A., Budhram, R., Dougan, G., Farrar, J., Nguyen, V. V., and Dolecek, C.(2010), A randomised trial evaluating the safety and immunogenicity of the novel single oral dose typhoid vaccine M01ZH09 in healthy Vietnamese children.*PLoS.One.*5,7,e11778-

Vega, M. I., Santos-Argumedo, L., Huerta-Yepez, S., Luria-Perez, R., Ortiz-Navarrete, V., Isibasi, A., and Gonzalez-Bonilla, C. R.(2003), A Salmonella typhi OmpC fusion protein expressing the CD154 Trp140-Ser149 amino acid strand binds CD40 and activates a lymphoma B-cell line. *Immunology.* 110,2,206-216.

Vermeulen, A. N.(1998), Progress in recombinant vaccine development against coccidiosis. A review and prospects into the next millennium.*Int.J.Parasitol.*28,7,1121-1130.

Wong, R. S., Wirtz, R. A., and Hancock, R. E.(1995), Pseudomonas aeruginosa outer membrane protein OprF as an expression vector for foreign epitopes: the effects of positioning and length on the antigenicity of the epitope.*Gene.*158,1,55-60.

Yepez, S. H., Pando, R. H., Argumedo, L. S., Paredes, M. V., Cueto, A. H., Isibasi, A., and Bonilla, C. R.(2003), Therapeutic efficacy of an E coli strain carrying an ovalbumin allergenic peptide as a fused protein to OMPC in a murine model of allergic airway inflammation.*Vaccine.*21,5-6,566-578.

Zenteno-Cuevas, R., Huerta-Yepez, S., Reyes-Leyva, J., Hernandez-Jauregui, P., Gonzalez-Bonilla, C., Ramirez-Mendoza, H., Agundis, C., and Zenteno, E.(2007), Identification of potential B cell epitope determinants by computer techniques, in hemagglutinin-neuraminidase from the porcine rubulavirus La Piedad Michoacan. *Viral Immunol.*20,2,250-260.

Salmonella enterica Serovar Enteritidis (SE) Infection in Chickens and Its Public-Health-Risk Control Using an SE Vaccine in Layer Flocks

Hiroaki Ohta[1] and Yukiko Toyota-Hanatani[2]

[1]*CAF Laboratories, Fukuyama, Hiroshima*
[2]*Laboratory of Veterinary Internal Medicine, School of Veterinary Science*
Osaka Prefectural University, Izumisano, Osaka
Japan

1. Introduction

Food poisoning caused by *Salmonella enterica* serovar enteritidis (SE) became a major public health problem in the middle of the 1980s, and several years were required to identify that the main causative food material was chicken eggs (Altekruse S. et al. 1993, [a),b)]CDC 1990, Cogan TA et al., Cowden JM et al. 1989, Henzeler DJ et al. 1994, Humphrey TJ 1994, Kusunoki J et al. 1996, Lin FY et al. 1988, Shivaprad HL et al. 1990, St Louis ME et al. 1988). Since CDC had firstly-reported the main causative origin of SE food born disease being shell eggs (CDC. 1987), shell eggs as a causative food have attended (Hogue A. et al. 1997, Humphrey TJ et al. 1991, Rodrigue DC et al. 1990,). World status of SE outbreaks at around 1999 is well-reviewed in the book of "*Salmonella enterica* serovar enteritidis in human and animals". (Saeed AM. Ed. 1999. Iowa State University Press). SE-contaminated chicken eggs are indistinguishable from non-contaminated eggs in appearance. As the sensory elimination of SE-contaminated chicken eggs was shown to be impossible, greater importance has been attached to the control of SE contamination in the egg production step [(a),b)]CDC 1990, Okamura M et al. 2001, Rodrigue DC et al. 1990, and Stevens A et al. 1989, Thomas RD 1989). The development of live and inactivated SE vaccines has been investigated because SE contamination of chicken eggs remained even after various hygienic countermeasures were taken on layer farms, and SE vaccine administration was started in the 1990s in Western countries. However, SE vaccines have only recently been recognized as an important tool to reduce SE-contaminated chicken egg production on layer farms.

Regarding efficacy evaluation of SE vaccines, the effect of live SE vaccine was understood as competitive elimination (Barrow PA et al. 1991, Hassan JO et al. 1997, Nasser TJ et al. 1994, and Parker C 2001), but many questions remained regarding inactivated SE vaccine (Davies R et al. 2003, Gast RK et al. 1992, Okamura M et al. 2007). The question concerning efficacy was whether the vaccine can eliminate SE which colonizes the digestive tract and reproductive organs even after elevating resistance in blood and parenchymal organs by SE antigen inoculation. For example, 100% elimination of gastrointestinal SE could not be achieved immediately after challenge in the chickens administrated with an inactivated SE vaccine, and also orally ingested bacteria proliferated in the gastrointestine. Moreover, concerning the mode of infection of pullorum disease (*Salmonella enterica* serovar pullorum infection) in chickens as a model (Gwatkin R. 1948, Shivaprasad HL. 2000), SE infection normally manifests no clinical symptoms and natural resistance levels rise in chickens aged 3 weeks or older, at which time inactivated SE vaccine becomes administrable, and SE colonization in the gastrointestine is very limited. Accordingly, it was considered that inactivated SE vaccine is unnecessary for chickens aged 3 weeks or older because of enhanced resistance to SE infection and the effect of inactivated SE vaccine is not useful. However, considering that the ultimate objective of inactivated SE vaccine administration to chicken flocks is to reduce SE-contaminated chicken egg production, we performed studies assuming that SE contamination of chicken eggs can be prevented by employing inactivated SE vaccine through a mechanism different from those of vaccines preventing chicken diseases, and epidemiologically clarified that inactivated SE vaccine administration in layer farms reduced the number of SE-contaminated chicken eggs to 1/260 and the isolation frequency to 1/10 as shown in Table 1 (Summary of field study on SE isolation incidence and bacterial No. in the presence and absence of SE bacterin application in four layer farms. [a]Toyota Hanatani Y et al. 2009). We also confirmed a high epidemiological risk-reducing effect of inactivated SE vaccine administration to flocks. Furthermore, we investigated the active component of inactivated SE vaccine and identified that the important activity is located at the flagellar g.m. antigen site (SEp 9) and the other components do not induce potent specific antibody production.

Incidence/Percentage	Vaccination status with inactivated SE vaccine	Result	bacterial No. or Mean (MPN/100mL) ±SE a)
Bacterial No.	Yes	<2 to <8	2.5±0.1**
	No	<2 to >1,600	674±4.1
Incidence	Yes	14/571	2.45%**
	No	10/40	25%

**$p<0.01$ a) Mean ± Standard error

Table 1. Summary of field study on SE isolation incidence and bacterial No. in the presence and absence of SE bacterin application in four layer farms. ([a]Toyota-Hanatani Y 2009. et al.) This table summarizes SE detection using more than 20 kg of liquid egg in four laying houses where inactivated SE-vaccinated and -unvaccinated chickens were housed for three years. As shown in this table, SE was isolated at up to 8 MPN (Most Probable Number) per 100 ml from inactivated SE-vaccinated chickens and over 1,600 per 100 ml from some non-vaccinated flocks. The inactivated SE vaccine significantly reduced public health risks. Material and Method; Four layer farms were monitored using over 20 kg liquid eggs for 3 years according to the method described in Fig. 1. In this duration, the vaccination status with SE bacterin was mixed (Vaccinated and un-vaccinated flocks were there in each farms).

In previous studies on the mode of SE infection in flocks, the mode of infection of pullorum disease (PD) bacteria in chickens (Table 2. Mode of SE infection in chickens and chicken flocks) was referred to and investigated as a model in the SE vaccine development. However, a recent study and our survey results suggested that SE contamination/infection of chickens disseminates through mouse-mediated transmission between hen houses (Davies RH et al. 1995, Henzler DJ et al. 1992), and not by vertical infection as in PD infection model (Yamane Y et al. 2000). Unlike SE infection of mice and humans (Guiney DG et al. 1995), SE infection of individual chickens occurs as opportunistic infection excluding that immediately after hatching ([a),b)]Bohez L et al. 2007, and 2008, Dhillion AS et al. 2001, and Roy TP et al. 2001), and the infection is not systemic and manifests no symptoms in infected chickens. Based on previous study results reported, it is suggestive that SE ingested by a chicken evades the chicken's immune system by changing substances expressed on its surface, which may be the essence of Salmonella infection

Mode of infection	Concept of mode	Examples and explanations
1. Vertical transmission	In chicks SE Infection occurred from infected parent chickens via embryonated eggs. PEQAP research data said this mode infection might be positioned at less than 5 % in all the infection cases in chicken flocks.	Infection mode of pullorum disease(PD) in chickens (PD is asymptomatic in adult chickens, but is highly lethal in chicks.) Thus, an antibody test was conducted in adult chickens to successfully eradicate positive ones.
2. Horizontal infection	Between chicken flocks (This route is more common according to the PEQAP survey)	Common route (SE may be transmitted by mice.)

Table 2. Mode of SE infection in chickens and chicken flocks. (Summarized by Ohta H)At first, many poultry farmers considered that SE was vertically transmitted. This is because both Salmonella pullorum and SE carry O9, O12, and O1 (belong to Salmonella D group and have the same antigenicity except for the presence of flagellar antigen) as bacterial antigens. Thus, many poultry farmers had tried that they could take measures by conducting a SE contamination (antibody) survey in breeding chickens to remove positive chickens. However, the PEQAP and our surveys demonstrated that SE infection was transmitted among laying flocks and that contamination was limited in growing and breeding farms. Thus, we considered that the inactivated SE vaccine might be effective.

2. Historical changes in the concept of 'food safety' with the recent emergence of SE food poisoning

2.1 History of occurrence of SE-contaminated chicken egg-induced SE food poisoning

Outbreaks of SE food poisoning, not previously noted, frequently occurred worldwide after 1980, mainly in Western countries, and became a social problem (Davison S et al. 2003, Stevens A et al. 1989). Before 1980, Salmonella food poisoning was mainly caused by *Salmonella enterica* serovar Typhimurium (ST), and so this species was mainly studied. Countermeasures against ST-induced food poisoning were taken to avoid hygiene problems of cooking facilities in many cases, and actions rarely reached the management of food material production. However, outbreaks of SE-induced food poisoning occurred in the middle of the 1980s in Western countries, and studies and study result-based countermeasures for not only ST but also SE food poisoning became necessary. The most surprising evidence with SE outbreaks for epidemiologists is that the outbreaks are sometimes caused with several number of SE not caused with numerous SE (Foley SL et al. 2008). At the beginning, contaminated food products and materials in SE food poisoning were unclear, and actions were mainly taken involving only cooking facilities, similarly to measures taken for ST. Many cooking facilities tried to eliminate 'inappropriate food materials (food materials detectable by sensory evaluation, such as those which had started decomposition and color change)', and complete hygiene measures were taken at cooking facilities. However, the occurrence of SE food poisoning did not stop. Around the end of the 1980s, researchers started to point out that chicken eggs were very likely to be contaminated by SE [a),b)]CDC 1990, Rodrigue DC et al. 1990, Steven A et al. 1989, Thomas RD 1989). In poultry industry in United Kingdom, a poultry association consisting of eggs producers, feed suppliers, eggs traders, egg-packing sections joined together to establish an egg sanitary standard like as HACCP (Hazard analysis and critical control point) system, so called Red Lion Code. On the other hand, SE-contaminated chicken eggs were vigorously studied, and several tens of thousands of chicken eggs were individually tested for SE contamination, and the incident of SE contamination eggs originated from SE infected poultry flocks also reported to be about 10 folds increasing compared with those of ordinary shell eggs (Humphrey TJ et al. 1994). In the U.S., it has been said that several million chicken eggs were individually tested for SE. These are unbelievable numbers compared to that in the current SE test, and our studies are based on the efforts of researchers at that time, to which I express my respect.

2.2 Why did SE-contaminated chicken eggs become a public health issue?

Since very few chicken eggs on layer farms are contaminated, reportedly, several in 10,000 eggs, SE-contaminated chicken eggs were not recognized as the cause of SE food poisoning earlier, as described above. However, many retroactive surveys suggested that the main cause was chicken egg-mediated SE contamination. On the other hand, layer farm-related test facilities performed SE tests based on the mode of PD infection as the model, and this was mainly performed for breeding flocks and chicks. Nearly 100% of samples from layer breeder farms were negative, and the isolation rate from breeding flocks and just hatched chicks was not high enough to explain the occurrence of SE food poisoning in humans. It was revealed that growing chickens and chicks are rarely contaminated with SE, which was markedly different from the retroactive survey results of food poisoning. Accordingly, chicken egg producers and chicken salmonellosis researchers believed that the frequency of SE-contaminated chicken eggs is very low, SE contamination of eggs on layer farms is not the main cause, and food poisoning occurs due to SE contamination of chicken eggs during distribution or at the consumer level due to inappropriate classic hygiene management. The detection and removal of SE-contaminated chicken eggs by employing sensory tests were considered possible at that time, and some people strongly believed that detection and removal at cooking facilities using the conventional method were possible. However, many retroactive surveys (Fris C et al. 1995, Henzler DJ et al. 1994, Kusunoki J et al. 1996) revealed that SE contamination was present in eggs that appeared normal as well as on layer farms, which led to recognizing that measures to reduce SE-contaminated chicken eggs are necessary and the safety control of food products and materials should be facilitated by producers, distributors, and consumers in unity. However, considerable time was necessary to spread the necessity of taking actions to reduce SE-contaminated chicken egg production to people related to chicken egg production.

1. Characteristics of farms with frequent SE contamination
1. Large scale (many chickens)
2. Many mice
3. Multiage flocks
4. Many poultry houses
5. EP center established side by side

2. Age (days) when SE infection occurs more commonly
1. After transfer to egg collection poultry farm and during laying peak (170-200 days old)
2. After laying peak
3. Newborn to 10 days old
4. 10 days old to transfer to egg collection poultry farm (in rare cases)

*Most SE infections occur in 1) and 2).

Table 3. Characteristics of SE infection in layer flocks (Biomune's seminar on SE vaccine in Japan in 1996, based on PEQAP research data)

In the northeastern U.S., severe SE contamination occurred in the latter half of the 1980s. Thus, egg consumption markedly decreased because of harmful rumors among consumers. Poultry farm associations, Pennsylvania government, U.S. federal government, etc. established the Pennsylvania Egg Quality Assurance Program (PEQAP) to survey SE contamination in egg-collecting farms in Pennsylvania. In addition, SE contamination rates of eggs and measures, such as inactivated SE vaccine, were evaluated. As shown in this table, SE infection spread in egg-collecting farms, in which mice played an important role. In addition, the inactivated SE vaccine reduced SE-contaminated eggs, but did not effectively eradicate them.

In the 1980s to early 1990s, chicken eggs were reported to be the main cause of SE food poisoning by various media, resulting in a marked reduction of chicken egg consumption. The SE issue put chicken eggs in a disadvantageous sales situation: egg consumption decreased not due to the reports of scientists, but because consumers and egg sellers (supermarkets) considered that eggs were produced by large layer farms performing no hygiene control. Layer farms in North East America, and England and Wales specified as SE-contaminated regions instantaneously lost more than 50% of chicken egg consumption, so-called 'damage caused by harmful rumors' (Davison S et al. 2003). The layer farm industry and instructing administrative agency were surprised and started joint studies involving layer farms, universities, and related administrative agencies. Based on the study results, the administrative agency adopted various measures for chicken egg production and distribution including legal action. For example, in response to this situation, Pennsylvania State in the U.S. vigorously surveyed the actual state of SE contamination on layer farms, investigated the mode of SE contamination, and evaluated various countermeasures. The survey results of the Pennsylvania Project (PEQAP) concerning the mode of SE infection of chickens (Hogue A et al. 1997) revealed that no SE infection was observed in chickens in the growing period, and most cases of infection occurred after chicks were introduced into layer hen houses (Table 3. Characteristics of SE infection in layer flocks unpublished data). The point emphasized was the presence of mice playing an important role in SE transmission between hen houses as above mentioned. Based on these epidemiological study findings, the U.S. government took administrative action by establishing a law which allows selling only sterilized egg liquid prepared from eggs collected from SE-contaminated hen houses (Table 4 US-FDA egg safety rule; US Federal Register 2009. 74: 33030-330101.). In Japan, labeling chicken eggs sold in packages with a date (laid, packaged, or sell-by date) is required. Administrative actions have been taken against SE food poisoning in many countries. In which "Red Lion Code" is involved. The history of the recent emergence of SE food poisoning emphasized the necessity of analyzing the cause of food poisoning in the processes of chicken egg production through distribution and consumption (Schroeder CM et al. 2006). For analysis and the control of health damage risks of not only chicken eggs but also all food products, identification and evaluation of possible risks in each step of production, distribution, and consumption and investigation of countermeasures while considering the cost-effectiveness have been established as " risk analysis concept" and applied to the problem of SE-contaminated chicken eggs as one case. However, no basic concept for the control of bacterial food poisoning has been established, and many epidemiological studies are still necessary.

Testing or Procedure	FDA
Chicks	NPIP SE Clean breeders
Pullet testing	14 to 16 weeks
Requirement for pullet + manure	Egg testing of 4 sets of 1000 eggs at 2 week intervals
Layer testing	40w and 4-6 weeks after molt completion
Egg testing if manure positive	1000 eggs at 2 week intervals, 4 submissions
Diversion to pasteurization required for egg+ flocks	Yes
Return to shell market allowed	Yes after a completed set of 4 submissions of 1000 eggs every 2 weeks
Egg testing after initial egg test set	None if negative first set; once a month if were previously egg positive
C&D of manure or egg + houses	Wet or dry cleaning
Vaccination required	None
Biosecurity plan	Required
Rodent Control Plan and Records	Required
Fly Control Plan and Records	Required

(personal information from Dr. Lozano F)

Table 4. US-FDA egg safety rule (Established by USDA, 2009)

2.3 History of SE vaccine

Live ST vaccine was used as SE control in large-scale state layer farms in Former Eastern Europe before the reunification of East and West Germany as described below. The safety and efficacy of this live ST vaccine for SE control were investigated, and several preparations are still used now.

Regarding inactivated SE vaccine, I would like to introduce the history of its first appearance in the world. The in-house vaccine system was established in the U.S. in the 1980s. In this system, farms which isolated the pathogen were allowed to use an inactivated vaccine for the infectious disease not included in highly pathogenic infectious diseases of animals, such as legal infectious diseases, and approval was granted to in-houses vaccine manufacturers. An in-house vaccine manufacturer produced a vaccine using an SE strain isolated from a layer farm, and the vaccine reduced the SE isolation frequency on the farm. This was the first preparation of inactivated SE vaccine. The world's first state approval was granted for a vaccine which showed efficacy in the field, not prepared through establishing an evaluation method in a laboratory and then confirming the efficacy in the field. Subsequently developed inactivated SE vaccines were produced following the first inactivated SE vaccine as the standard.

3. Discussion on the usefulness of SE vaccine administration to chickens

3.1 Situation at the time of early approval of inactivated SE vaccine

The world's first approval of inactivated SE vaccine by the administrative authority was granted to Layermune SE (Biomune Co., Kansas) in the U.S.A. in 1992. Since then, inactivated SE vaccine has been discussed with regard to not only the efficacy but also many other aspects. Discussions have mainly concerned doubt regarding the efficacy, and, secondly, vagueness of the objective of use. Generally, the objective of animal and human vaccines is the prevention of clinical problems of vaccinated animals and humans, but SE infection manifests no clinical symptoms in chickens excluding newborn chicks, causing no economic damage. For newborn chicks, there is no time for vaccination because infection occurs before inactivated SE vaccine exhibits an immunological effect. Accordingly, inactivated SE vaccine is administered to chickens developing no clinical problems, and the objective is only to reduce the public health risk (reduction of SE-infected chicken egg production). Chickens are vaccinated for a disease manifesting no clinical symptoms, but the effect of the vaccine has to be investigated in these chickens. Economically, inactivated SE vaccine has to be administered to individual chickens, requiring considerable human labor and expense for purchasing the vaccine. Since SE infection causes no direct economic damage, vaccine administration to chickens on farms requires the high-level motivation of vaccine users. The first inactivated SE vaccine was a new type of vaccine, i.e., emergence of a high-cost vaccine slightly stressful to vaccinated animals and not preventing disease in the animals white leghorn chickens (Mizumoto N et al. 2004).

We also investigated the efficacy of inactivated SE vaccine employing various challenge tests. In one of the tests, SE was orally challenged 3 or 4 weeks after inactivated SE vaccine administration, and SE was re-isolated from the gastrointestine and parenchymal organs. Concretely, 3-week-old SPF chickens were vaccinated at the normal dose and orally challenged with food poisoning-derived SE at a high bacterial count after 4 weeks (at 7 weeks of age), and the bacteria were re-isolated from the cecum after 1-7 days. SE was isolated from nearly 100% of chickens despite the vaccine having been administered. When the number of challenged bacteria was reduced to a moderate count, the number of isolated bacteria was significantly decreased in many animals in the vaccinated group, but the results were not stable. In chickens subjected to the test at 5 or 7 weeks of age, the isolation frequency after challenge (at 11 weeks of age) was markedly lower in the control non-vaccinated group. Accordingly, a large number of chickens are necessary to perform the challenge test at this age, which is not routinely possible. In this laboratory test, a significant reduction of the intestinal bacterial count was observed, but complete disappearance of the bacteria from the gastrointestine has not been confirmed within a couple of weeks.

3.2 Situation at the time of the initiation of our study

In 1990, we were informed of SE-contaminated chicken egg production on layer farms covered by our veterinary care activity. We administered bacteriostatics and organic acids on large-scale layer farms, hoping to avoid a decline in consumption, which occurred in America and England, but no effect was obtained. Thus, we performed a field epidemiological study of the efficacy of inactivated SE vaccine (Yamane Y et al. 2000). Inactivated SE vaccine was administered to flocks on a large-scale layer farm with apparent SE contamination. Eggs (500 kg) were broken in a liquid egg plant, 1,000 ml was sampled from the liquid egg batch, and 400 ml was subjected to SE isolation. The isolation rate was compared between the vaccinated and non-vaccinated groups. The results are shown in Fig. 1. (Fig. 1 Number of SE isolates and SE isolation frequency of SE-contaminated chickens and inactivated SE-vaccinated chickens in the same poultry house). Furthermore, our study confirmed horizontal infection of 4 industrial poultry farms (Table 5. SE samples monitored and their results with 4 integrated layer companies (1996-1998)). Based on those results, it was considered necessary for the positivity rate on plate agglutination with pullorum disease-diagnostic antigen to be 90% or higher in the SE-inoculated group (0% in the non-inoculated control group), while vaccination of SE-contaminated farms significantly reduced the number of bacteria isolated from liquid egg samples from 500 kg or more of eggs compared to that from non-vaccinated chickens (Table 6 Evaluation criteria for the inactivated SE vaccine (Layermune SE) in field chickens). In addition, the requirement of the number of isolated bacteria from chicken feces was set at 1 CFU or lower per 1 g in the inactivated SE vaccine-treated group. Later, similar results were obtained in the test using 20 kg of eggs (about 320 eggs). We partially demonstrated these established values epidemiologically after more than 10 years ([a]Toyota-Hanatani Y et al 2009).

Vertical section for SE monitor	Sample materials	Result monitored	Memo
Breeder farms	1) Several swabs	No detection	
	2) Manure	at all	
	3) Sera to detect antibody		Using SE cell antigen coated
	4) Workers feces		ELISA
(Hatchery)	1) Swab	No detection	
	2) Worker feces		
Feed mile	Any protein source	No detection	
Growing	Like as breeder	No detection	
Laying	1) Swabs	No detection	
	2) Manure	A few positive	
	3) Dusts	A few positive	
	4) Liquid eggs	Several positive	
	5) Workers feces	No detection	
	6) Water in EP center	Detectable	

Table 5. SE samples monitored and their results with 4 integrated layer companies (1996-1998) (Yamane Y et al. 2000. modified). Our severance studies (Yamane Y et al. 2000) summarizes the results of SE tests conducted in breeding farms, feed mills, and EP centers for three years. As shown in this table, no vertical transmission (infection from laying to adult chickens) occurred. The infection was repeated within a laying poultry houses. Materials and Methods; See Fig. 1.

Salmonella enterica Serovar Enteritidis (SE) Infection in Chickens and Its Public-Health-Risk Control Using an SE Vaccine in Layer Flocks

29

		Case 1			Case 2		
		Age	SE(MPN)	ELISA[c]	Age	SE(MPN)	ELISA
Year1	Jan.	237	○	0 %	181		
		260	○		252	●(>1600)	0 %
				0 %			
		350	○				
				0 %			45 %
	Jul.			20 %	349	●(NT)	
		455	●(NT)	20 %			28 %
							35 %
					447	●(NT)	
		552	●(39)	15 %			20 %
Year2	Jan.			35 %			25 %
				25 %			10 %
				25 %			
				40 %			
	Jul.				729	Replaced	
		881	△				
		911	▲	35 %			
					218	△	
		937	Replaced		245	(<2)	
					259	△	
Year3	Jan.	168	△		316	△	
		239	△		386	△	
		302	△				
					435	△	

Fig. 1. Number of SE isolates and SE isolation frequency of SE-contaminated chickens and inactivated SE-vaccinated chickens in the same poultry house (Yamane Y et al. 2000) SE isolation in field chickens before and after inoculation of the inactivated SE vaccine (Layermune SE) (Four cases are shown in the reference. Two cases are shown here.) (Filled circles indicate SE isolation). The number indicates the number in ( ) of SE isolates. The detection rates of SE antibodies in unvaccinated chickens by ELISA coated with SE cell antigen was 0-40% for Case 1 and 0-45% for Case 2. To our experiences, the antibody positive rate of 400-500-day-old chickens inoculated with the inactivated SE vaccine (at 300-400 days after vaccination) was about 70-100%. Inaccurate administration of SE bacterin may induce the antibody positive rate of inactivated vaccine to be further decreased. Thus, vaccination and field infection cannot be distinguished at antibody level. The number and frequency of SE isolates decreased in the vaccinated group.

Material &Methods: An industrial layer farm was monitored. SE isolation was done using liquid eggs samples originating from 500 kg of shell eggs. And then most probable number (MPN) per 100 m was determined. For detection of specific antibody in the sera of the flocks, an ELISA coated with SE cell antigen was used.

Test item	Method	Procedures	Criteria
Antibody response	RPA	Twenty chickens were examined at 4 weeks after vaccination.	≥90%: Markedly effective <90%~≥80%: Effective <80%: Non effective
Antibody response	ELISA	Same as above	Same as above
Bacterial isolation	Bacterial isolation	500 kg of eggs are collected from the vaccinated group. The eggs are broken and cultured within 48 hours after collection.	≤10MPN/100mL: Markedly effective (if materials from the unvaccinated group of the same farm showed ≥1,600 MPN/100 mL)

Table 6. Evaluation criteria for the inactivated SE vaccine (Layermune SE) in field chickens (application form for the reexamination of this formulation in Japan, provided by CAF Laboratories)
The effectiveness of the formulation (Layermune SE) in Japan is evaluated based on this table. The formulation was effective in all the 12 chicken groups by an antibody test. However, SE-contaminated farms could not be surveyed by bacterial isolation.

3.3 Risk of misjudging inactivated SE vaccine-treated chickens as SE-infected chickens

We had a problem in handling inactivated SE vaccine in our field facilities: inactivated SE vaccine-treated chickens and SE-infected chickens showed the same serological reaction (Table 7. Production of antibodies against SE bacterial antigens in inactivated SE-vaccinated and -unvaccinated chickens). Inactivated SE vaccine is generally administered at about 80 days of age. In chickens treated with a commercial inactivated SE vaccine, the anti-bacterial cell antigen-antibody positive rate determined using commercial antigen solution for the diagnosis of PD, or SE cell antigen coated ELISA reaches nearly 100% within about 120 days of age and then slowly decreases and reaches 20-60% at about 300 days of age, whereas the positive rate in SE-infected chickens is about 5-70%. We attempted to distinguish SE-infected from inactivated SE vaccine-treated chickens because eggs laid by inactivated SE vaccine-treated chickens are misjudged as those laid by SE-infected chickens, if the 2 chicken groups of SE infected and vaccinated cannot be distinguished. Thus, we investigated specific antibodies present only in chickens with 'inactivated SE vaccine treatment' described below (Fig. 4. Detection of specific antibodies in sera against SE cell antigen and SEp9 on oral SE administration to field white leghorn chickens) (Mizumoto N et al. 2004).

Group	Positive rate (references)	Test methods**	(References)
Inactivated SE vaccine In the laboratory	Vaccination At 30-40 dpv: 95-100% : ≥ 90%	ELISA RPA	
In field	300~400 days old: 70~100%	ELISA	
SE infected group (Field group)	Shipping to slaughterhouse (about 700 days old): 0-15% Induced molting (400-500 days old): 0~45%	ELISA ELISA	(Mizumoto N et al. 2004, Sunagawa H et al. 1997, Yamane Y et al. 2000)

* Age of vaccination: around 80 days old
** ELISA: Indirect method with SE cell antigen coated.
RPA: rapid plate agglutination with diagnostic for pullorum disease antigen.

Table 7. Production of antibodies against SE bacterial antigens in inactivated SE-vaccinated and -unvaccinated chickens (summarized by our research group)
Almost all the 3-week-old or older chickens inoculated with the inactivated SE vaccine were positive at around four weeks by both ELISA (coated with SE cell antigen) and RPA. Subsequently, the positive rate decreased at 250 days or later after inoculation. The positive rate in the ELISA coated with the g.m. antigen of SE was shown above 80% up to about 700 days old. On the other hand, SE-contaminated chickens showed the similar positive rates as those of inactivated SE-vaccinated chickens in ELISA coated with SE bacterial antigen and RPA. Generally, the positive rate of SE-contaminated chickens is lower than that of inactivated SE-vaccinated chickens. However, an antibody test cannot distinguish these 2 groups, because some SE-contaminated chickens show higher positive rate.

3.4 Active component of inactivated SE vaccine (main Fli C antigen: SEp 9)

Using sera from inactivated SE vaccine-treated and SE-infected chickens, we compared the production of antibodies against the SE cell antigen to investigate differences between the sera. A strong reaction with a 53-kDa polypeptide (Fli C) (Namba K et al. 1997) was observed in all serum samples from inactivated SE vaccine-treated chickens, but rare reaction with a specific antigen was noted in SE-infected chicken-derived serum samples (Fig. 2. Western blotting with sera from SE-infected and inactivated SE-vaccinated chickens using formalin-treated SE antigens (surface antigens)). Fli C is considered to be strongly antigenic as inactivated SE vaccine. When the SE-specific polypeptide (g.m. antigen) in Fli C (Van Asten AJ et al. 1995, and Yap LF et al. 2001) was prepared by genetic engineering and reacted with serum from inactivated SE vaccine (Layermune SE)-treated chickens, strong reactivity was noted, but SE-infected chicken-derived serum did not react with g.m. antigen. When the specific antibody reaction was investigated in sera from chickens treated with other vaccines sold in Japan (oil adjuvant vaccine 3 and aluminum hydroxide gel vaccine 1), a specific antibody reaction with g.m. antigen was noted in the serum of oil adjuvant vaccine-treated chickens (Fig. 3.

Production of specific antibodies against commercial inactivated SE vaccines SE cell and SEp9 antigens). In an experiment, the inoculated chickens with SE induces antibody against SE cell antigen but not SEp 9. In field poultry flocks, inactivated SE vaccine administration was confirmed a long period persistency of specific antibody level against SEp 9 until 700 days of age (Fig. 5. Positive rates of g.m.-specific antibodies in the yolks derived from field chickens inoculated with the inactivated SE vaccine).

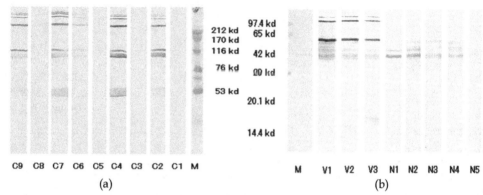

Fig. 2. Western blotting with sera from SE-infected and inactivated SE-vaccinated chickens using formalin-treated SE antigen (surface antigen) (Nakagawa Y et al. reported by Japanese) Figure 2a shows the reactivity of sera from 3-week-old SPF chickens which received oral SE administration (C1~9, M: marker protein), examined by Western blotting (SDS-PAGE) with SE surface antigen. Fig. 2b shows Western blotting with the same antigen using sera from 3-week-old SPF chickens inoculated with the inactivated SE vaccine (Layermune SE) (at 4 weeks after inoculation) (V1~3) or from those from which SE was isolated from naturally-infected-field flocks (N1~5; 710 days old).

Fig. 2a shows that light antibody response against 53 kDa (Fli C of SE) was noted in two chickens (one chicken at 2 weeks) and no band against Fli C (53 kDa polypeptide) was noted in all the nine SE-intraoral inoculated chickens. As shown this figure, one of 2 responded band at week post inoculation (wpi) was continued by 2 wpi but not by 4 wpi. Thus, the responsive antibody was considered to be IgM antibody. In another our report, a 53 kDa band was not detected in 4-week-old SPF chickens and 300-day-old field chickens, which received SE administration, but was detected in molting-induced chickens (Mizumoto N et al. 2004, Piao Z et al. 2007). Thus, the antibody against the 53 kDa polypeptide after SE inoculation is suspected no invasion into the internal organs.

Fig. 2b shows strong bands against the 53 kDa polypeptides and its dimer (98 kDa) in inactivate SE-vaccinated chickens. However, in chickens from which SE could be isolated, a weak band could be detected at around 42 kDa, but no band could be detected at 53 kDa.

Materials and Methods: For antigen preparation, SE was treated with formalin and centrifuged at 2000 g for 20 min. Then the supernatant was further centrifuged at 10,000g for 60 min and the precipitate dissolved in a buffered saline. The antigen was used in this analysis. The sera for SE infected chickens were prepared from the chickens inoculated with SE at the age of 3 weeks, and were weekly bled individually for this study. To the "vaccine sera", SPF chickens were injected with Layermune SE at the age of 3 weeks and bled 4 weeks post injection. The sera were designed as vaccine sera.

Salmonella enterica Serovar Enteritidis (SE) Infection in Chickens and Its Public-Health-Risk Control Using an SE Vaccine in Layer Flocks

33

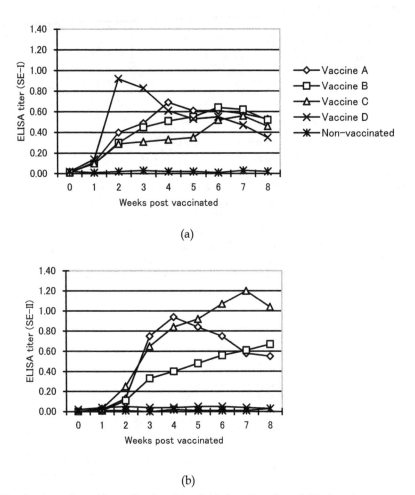

(a)

(b)

Fig. 3. Production of specific antibodies SE cell (deflagellated) and SEp9 antigens (Nakagawa Y et al. Japanese report)
(a; Antibody response to SE cell antigen, b; Antibody response to SEp9)
Four commercial inactivated SE vaccines (Vaccine A to D) were used to inoculate five 3-week-old SPF chickens/group to examine the responsiveness to SE cell antigen and SEp9. Results shown in Fig. 3A and 3B were obtained. No response was noted in unvaccinated chickens. The inactivated SE vaccine responded to SE cell antigen in all the chickens. Notably, the antibody response of the formulation with aluminum gel used as adjuvant rapidly increased and then decreased. On the other hand, the antibody response to SEp9 was specific to each vaccine. However, this may have resulted from vaccine lot-variation. Further studies are needed to make a conclusion. Notably, there was no response to the formulation with aluminum gel used as adjuvant.

When the levels of antibodies against inactivated SE vaccine-induced SE cell antigen and flagella were compared, as shown in Table 8. (Table 8. Detection of SE-specific antibodies by

Age (day)	SE I		SE II	
	Mean E value±2SD	Positive (%)	Mean E value±2SD	Positive (%)
125	0.66 ± 0.36	100	1.37 ± 0.76	100
330	0.26 ± 0.18	60	0.77 ± 0.41	100
550	0.29 ± 0.16	65	0.49 ± 0.34	100
650	0.35 ± 0.23	65	0.47 ± 0.39	95

n=20/group (chicken groups in a farm where molting is induced once at 450 days old)

Table 8. Detection of SE-specific antibodies by ELISA coated with SE cell antigen (SE-I) or the g.m. antigen (SEp 9; SE-II) in inactivated SE-vaccinated chickens (Nakagawa Y et al. Japanese report)
In this survey, 20 chickens were randomly extracted from inactivated SE (Layermune SE)-vaccinated chickens (applied at about 80 days old). The positive rates of specific antibodies against serum SE-I and -II and mean antibody titer (E value) were examined in these flocks. As a result, the positive rates of specific antibodies against bacterial antigen (SE-I antigen) were 100% in 125-day-old chickens and 60% in 330-day-old chickens. Subsequently, these positive rates remained at the same levels. The positive rate of specific antibodies against the g.m. antigen (SEp9) gradually decreased, but remained at a high level of positive ratio.

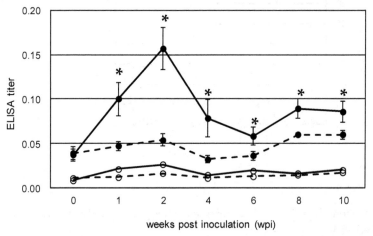

weeks post inoculation (wpi)

Fig. 4. Detection of specific antibodies in sera against SE cell antigen and SEp9 on oral SE administration to field white leghorn chickens (Mizumoto N et al. 2004)
SE was orally administered to 500-day-old laying chickens. Specific antibodies against SE bacterial antigen were produced in the blood. However, no specific antibody against SEp9 was produced. Specific antibodies were similarly detected in yolks. (The dashed line indicates an antibody level against SEp9, and the solid line indicates an antibody level against an SE cell antigen.) The symbol of closed circles means the antibody level in sera obtained from inoculated chickens. The open circle means the ones from not-inoculated chickens. The yolk antibody responses obtained from same birds were shown similar pattern as this figure.

ELISA coated with SE cell antigen (SE-I) or the g.m. antigen (SEp 9; SE-II) in inactivated SE-vaccinated chickens), the anti-CE cell antigen antibody level was high at 120 days of age about 50 days after vaccination, the antibody positivity rate was 50-60% at 300 days of age

(220 days after vaccination), and the rate was retained thereafter. In contrast, g.m. antigen
(SEp 9)-antibody level was maintained at a high level until 700 days of age (about 620 days
after vaccination). An experimental inoculation with SE in SPF chickens showed similar
response (Fig. 4. Detection of specific antibodies in sera against SE cell antigen and SEp9 on
oral SE administration to field white leghorn chickens). This tendency of the presence of
specific antibody in egg yolk was observed (date not shown).

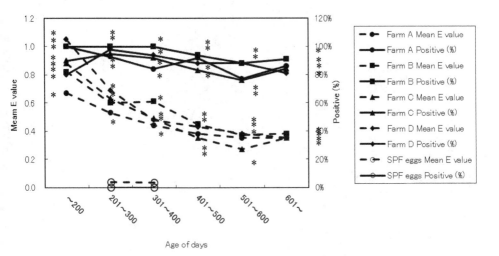

Fig. 5. Positive rates of g.m.-specific antibodies in the yolks derived from field chickens
inoculated with the inactivated SE vaccine (Publishing elsewhere by Nakagawa Y et al.)
In this study, the inactivated SE vaccine (Layermune SE) was used to inoculate about 80-
day-old chickens in six farms. Ten eggs were randomly collected once in two months from
150-700-day-old chickens of each farm. Mean antibody titers (positive: ≥ 0.1 E values)
against the g.m. antigen (SEp9) in yolks were determined. These chickens were giving an
induced molting for about 40 days after day 450. During this period, eggs were not sampled.
The determination with specific antibody to g.m. antigen was done according to the method
described by Mizumoto N et al. 2004.
The mean positive rate of the farms was about 88%. The positive rates were above 80% in all
the farms. Thus, about 700-day-old chickens carried antibodies against SEp9. Antibodies
against SEp9 markedly decreased the number of SE isolates in the gastrointestinal tract). In
addition, antibodies against SEp9 inhibited SE isolation from eggs in the report. Proper
vaccination prevented SE infection for a long time.
Thus, specific antibodies remained in chickens inoculated with the inactivated SE vaccine,
even after molting was induced once, as examined by SEp9-coated ELISA. The specific
antibodies could be detected also in yolks.

3.5 Immunogenicity of SEp 9

A high specific antibody production level was noted in antibodies against a flagellar
component, Fli C, in inactivated SE vaccine-treated chickens, as described above. The SE-

specific region in Fli C is g.m. antigen (SEp 9), and the antigen was assumed to be effective as the antigenic site of inactivated SE vaccine (Toyota-Hanatani Y et al. 2008, and [b)]Toyota-Hanatani Y et al. 2009), for which we prepared SEp 9 antigen by genetic engineering and investigated the efficacy of SEp 9 vaccine. Since no international method (challenge test model) has been established for efficacy evaluation of inactivated SE vaccine, we analyzed tissue reactions at the vaccine administration site in vaccinated chickens.

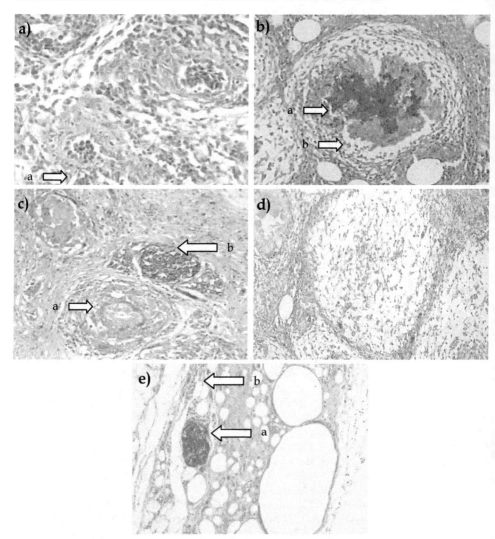

Fig. 6. Histological reactions at the inoculation with f the inactivated SE vaccine or the g.m site of Fli C ([b)]Toyota-Hanatani H et al. 2008)
We investigated a kinetic of histological reactions at the inoculation site of commercial inactivated SE vaccine or SEp 9 antigen. In the inoculation site (7a) at one week post vaccination (wpv), many histocytes were infiltrating, and hyperplastic connective issues are

shown (arrow a). However, tissue images, such as oil cyst, were not observed. In (7b) at 2wpv, necrosis (arrow a), surrounded by granulomatous structures (arrow b), was observed in the middle of inflammatory response. Polynuclear cells appeared in some granulomatous structures. Oil cyst was also observed. These images indicate that the antigen and oil ingredients were actively excluded from the vaccine, suggesting the establishment of specific immunity. At 4 wpv (7c), severe necrosis at 2 weeks became smaller, and the inflammatory response resolved (arrow a). In addition, peripheral lymphoid node structures (arrow b) appeared near the disappearing necrosis, suggesting active antibody production. At 6 wpv (7d), hyperplastic connective tissues also disappeared. Of the tissue reactions in the vaccination site, the characteristic responses during specific immune reaction are the emergence of polynuclear, which surrounded the granulomatous structure, and peripheral lymphoid node like structure. Thus, the inoculation site of SEp9 antigen was histologically examined at four weeks. As shown in Figure 7e, a lymphoid node like structure (arrow a) and a small number of polynuclear cells (arrow b) appeared in the SEp9 inoculation site. Thus, we concluded that SEp9 could induce specific immunity in chickens.
Materials and Methods; A commercial inactivated SE vaccine was injected and weekly taken tissue sample at the injected site, and then fixed and stained as usual (HE staining, X50).

The general time course of histological changes at the inoculation site with inactivated SE vaccine (oil-adjuvant-type) is shown in Fig. 6 (Histological reactions at the inoculation with f the inactivated SE vaccine or the g.m site of Fli C); nonspecific inflammation characterized by marked monocyte infiltration was noted after 1 week, and perivascular granulomatous changes were noted at 2 weeks including the appearance of multinucleated giant cells. At 3 weeks after vaccination, lymphocyte clustering showing a lymph node-like structure, considered to be an antibody production site, was noted. These reactions then slowly disappeared. In granulomatous changes accompanied by multinucleated giant cell infiltration observed after 2 weeks, cellular reactions of delayed hypersensitivity were noted (Table10. Characteristics of histological lesions at the inoculation site in the chicken applied with commercial SE vaccine (4wpi)). The tissue reactions at the SEp 9-administered site were similar to those induced by commercial inactivated SE vaccine, confirming anti-SEp 9-specific antibody production (Table 11 Production of specific antibodies in chickens inoculated with the inactivated SE vaccine or the g.m. site of Fli C).

When SEp 9-treated and non-treated chickens were orally challenged with SE, gastrointestinal SE was significantly decreased in the SEp 9-treated group compared to that in the non-vaccinated group, and the number of isolated bacteria was decreased similarly to that in the commercial inactivated SE vaccine-treated group (Fig. 7. Challenge test in chickens inoculated with the inactivated SE vaccine or the g.m site of Fli C). Although it is not clarified why the specific immunity induced by SEp 9 injection in chickens is able to reduce SE colonization in gastrointestinal organs, we have suspected that the induced immunity may affect SE yielding lower colonization ability SE. For example, the amount of a fibrin molecular, 21 kDa polypeptide, might be reduced on surface resulting from the induced specific immunity without SE-proliferation reduction. This is because the isolation level at 1 week post challenge in Fig. 7 (Challenge test in chickens inoculated with the inactivated SE vaccine or the g.m site of Fli C) does not show different bacterial level between SEp 9 injection and non-injection groups, even though statistical difference is observed. To this point, we will attempt to further clarify the mechanisms of lower SE-colonization in SEp 9-injected birds.

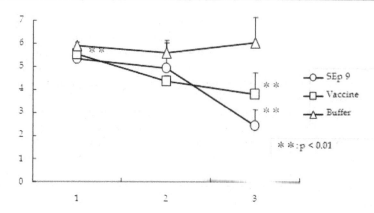

weeks post challenge (wpc)

Fig. 7. Challenge test in chickens inoculated with the inactivated SE vaccine or the g.m site of Fli C (Toyota-Hanatani Y et al. 2009)

SEp9 induced specific immunity. Subsequently, a challenge test was conducted in SEp9-inoculated chickens. The results are shown in this figure. When buffer alone was used for inoculation, the number of SE isolates did not decrease, but remained constant. The number of SE isolates decreased in chickens, inoculated with a commercial inactivated SE vaccine or SEp9, with aging.

Material and Methods; a buffered saline, SEp 9 and Layermune SE were twice-injected with mixture with an oil adjuvant, respectively. Four weeks post injection from final application, those chickens were orally challenged with SE Y 24 strain, and SE isolation was performed from intestinal samples.

Type of Bird	Type of Vaccine	Vaccination Age	Route of Administration	Program Advantages
Breeders	Live	1 day old 7 Weeks old	Coarse Spray Drinking Water	Broad Protection Selective Competitive Exclusion
	Killed	12-14 Weeks of Age 18- 20 Weeks of Age	Subcutaneous Subcutaneous	Strong Maternal Immunity
Layers	Live	1 day old 7 Weeks old	Coarse Spray Drinking Water	Broad Protection Selective Competitive Exclusion
	Killed	10-12 Weeks of Age	Subcutaneous	Strong Immunity
Broilers	Live	1 Day old Coarse Spray or Drinking Water	Coarse Spray or Drinking Water	Strong Immunity

Table 9. Recommended Salmonella vaccination programs in poultry.

Indicator No.	Activation result of	Characteristics in histological observations	Immunological properties
1	Cellular immunity	Granular formation (lumps) with epithelioid cells	Type 4 hypersensitivity (Pellertier M et al. 1984. Uthoaisangssok S et al. 2002)
2	Humoral immunity	Perivascular accumulation with lymphocytes	Activated B-lymphocytes
3	Non-specific immunity	Hyperplastic connective tissue, infiltration of non-specific immune cells	Early or late non-specific immune reaction

Table 10. Characteristics of histological lesions at the inoculation site in the chicken applied with commercial SE vaccine (4wpi) (supplementary data by Toyota-Hanatani Y et al. 2008). This table shows three categories of reactions characteristic of tissue images on inoculation of a commercial inactivated SE vaccine: (1) cellular immunity, (2) hormonal immunity, and (3) nonspecific reaction. All the above reactions were observed in SEp9-inoculated chickens at 4 weeks. Supplementary data from other studies by Toyota-Hanatani Y et al. were also discussed in this table.

Granulomatous reaction, observed at 2-4 weeks after inoculation of the inactivated SE vaccine, was considered to be the same tissue reaction as tuberculin reaction. We considered it to be cellular type IV (delayed) hypersensitive reaction. This might be a process of developing cellular immunity.

Immunizing antigen	Tested chickens	Production of antibodies against g.m. (SEp 9)	
		Antibody positive conversion	Mean value in ELISA
g.m.(SEp 9) antigen	4	4	0.84
De-flagellated SE antigen	4	0	0.00
Buffer	4	0	0.00
Inactivated SE vaccine	4	4	0.83

Table 11. Production of specific antibodies in chickens inoculated with the inactivated SE vaccine or the g.m. site of Fli C (supplementary data by Toyota-Hanatani Y et al. 2008) This table shows specific humoral immunity induced by the g.m. antigen site (SEp9). As shown in the table, specific antibodies were produced when SEp9 was used to inoculate chickens with adjuvant. However, no specific antibodies against SEp9 were produced when SE cell antigen was used. Importantly, the g.m. antigen site has high immune induction capacity in chickens because a small amount of antigen (100 µg/bird; about 30 µg/bird of not involving GST) induces specific immunity.

In another study where SEp9 in buffer was used to inoculate chickens (un-published data), specific antibodies were produced. Thus, specific immunity can be induced even without adjuvant.

Materials and Method; See Fig. 7

3.6 The details of attenuated live *Salmonella* vaccines for poultry

The first live *Salmonella* vaccine for poultry was a *Salmonella enterica* Serovar Gallinarum (SG) developed in the early 1950's (Williams SH.et al. 1956). This attenuated SG rough strain called 9R has been used in many countries around the world for the control of fowl typhoid. However, interference with official *Salmonella* control and eradication programs using serological methods has limited the wider use of this attenuated strain in addition to scattered field reports of excessive attenuation and reversion to virulence. The development of paratyphoid live attenuated *Salmonella* vaccines is an advancement and reinforcement to the use of inactivated vaccines for *Salmonella* control programs in the poultry industry. These new attenuated live *Salmonella* vaccines elicit cell-mediated, mucosal and humoral immune responses (Gomez-Duarte. et al. 1999, Roy Curtiss R 3rd et al. 1996, Kulkarni KK et al. 2008, Ashraf S et al. 2011). In addition, new recombinant DNA technology permits the expression in *Salmonella* serovar strains of protective antigens from unrelated bacterial, viral or parasitic pathogens.

There are two common approaches which have been applied in the development of the new paratyphoid live *Salmonella* vaccines. One of them is the genetic manipulation through recombinant technology selecting virulence genes to be deleted in selected *Salmonella* serovars. The other approach is the manipulation of the media used for *Salmonella* propagation resulting in a metabolic drift mutation reducing the activity of essential enzymes and the bacterial metabolic regulatory systems resulting in slower propagation cycles under natural infection conditions and this prolonged generation time cause reduced bacterial multiplication within the host at a significant rate. Consequently, when the genetically or chemically attenuated *Salmonella* strain is administered to the birds, the modified bacteria lives long enough to stimulate an immune response in chickens before to be eliminated within few weeks after administration of the vaccine. Currently, two paratyphoid serovars are commercially available as live attenuated vaccines: ST and SE.

It is considered that the genetic deletion of selected virulent genes induced a more attenuated recombinant *Salmonella* serovar strains compared with the chemically induced metabolic mutants, which still have residual enzymatic activity and more invasivity inducing a stronger immune response.

Epidemiological markers (Specific antibiotic resistance or sensitivity patterns) are included in the development process of these live vaccines to be able to differentiate the new construct or mutant from similar wild bacterial serovars in case of a field combined infection.

The field use of these new live attenuated *Salmonella* vaccines has advantages and precautions to observe when administered to the chickens. The advantages of these live vaccines are: mass administration, different routes of administration (Drinking water, coarse spray), selective competitive exclusion and broader spectrum of immunity. Among the precautions to be observed are: Not compatible with antimicrobials, no water chlorination when administered in the drinking water, careful handling by the operator to protect the worker from self-infection. Different recommendations on the use of the attenuated live *Salmonella* vaccines may be found in the literature to obtain the best protection against field challenge in a specific environment. Short duration of immunity of the live attenuated vaccines may require 2 to 3 applications every 6 to 10 weeks to obtain a more solid protection. The combined administration of live and inactivated *Salmonella* vaccines provides broader and long lasting immunity, especially in breeders to transfer strong maternal immunity to the progeny. (Table 9. Recommended *Salmonella* vaccination programs in poultry).

3.7 SE vaccine in the future

The current live and inactivated SE vaccines have advantages and disadvantages. Live vaccine is readily administrable to newborn chicks, but inactivated SE vaccine cannot be administered before 3 weeks of age. The detail potency mechanisms with live vaccine has not been clarified yet, and concerns over causing public health problems are always present: the possibility of back mutation of the vaccine production strains of SE and ST (such as reversal of pathogenicity) or mutation to a pathogenic strain cannot be completely ruled out, and, accordingly, live vaccine is not applicable for laying chickens as described above. Currently, inactivated SE vaccine is manufactured using the whole cell body containing endotoxin, which may induce stress in chickens, although this is slight.

To overcome these problems, the development of a subunit or vector vaccine comprised of active components of SE is awaited, and many researchers may have started research and development.

4. Marked usefulness of inactivated SE vaccine administration to flocks for reducing the human health risk

4.1 Reduction of SE contamination risk of chicken eggs by inactivated SE vaccine

We have surveyed the reduction of the SE contamination risk of chicken eggs by employing inactivated SE vaccine on field layer farms for a prolonged period. Herein, we report the study results.

Four-year surveys were performed on 4 field layer farms (a total of 2,300,000 chickens maintained in 37 hen-houses). Records of SE isolation from liquid eggs were analyzed. Some chickens in these layer farms were treated with inactivated SE vaccine as a trial before analysis, and all chickens were vaccinated in the 4th year of analysis.

The mean numbers of SE isolated from liquid eggs (MPN/100 mL) in the vaccinated and non-vaccinated groups were 2.5±0.1 and 674.8±162.9, respectively, and the isolation frequencies were 2.45 and 25%, respectively, showing that the isolation frequency was reduced to 1/10 in the vaccinated group. In addition, no SE was isolated after vaccination of all chickens in the 4th year (0 of 257 samples), as described above.

It was clarified that the use of inactivated SE vaccine on layer farms significantly reduced the number of SE isolated from SE-contaminated eggs and the isolation frequency.

4.2 Risk reduction by inactivated SE vaccine on risk analysis

As described above, inactivated SE vaccine decreased the mean number of SE contaminating eggs as a food product to about 1/260 and the isolation frequency to 1/10. These occurred on SE-contaminated farms when vaccinated and non-vaccinated chickens were mixed. When these were simply compared with the number of orally ingested SE and the incidence of patients reported by the [a], [b]WHO and FAO-US, the incidence of SE patients in healthy subjects was estimated to be decreased to 1/100 or lower.

The 4 farms involved in our study on the reduction of SE contamination of liquid eggs by inactivated SE vaccine were large-scale farms maintaining 350,000-950,000 chickens. These were windowless farms and high-level general hygiene control was also performed.

Accordingly, similar surveys should be conducted on floor feeding and loose housing layer farms, and the risk-reducing effect of SE vaccine should be investigated based on the combined results at national and community levels. In previous reports, the frequency of SE isolation from feces was reduced by about 70% in regions which applied live and inactivated SE vaccines individually or in combination [a),b)] WHO FAO-US, 2002). The accumulation of individual epidemiological surveys and studies may lead to the effective control of SE food poisoning.

5. Re-consideration of the mode of SE infection in chickens

5 1 Mode of SE infection on farms and in flocks

Many points regarding the mode of SE infection on layer farms were unclear around 1990. Layer farm veterinarians referred to the mode of infection of PD (vertical infection), considering that SE also infects in this mode, and prepared an SE detection and monitoring system. Briefly, the mode of SE infection was considered as follows: SE infects breeding chickens and the infection transmits to chicks through breeding eggs (eggs raised to chickens). Some chicks die, but latent infection occurs in survivors and these chicks grow and lay SE-contaminated eggs. Accordingly, they considered that the antibody test in breeding chickens and SE test in chicks after hatching are important, and did not attach greater importance to SE tests of grown chickens, especially laying hens. Moreover, they considered that inactivated SE vaccine is ineffective for chicks after hatching, and only bacteriostatics and analogous agents are effective. The Pennsylvania Egg Quality Assurance Project (PEQAP) of the U.S.A. actively performed field SE contamination surveys to investigate this hypothesis, and found several new facts, as described above (Davison S et al. 2003, Henzler DJ et al. 1998), Hogue A et al. 1997, Lin FY et al. 1988, Stevens A et al. 1989). The points particularly attracting attention in the PEQAP report are a very low infection frequency in newborn chicks, although contamination occurred, and the absence of SE contamination in raising houses. However, SE contamination was observed most frequently after transfer to layer hen houses over 180 days of age. Even though new episodes of SE contamination occurred thereafter (after the laying peak), the frequency was very low.

SE sensitivity of chickens is schematically presented based on the study results reported by PEQAP and our experience in Fig. 8 (Age-dependent susceptibility of chickens against SE colonization). Chicks are very sensitive to SE infection immediately after hatching, but the sensitivity rapidly decreases. No clinical symptoms develop over the growth and egg-laying periods, but the sensitivity rises around the initiation of sexual maturation (100-120 days of age). In layer hen houses, the frequency of SE contamination is high, elevating the infection risk of chickens. It is considered that most SE infection of chickens occurs after transfer to layer hen houses (around 115 days of age) over the peak laying period (around 180 days of age). The sensitivity of layer hens slightly decreases thereafter but then slowly rises with aging. SE sensitivity may be enhanced when induced molting is performed during this period, but these chickens are already infected immediately after transfer to layer hen houses. Therefore, the infection rate is not actually elevated by induced molting, although the sensitivity is high. Considering SE sensitivity and SE control of layer flocks and economic damage, chicks infected immediately after hatching may be culled because they develop clinical symptoms. The period after transfer to layer hen houses over the egg-laying peak is the most important for hygienic SE control because chickens are highly sensitive to

SE but infection is unclear. The survey results of PEQAP well reflected this condition. Therefore, how hygiene control is performed during this period (after transfer to layer hen houses over the egg-laying peak) is important, and inactivated SE vaccine can be administered corresponding to this high contamination risk period.

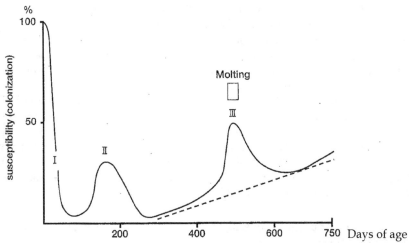

Fig. 8. Age-dependent susceptibility of chickens against SE colonization (Ohta H et al. presented in 2nd Symposium of the Germany-Japan veterinary association 1998)
The susceptibility of chickens to SE infection changed with aging. Oral SE administration killed almost 100% of chicks before feeding. However, the death rate rapidly decreased after feeding. No death was usually noted in 2-3-week-old chickens even after administering 10^9 FFU/bird. However, SE colonized in the intestine only for a short time. However, chickens became more susceptible to SE infection at around 100 days old when the reproductive organ developed. SE infected and colonized in the intestine for a long time in chickens of 50-100% laying age (145-180 days old). Subsequently, chickens gradually became susceptible with aging. Molting chickens were more susceptible to SE infection. More susceptible chickens were not necessarily more vulnerable to SE infection. SE infection risks became higher during the stage II (145-180 days old) in the figure, causing environmental SE contamination condition. Thus, chickens of this age group were more susceptible to SE infection. These facts were taken into consideration in the measures taken in the U.S.

5.2 SE infection of chickens

The epidemiological mode of infection of chickens is described above, but how does it occur in individual chickens? Generally, SE is orally ingested. Regarding experimental SE infection of chickens, Bohez et al. and other study-groups actively investigated pathogenicity in young chickens as above mentioned, and observed that the pathogenicity manifestation mechanism was similar to that in mice and systemic sepsis occurred and resulted in death at a high rate. We also obtained similar results (data not shown). In contrast, pathogenicity was rarely observed and the course was asymptomatic when grown chickens and layer hens were infected. Weakened chickens were observed in very rare cases, but the presence of other factors, such as stress, is generally considered for these cases, and SE infection alone is considered to induce no morbidity. It has been considered most SE strains are not actually pathogenic for chickens. Therefore, it is unclear what roles are played in chickens by the genes, components, and molecules reported to exhibit pathogenicity in mice.

However, unlike other *Salmonella* species, such as ST, SE infection shows high tropism for intestinal and reproductive organ epithelial cells in chickens, and the colonization rate in the chicken intestine is high (Mizumoto N et al. 2005, Okamura M et al. 2007). Regarding tropism, there has been no report on differences in tropism for epithelial cells of SE and other *Salmonella* species in other animal species, but SE shows high species specificity for chickens. When the oviduct surface was contacted and colonized by various *Salmonella* species, the number of colonizing bacteria of SE strains was the highest and the number decreased in the other of *S.* Agona, *S.* Typhimurium, *S.* Heidelberg, *S.* Harder, *S.* infantis, and *S.* Montevideo. The high tropism of SE for chickens is an interesting study subject. For example, if SE-contaminated chicken eggs serve as the main cause of SE food poisoning resulting from the acquisition of species specificity for chickens by many currently isolated SE strains, this property of SE will be a major epidemiological study subject, i.e., it explains the sudden emergence of appearance of chicken egg-mediated SE food poisoning caused by SE contaminating chicken farms in the 1980s.

6. Proposal for food safety based on the history of emergence and decline of SE food poisoning

We selected 2 topics concerning chickens and SE infection in this chapter. One was the usefulness of inactivated SE vaccine administration to chickens to reduce the public health risk. The other was the introduction of some of our studies on SE infection of chickens. In the first topic, the history was described in some detail because a description of the historical background is necessary to understand why we wanted to describe the history of the emergence of chicken egg-induced food poisoning. In the 1980s, the production, distribution, and consumption of food products and materials became global. Safety standards became necessary for mass production, international distribution, and the selling of food products and materials, with which SE food poisoning occurred and rapidly spread in Western countries and then declined. However, this declined incidence has recently tended to slightly re-increase in some countries, suggesting that it is time to review SE control from the basics. Together with the history of overcoming the BSE problem of beef, the history of emergence and control of SE food poisoning contributes to establishing the concept of 'risk analysis of health damage by foods'.

In the control of food poisoning before 1980, hygienic measures were mainly taken in the steps after cooking, but analysis of SE-contaminated chicken eggs led to a new concept of food poisoning control: SE infection of chickens should be prevented although no clinical symptoms develop in chickens excluding chicks immediately after hatching. The problem of food poisoning seems to widely extend over the world. This is of course due to large-scale distribution and consumption of food products and materials, but it may also be due to failure of inheriting food culture in countries throughout the world. Previously, sensory elimination of problematic food products and materials was performed in each home as 'food culture', but this may not have been passed on in modern society in many countries. Studies on food poisoning are required to closely investigate the safety of the globalized production, distribution, and consumption of food products and materials. These worldwide changes in food culture are a background to the emergence and decline of SE-contaminated chicken egg-induced food poisoning.

A large part of the text was also devoted to the usefulness of inactivated SE vaccine in this chapter. The first vaccine approved by the US government does not completely stop SE proliferation in the gastrointestine after SE challenge, and the bacterial count rather increases transiently. The previous concept of vaccine for chicken diseases was the inhibition of clinical symptoms and bacterial proliferation after challenging the pathogen, but SE infection does not induce clinical symptoms in chickens excluding chicks immediately after hatching. In other words, inactivated SE vaccine is administered for asymptomatic SE infection, and stress load of vaccination gives no advantage to farms. The use of inactivated SE vaccine was initiated in response to demands from consumers, to which layer farms had strong resistance, and the usefulness was frequently questioned. In this situation, we investigated the usefulness of inactivated SE vaccine.

Our study demonstrated that inactivated SE vaccine is very useful with regard to the inhibition of SE-contaminated chicken egg production, unlike conventional vaccine for chicken diseases. Although SE temporarily proliferated in the chicken gastrointestine on the SE challenge test, the production of SE-contaminated chicken eggs was markedly inhibited. Specific immunity against flagellar components plays a central role in the inhibition, and, particularly, specific immunity against g.m. antigen is assumed to play a major role. Unfortunately, the mechanism of the effect of flagellar component-specific immunity has not been clarified, and so remains to be investigated.

Reportedly, the current inactivated SE vaccine may induce stress in some cases. The development of vector vaccine with the insertion of flagellar components inducing no stress is underway, and may be realized in the near future.

Some of our study results on SE infectious disease in chickens were introduced in this chapter. SE infection of chickens may be opportunistic infection, unlike infections of mice and humans. However, SE infection of chickens was not regarded as opportunistic infection in previous studies. SE infection of chickens has been investigated employing the mode of PD infection or partially employing the mode of SE infection in mice and humans as a model, but we have been considering that it is appropriate to basically regard *Salmonella*-induced infectious disease as 'opportunistic infection' or less pathogenic 'indigenous bacteria'. Although it causes food poisoning in humans and may result in death, it is very rarely fatal in chickens. For fatal cases, other factors may be the major cause, such as hot conditions in summer. SE infection causes no damage to chickens, but there is no doubt that SE-contaminated chicken eggs cause food poisoning in humans, although the mode of SE infection in chickens cannot be fully explained.

We attempted to describe live SE vaccine. However, I could not draw the efficacy of the live vaccine for the applied flocks in their whole the life.

Responses of SE to stimulation by chickens were confirmed as the lacy phase changed to the colonial phase when the bacteria entered the intestine, but other responses are slightly unclear. In our study, marked colonization (tropism) of the reproductive organs by SE was noted, compared to that by other *Salmonella* species, but enumeration of these facts will not lead to studies in the future. Thus, we selected 2 topics in this field as study subjects (working hypotheses). Various conversions occur in SE and these are important to evade the chicken's immune system and become indigenous. We selected these conversions as one topic. The other is the colonization mechanism of SE on layer farms which should be deduced based on the history. The colonization may have resulted from the facts that flagellar antigens are not expressed in chickens and chickens do not respond to the fimbrial components. These two mechanisms have been well-known for a long time, but it may be essential to analyze the mode of SE infection of chickens. SE infection of chickens is a never-ending study area.

7. Acknowledgement

First of all, we apologize that this chapter mainly describes our field experiences, with little reference to literature. This chapter is comprised of not only studies performed mainly by us but also joint studies with many researchers, and contains fewer descriptions referring to the literature. Many contents of this chapter were those of cooperative studies performed by Osaka Prefectural University and CAF Laboratories. We would like to thank for the persons related to this publishing program, to be given a chance to write this chapter.

We express our respect to Dr. Joan D. Leonard of Biomune Co., U.S.A., who launched the world's first inactivated SE vaccine, for her courage and fine studies, and we deeply appreciate having an opportunity to perform studies using this vaccine. We are also grateful to Dr. Fernando Lozano of Ceva/Biomune Co. for providing us with global information concerning SE vaccine.

We are grateful to the many researchers who performed cooperative studies with us, especially the following persons:

Dr. Shizunobu Igimi, National Institute of Health Sciences Japan
Dr. Eiichiroh Baba, formerly of Osaka Prefectural University
Dr. Kazumi Sasai, Osaka Prefectural University

8. References

Altekruse S, Koehler J, Hickman-Brenner F, Tauxe RV, Ferris K, (1993) A comparison of Salmonella enteritidis phage types from egg-associated outbreaks and implicated laying flocks. *Epidemiol. Infect.* 110: 17-22.

Ashraf S, Kong W, Wang S, Yang WJ, Curtiss R. (2011) Protective cellular responses elicited by vaccination with influenza nucleoprotein delivered by a live recombinant attenuated Salmonella vaccine. *Vaccine.* 29: 3990-4002.

Barrow PA, Lovell MA, Berchieri A. (1991) The use of two live attenuated vaccines to immunize egg-laying hens against Salmonella enteritidis phage type 4. *Avian. pathol.* 20: 681-692.

a)Bonhez L, Dewulf J, Ducatelle R, Pasmans F, Haesebrouck F, Van Immerseel F. (2008). The effect of oral administration of homologous HilA mutant strain on the long-term colonization and transmission of Salmonella enteritidis in brolar chickens. *Vaccine.* 17:372-378.

Bonhez L, Ducatelle R, Pasmans F, Haesebrouck F, Van Immerseel F. (2007) Long-term colonisation inhibition studies to protect broiler against colonization with

Salmonella enteritidis, using Salmonella pathogenicity island 1 and 2 mutants. *Vaccine.* 22: 4235-4243.

b)Bonhez L, Gantois I, Ducatelle R, Pasmans F, Dewulf J. Haesebrouck F, Van Immerseel F.(2008) The Salmonella pathogenicity island 2 regulator ssrA promoter reproductive tract but not intestinal colonization in chickens. *Vet. Microbiol.* 126: 216-224.

Bumann D, Meyer TF, Gomez-Duarte OG,(1999). The attenuated Salmonella vaccine approach for the control of Helicobacter pylori-related diseases. *Vaccine.* 17: 1667-1673

Center for Disease Control (CDC). (1987). Update: Salmonella enteritidis infections in northeastern United State. *MMWR Morb Mortal wkly Rep.* 36: 204-205

a)Center for Disease control (CDC) (1990). Update: Salmonella enteritidis infections and grade A shell eggs-United States 1989. *MMWR Morb Mortal Wkly Rep.* 38: 877-880.

b)Center of Disease Control (CDC). (1990). Update: Salmonella enteritidis infections and shell eggs-United States,1990. *MMWR Morb. Mortal. Wkly. Rep.* 39: 909-912.

Cogan TA, Humphrey JM, (2003). The rise and fall of Salmonella enteritidis in the UK. *J. Appl. Microbiol.* 94: 114s-119s.

Cowden JM, Chisholm D, O'Manhony M, Lynch D, Mawer SL, Spain GE, Ward L, Rowe B. (1989) Two outbreaks of Salmonella enteritidis phage type 4 infection associated with the consumption of fresh shell-egg products. *Epidemiol. Infect.* 103: 47-52.

Curtiss R, Hassan JO. (1996). Nonrecombinant and recombinant avirulent Salmonella vaccines for poultry. *Vet Immunol. Immunopathol.* 54: 365-372.

Davies RH, Wray C. (1995) Mice as carriers of Salmonella enteritidis on persistently infected poultry units. *Vet. Rec.* 137: 337-341.

Davies R, Breslin M. (2003) Effects of vaccination and other preventive methods for Salmonella enteritidis on commercial laying chicken farms. *Vet. Rec.* 153: 673-677.

Davison S, Benson CE, Munro DS, Rankin SC, Ziegler AE, Eckroade RJ. (2003) The role of disinfectant resistance of *Salmonella enterica* serovar enteritidis recurring infections in Pennsylvania egg quality assurance program monitored flocks. *Avian Dis.* 47: 143-148.

Dhillion AS, Shivaprasad HL, Roy TP, Alisantosa B, Schaberg D, Bandli D, Johnson S.(2001) Pathogenicity of environmental origin Salmonellas in specific pathogen free chicks. *Poul.Sci.* 80: 1323-1328.

Fris C, Bos J.(1995) A retrospective case-control study of risk factors associated with *Salmonella enterica* subsp. enterica serovar enteritidis infections on Dutch broiler breeder farms. *Avian Pathol.* 24: 255-272.

Foley SL, Lynne AM. (2008) Food animal-associated Salmonella challenge: pathogenicity and antimicrobial resistance. *J. Anim. Sci.* Apr;86 (14 suppl): E137-187.

Gast RK, Stone HD, Holt PS, Beard CW, (1992). Evaluation of the efficacy of an oil-emulsion bacterin for protecting chickens against Salmonella enteritidis. *Avian Dis.* 36: 992-999.

Guard-Petter J. (1997) Induction of flagellation and a novel agar-penetrating flagellar structure in *Salmonella enterica* grown on solid media: possible consequences for serological identification., *FEMS Microbiol Lett.,* 149: 173-180

Guiney DG, Fang FC, Krause M, Libby S, Buchmeier NA, Fierer J. (1995). Biology and clinical significance of virulence plasmids in Salmonella serovars. *Clin Infect Dis.* Oct;21 Suppl 2:S146-151

Gwatkin R. 1948. Studies in Pullorum Disease: XXI Response to oral infection with Salmonella pullorum in comparable groups of turkeys and chickens.*Can J Comp Med Vet Sci.* 12(2):47-55.

Hassan JO, Curtiss R 3rd. (1997) Efficacy of a live avirulent Salmonella typhimurium vaccine in preventing colonization and invasion of laying hens by Salmonella typhimurium and Salmonella enteritidis. *Avian. Dis.* 41: 783-791.

Henzler DJ, Ebel E, Sanders J, Kradel D, Mason J, (1994) Salmonella enteritidis in eggs from commercial chicken layer flocks implicated in human outbreaks. *Avian Dis.* 38: 37-43.

Henzler DJ and Opitz HM. (1992). The role of mice in the epizootiology of Salmonella enteritidis infection on chicken layer farms. *Avian Dis.* 36: 625-631.

Henzler DJ, Kradel DC, Sischo WM. (1998). Management and environmental risk factors for Salmonella enteritidis contamination of eggs. *Am. J. Vet. Res.* 59: 821 829.

Hogue A, White P, Guard-Petter J, Schlosser W, Gast R, Ebel E, Farrar J, Gomez T, Madden J, Madison M, McNamara AM, Morales R. Parham D, Sparling P, Sutherlin W, Swerdlow D. (1997) Epidemiology and ontrol of eggs-associated Salmonella enteritidis in United State of America. *Rev. Sci. Tech.* 16: 542-553.

Humphrey TJ. (1994) Contamination of egg shell and contents with Salmonella enteritidis: a review. *Int. J. Food Microbiol.* 21:31-40.

Humphrey TJ, Whitehead A, Gawler Ah, Henley A, Rowe B. (1991) Number of Salmonella enteritidis in the contents of naturally contaminated hens' eggs. *Epidemiol. Infect.* 106: 489-496.

Kulkarni RR, Parreira VR, Sharif S, Prescott JF. (2008) Oral immunization of broiler chickens against necrotic enteritis with an attenuated Salmonella vaccine vector expressing Clostridium perfringens antigens. *Vaccine.* 26: 4194-4203.

Kusunoki J, Kai A, Yanagawa Y, Takahashi M, Shingaki M, Obata H, Itho T, Ohta K, Kudoh Y, Nakamura A. (1996) Characterization of Salmonella ser. enteritidis phage type 34 isolated from food poisoning outbreaks in Tokyo by epidemiological markers. *Kansenshogaku Zasshi.* 70: 702-709.

Lin FY, Morris JG Jr, Trump D, Tilghman D, Wood PK, Jackman N, Israel E, Libonati JP. (1988) Investigation of an outbreak of Salmonella enteritidis gastroenteritidis associated with consumption of eggs in a restaurant chain in Maryland. *Am. J. Epidemiol.* 128: 839-844.

Michella SM, Slaugh BT, (2000) Producing and marketing a specialty egg. *Poult. Sci.* 79: 975-976.

Mizumoto N, Sasai K, Tani H, Baba E. (2005) Specific adhesion and invasion of Salmonella enteritidis in the vagina of laying hens. *Vet. Microbiol.* 111: 99-105.

Mizumoto N, Toyota-Hanatani Y, Sasai K, Tani H, Ekawa T, Ohta H, Baba E. (2004) Detection of specific antibodies against deflagellated Salmonella enteritidis and S. enteritidis Fli C-specific 9kDa polypeptide. *Vet. Microbiol.* 99: 113-120.

Namba K, Vonderviszt F. (1997) Molecular architecture of bacterial flagellum. *Q. Rev. Biophys.* 30:1-65.

Nassar TJ, al-Nakhli HM, al-Ogaliy Z H. (1994) Use of live and inactivated Salmonella enteritidis phage type 4 vaccines to immunise laying hens against experimental infection. *Rev. Sci. Tech.* 13:855-867.

Okamura M, Tachizaki H, Kubo T, Kikuchi S, Suzuki A, Takehara K, Nakamura M. (2007) Comparative evaluation of a bivalent killed Salmonella vaccine to prevent egg contamination with *Salmonella enterica* serovars Enteritidis, Typhimurium, and Gallinarum biovar Pullorum, using 4 different challenge models. *Vaccine.* 25: 4837-4844.

Okamura M, Miyamoto T, Kamijima Y, Tani H, Sasai K, Baba E. (2001) Differences in abilities to colonize reproductive organs and to contaminate eggs in intravaginally inoculated hens and in vitro adherences to vaginal explants between Salmonella enteritidis and other Salmonella serovars. *Avian Dis.* 45:962-971.

Piao Z, Toyota-Hanatani Y, Ohta H, Sasai K, Tani H, Baba E. (2007) Effects of *Salmonella enterica* subsp. enterica serovar enteritidis vaccination in layer hens subjected to S. Enteritidis challenge and various feed withdrawal regimens. *Vet Microbiol.* 125:111-119.

Parker C, Asokan K, Guard-Petter J. (2001) Egg contamination by Salmonella serovar enteritidis following vaccination with Delta-aroA Salmonella serovar typhimurium. *FEMS. Microbiol.lett.* 195 :73-78.

Pelletier M, Forget A, Bourassa D, Skamene E. (1984) Histological studies of delayed hypersensitivity reaction to tuberculin in mice. *Infect. Immun.* 46: 873-875.

Rodrigue DC, Tauxe RV, Rowe B. (1990) International increase in Salmonella enteritidis; a new pandemic? *Epidemiol. Infect.* 105: 21-27.

Roy TP, Dhillion AS, Shivaprasad HL, Schaberg DM, Bandli D, Johnson S. (2001) Pathogenicity of different serogroups of avian Salmonella in specific pathogen free chickens. *Avian Dis.* 45: 922-937.

Saeed AM. Ed. (1999). *Salmonella enterica* serovar enteritidis in human and animals. *Iowa State University Press.*

Schroeder CM, Latimer HK, Schlosser WD, Golden NJ, Marks HM, Coleman ME, Hogue AT, Ebel ED, Quiring NM, Kadry AR, Kause J. (2006) Overview and summary of the Food Safety and Inspection Service risk assessment for Salmonella enteritidis in shell eggs, October 2005. *Foodborne Pathog Dis.* 3: 403-412.

Shivaprasad HL. (2000). Fowl typhoid and pullorum disease. *Rev Sci Tech.* 19: 405-24.

Shivaprasad HL, Timoney JF, Molares S, Lucio B, Baker RC. (1990) Pathogenesis of Salmonella enteritidis infection in laying chickens. I. Studies on egg transmission, clinical signs, fecal shedding, and serologic responses. *Avian Dis.* 34: 548-557.

Stevens A, Joseph C, Bruce J, Fenton D, O'Mahony M, Cunningham D, O'Connor B, Rowe B. (1989) A large outbreak of Salmonella enteritis phage type 4 associated with eggs from overseas. *Epidemiol. Infect.* 103: 425-433.

St Louis ME, Morse DL, Potter ME, DeMelfi TM, Guzewich JJ, Tauxe RV, Blake PA. (1988). The emergence of grade A eggs as a major source of Salmonella enteritidis infections. New implications for the control of salmonellosis. *JAMA.* 259:2103-2107.

Sunagawa H, Ikeda T, Takeshi K, Takada T, Tsukamoto K, Fujii M, Kurokawa M, Watabe K, Yamane Y. Ohta H. (1997) A survey of Salmonella enteritidis in spend hens and its relation farming style in Hokkaido-Japan. *Inter. J. Food Microbiol.* 38: 95-102.

Thomas RD. (1989) Grade A eggs as a source of Salmonella enteritidis infections. *JAMA.* 261: 2064-2065.

[a)]Toyota-Hanatani Y, Ekawa T, Ohta H, Igimi S, Hara-kudo Y, Sasai K, Baba E. (2009) Public health assessment of *Salmonella enterica* serovar enteritidis inactivated-vaccine treatment in layer flocks. *Appl. Environ. Microbiol* . 75: 1005-1010.

[b]Toyota-Hanatani Y, Kyoumoto Y, Baba E, Ekawa T, Ohta H, Tani H, Sasai K (2009) Importance of subunit vaccine antigen of major Fli C antigenic site of Salmonella enteritidis II: a challenge trial. *Vaccine*. 27: 1680-1684.

Toyota-Hanatani Y, Inoue M, Ekawa T, Ohta H, Igimi S, Baba E. (2008) Importance of the major Fli C antigenic site of Salmonella enteritidis as a subunit vaccine antigen. *Vaccine*. 26: 4135-4137.

Uthaisangsook S, Day NK, Bahna SL, Good RA, Haraguchi S. (2002). Innate immunity and its role against infections. *Ann. Allergy. Asthma. Immunol* 88: 253-264.

Van ASTEN AJ, Zwaagstra KA, Baay MF, Kusters JG, Huis in't Velt JH, Van der Zeijst BA. (1995) Identification of the domain which determines the g,m serotype pf the flagellin of Salmonella enteritidis. *J. Bacteriol*. 177:1610-1613.

Williams SH. (1956) The use of live vaccines in experimental Salmonella gallinarum infection in chickens with observation on their interference effect. *J. Hyg*. 54:419-432.

[a]World Health Organization (WHO), Food and Agriculture Organization of the United States (FAO-US), (2002) Microbiological risk assessment series 1. Risk assessments of *Salmonella* in eggs and broiler chickens.
http://www.fao.org/docrep/fao/005/y4392e/y4392e00.pdf. 28-29.

[b]World Health Organization (WHO), Food and Agriculture Organization of the United States (FAO-US), (2002) Microbiological risk assessment series 2. Risk assessments of *Salmonella* in eggs and broiler chickens.
http://www.fao.org/docrep/fao/005/y4392e/y4392e00.pdf. 80-89.

Yamane Y, Leonard DJ, Kobatake R, Awamura N, Toyota Y, Ohta H, Otsuki K, Inoue T. (2000) A case study on Salmonella enteritidis (SE) origin at three egg-laying farms and its control with an S. enteritidis bacterin. *Avian Dis*. 44: 519-526.

Yap LF, Low S, Liu W, Loh H, Teo TP, Kwang J. (2001) Detection and screening of Salmonella enteritidis-infected chickens with recombinant flagellin. *Avian Dis*. 45:410-415.

3

Use of Isolation and Antibody Detection for *Salmonella* Assessment

Marina Štukelj[1], Vojka Bole-Hribovšek[2],
Jasna Mićunović[2] and Zdravko Valenčak[1]
[1]University of Ljubljana, Veterinary faculty
Institute for the Health Care of Pigs, Ljubljana
[2]University of Ljubljana, Veterinary faculty
Institute for Microbiology and Parasitology, Ljubljana
Slovenia

1. Introduction

1.1 *Salmonella* in pigs

Salmonella infections of swine are of concern for two major reasons. The first is the clinical disease (salmonellosis) in swine that may result, and the second is that swine can be infected with a broad range of *Salmonella* serovars that can be a source of contamination of pork products. The genus *Salmonella* is morphologically and biochemically homogeneous group of Gram-negative, motile, non-spore-forming, facultative anaerobic bacilli with peritrichous flagella (Griffith et al., 2006). According to their biochemical characteristics it is divided in two species *Salmonella enterica* and *Salmonella bongori*. *Salmonella enterica* is further divided in six subspecies. Regarding their antigenic structure of somatic (O), flagellar (H) and capsular (Vi) antigens they are divided in serovars. Traditionally the serovars of *subspecies enterica*, which account for more than 99.5% of isolated *Salmonella* strains, have names, while all the others are named by their antigenic formula only (Grimont and Weill, 2007). Final differentiation within serovars is carried out by phage typing, plasmid profiling, restriction endonuclease analysis and resistance patterns. Serovars Typhimurium, Derby, Saintpaul, Infantis, Heidelberg, Typhisuis and Choleraesuis may all occur in pigs (Taylor, 2006).

The reservoir for *Salmonellae* is the intestinal tract of warm-blooded and cold-blooded animals. *Salmonellae* are hardy and ubiquitous bacteria that multiply at 7-47° C; survive freezing and desiccation well; and persist for weeks, months, or even years in suitable organic substrates. The bacteria are readily inactivated by heat and sunlight as well as by common phenolic, chlorine, and iodine disinfectants. Ability to survive in the environment, as well as prolonged carrier states in innumerable hosts ensures the widespread distribution of this genus worldwide (Griffith et al., 2006).

Pigs usually get infected through oral intake of the organism. After infection, animals can become carriers in the tonsils, the intestines and gut-associated lymphoid tissue (Wood et al., 1989; Fedorka-Cray et al., 2000). Most of the time, carriers are not excreting the bacteria but under stressful conditions, re-shedding may occur. In this way, carriers are permanent potential source of infection for other animals and humans. Stress factors can occur during the fattening period, but also prior to slaughter, for instance during transport to the slaughterhouse or during the stay in the lairage (Seidler et al., 2001; Rostagno et al., 2010). Along the slaughter line, several steps can be critical for *Salmonella* contamination, removal of the pluck set and meat inspection procedures (De Busser et al. 2011). During these steps, the carcass can be contaminated with faeces and bacteria can be spread all over the carcass and to subsequent carcass.

After tracing the *Salmonella* data from the colon content isolated in the slaughterhouse back to the herd level, it was estimated that 40% of the herds were *Salmonella* positive at the moment of slaughter. A high level of herd contamination was also found in the Netherlands with 23% of the herds *Salmonella* positive sampled on the farm (van der Wolf et al., 1999) and in the UK with 63% positive farms (Davies et al., 1999). For interpretation of our data, it has to be kept in mind that the pigs with positive colon content and/or mesenteric lymph nodes in the slaughterhouse could have been infected on the farm and during transport or during the waiting period in the lairage before slaughtering. There are indeed indications that the contamination could already be detected in the faeces and the mesenteric lymph nodes as early as 3 h after infection (Fedorka-Cray et al., 1994). Especially the lairage and the high contamination level of the slaughterhouse environment are probably the major source for *Salmonella* infections prior to slaughter (Hurd et al., 2001; Swanenburg et al., 2001). Hurd et al. (2002) demonstrated that rapid infection during transport, and particularly during holding, is a major reason for increased *Salmonella* prevalence in swine: a sevenfold higher *Salmonella* isolation rate and twice as many different serovars were observed from pigs necropsied at the abattoir than from those necropsied on the farm.

There is currently an explosion of investigational activity related to issue of food safety, including *Salmonella* contamination of variety of foods. Salmonellosis is considered to be one of the most common food-borne illnesses in humans. There has been an increased public awareness of microbiological hazards of food and improved monitoring. Over the recent years, salmonellosis has been the second most commonly reported zoonoses in the European Union, accounting for 151,995 recorded human cases in 2007 (EFSA, 2009b) and 131,468 in 2008 (EFSA, 2010). Although *Salmonella* contamination of poultry and beef products exceeds that of pork, *Salmonella* control programs in swine will continue to be a primary focus of food safety initiatives. *Salmonella* reduction programs are becoming commonplace, with long-range goals to include the production and marketing of *Salmonella*-free pork products. Numerous dynamic programs are in place utilizing hazard analysis and critical control point (HACCP) principles (Griffith et al., 2006). Those programs, that have been in place for sufficient period of time, such as the Danish program, have significantly reduced the rate of *Salmonella* infection in pork products (Nielsen et al, 1995). Fortunately, most of the methods useful for pre-harvest *Salmonella* reduction in swine populations are related to sound management practices that also improve the overall health of swine operation.

Reduction of *Salmonella enterica subsp. enterica* (*Salmonella*) prevalence in the pig industry will be set as a target at the EU level and it is believed to significantly contribute to the protection of human health. The specific reduction target will be based upon the results of a quantitative microbiological risk assessment on *Salmonella* in slaughter and breeder pigs as well as cost-benefit analyses, all conducted at the EU level. According to the Regulation EC-2160/20032, protection of human health from food-borne zoonotic agents is an issue of paramount importance. Farm-to-fork control programs will probably be needed to ensure a reduction of the prevalence of specified zoonoses and zoonotic agents. Moreover, Member States will have the responsibility to establish effective national control programs adjusted for the country-specific characteristics, including the disease burden and the financial implications for stakeholders. Results of the EU baseline survey on the prevalence of *Salmonella* in lymph nodes of slaughter pigs showed a wide range of prevalences in EU countries, from 0% to 29% infected pigs (EFSA, 2008). These findings suggest that country tailored surveillance-and-control strategies should be designed aiming to achieve the targets in a cost-effective way, assuring human-health protection (Baptista et al., 2010).

Bacteriological isolation methods are used to detect *Salmonella* positive pigs and to identify the *Salmonella* serovars, but because of the low sensitivity of bacteriological faecal or intestinal examination *Salmonella* positive pigs can be missed (Bager et al., 1991). Another method to screen pigs for *Salmonella* is detection of *Salmonella* serum antibodies. The *Salmonella* –LPS-ELISA (*Salmonella*-ELISA) has been developed in Denmark (Nielsen et al., 1995) and in The Netherlands (Van der Heijden et al., 1998). The setup of the *Salmonella*-ELISA is based on a mixture of lipopolysaccharides (LPS) from two *Salmonella* serovars and should theoretically detect 95% of *Salmonella* serovars (Baggesen et al., 1997). From field studies it became clear that the *Salmonella*-ELISA detects antibodies against serovars Typhimurium and Infantis more effectively than other *Salmonella* serovares (Basggsen et al., 1997). Experimental studies to investigate the feasibility of this method for other *Salmonella* serovars have not been carried out yet (Van Winsen et al., 2001).

Results from direct diagnostic methods (bacteriology) and indirect diagnostic methods (serology) cannot be compared easily. The actual shedding of *Salmonella* indicates true infection and transmission, whereas the positive serology indicates also silent transmission within the herd (Van Winsen et al., 2001). The two *Salmonella* ELISA´s have been shown to be useful to screen herd or groups that are possibly infected with certain serovars but are of no use to judge individual animals (Nielsen et al., 1995; Van Winsen et al., 2001). The EU baseline study in fattening pigs showed that due to the diversity of tests and cut-off points, used by the 9 Member States (MSs) that chose to collect meat juice samples, no group level prevalence can be estimated. The sensitivity and specificity of these tests is not precisely known and in most MSs, some inconclusive results were reported. The sero-prevalence amongst these 9 MSs was estimated to have been from as low as 2.2% (lower boundary of 95% CI, classifying inconclusive results as negative) in Sweden to as high as 41.6% (upper boundary of 95% CI, classifying inconclusive results as positive) in Cyprus (EFSA, 2008). Community reference laboratory for *Salmonella* received from this study 60 meat juice samples per participating Member State and additionally tested them to evaluate possible comparison of results between member States. Four different ELISA kits were used by Member States and considerable discrepancies between Member States' results and the results of Community Reference Laboratory were found (Berk, 2008).

Danish *Salmonella* scheme categorised pig farms in four levels from 0 to 3. Once a month, all herds were assigned to official *Salmonella* level (1, 2 or 3) according to the results from the preceding 3 months. Level 1 included herds with low acceptable prevalence of *Salmonella*, Level 2 included herds with a moderate still acceptable prevalence of *Salmonella*, and Level 3 included herds with a high unacceptable prevalence (Alban et al., 2002). Farm category must be a result of several consequential serological testing (two or three) in different period (monthly or four times per year) which is for determination of "serological salmonella index" in monitoring schemes in EU members differently regulated. Number of samples from each farm is also important for estimation of seroprevalence for *Salmonellae*. In Danish *Salmonella* control program the sampling has been simplified into 60, 75 or 100 samples per herd per year depending on herd size after revision of their program in 2001. Also cut off for tested samples has been reduced from OD 40 % to OD 20 % which increases the number of seropositive samples approximately two times. Level 1 herds have an index of <40, Level 2 herds have an index between 40 and 70, and Level 3 herds have an index >70. A Level 0 category is currently being evaluated for herds in which the seroprevalence is 0 for 3 consecutive months. Three months results of the prevalence were weighed 0.2: 0.2: 0.6 where the immediate month is counting three times as much as the previous months. Producers are interested to be introduced in level 0 where herd is seronegative for *Salmonellae* in certain period (Alban et al. 2002; Benchop et al., 2008). Beginning in 2002, Germany initiated a voluntary *Salmonella* control program similar to the Danish one, and the United Kingdom introduced the Zoonoses Action Plan (ZAP) *Salmonella* monitoring program, also based on meat juice ELISA. The Netherlands and Belgium are considering similar programs (Nielsen, 2002). Presently, there is no national *Salmonella* monitoring program for pig producers in the United States or Canada. Sera collected as part of the National Animal Health Monitoring System (NAHMS) Swine 2000 Study being evaluated with the DME conducted at Iowa State University, Ames, Iowa (Turney, 2003). The Norwegian *Salmonella* surveillance and control programme (NSSCP) was launched in 1995 and has been approved by the EU (EFTA Surveillance Authority Decision No. 68/95/COL of 19 June 1995) as the background for accepting testing meat, meat products or live animals for *Salmonella* before it is allowed to enter Norway from EU member countries. The program covers activities directed towards both live animals (cattle, pig and poultry) and meat (cattle, pig, sheep and poultry) and is designed similarly to the Swedish and Finnish *Salmonella* control programmes (Hopp et al., 1999). The program includes systematic sampling in the breeding herds (BH) and random sampling of carcasses at the abattoirs in order to identify infected carcasses originating from BH, IH (integrated herds) and FH (finishing herds). The sample sizes have been calculated so that a prevalence of 5% in any breeding herd and 0.1% in the total population can be detected, assuming a diagnostic test sensitivity of 100% (Sandberg et al., 2002).

The control program was based on the assumption that there was an association between serological reaction and bacteriological *Salmonella* prevalence. This association has been described (Nielsen et al., 1995; Stege et al., 1997; Christensen et al., 1999; Sørsen et al., 2000). The general conclusion of these studies was that the serological test was effective mainly at herd-level and especially well suited to detect high prevalence herds. A central question is how to describe the association between serology and bacteriology, because the serological results from a herd may be interpreted differently (Alban et al. 2002).

In 2008 there were 43,124 breeding pigs and 432,011 fattening pigs in Slovenia, reared on 34,725 holdings. Pig production in 2010, which includes only pigs, slaughtered in slaughterhouses in Slovenia, was 241,332 for year 2010. Number of breeding pigs was 30,345 which were on 4,373 farms. From these farms there were 3,296 farms with five or less than five breeding sows. All these farms are one-site farms, which means, that all categories of pigs from breeding pigs till fatteners are located on one site. All pigs were raised indoor (Statistical office of the Republic of Slovenia, 2011).

Seroprevalence of *Salmonella* in Slovenia is low. Comparison of the seroprevalence between large and small farms shows that the number of positive breeding swine and fatteners are higher at the large farms than in small farms. The seroprevalence of fatteners from small farms was 0.1 and of breeding sows was 0.3. The seroprevalences of pigs from large farms were higher; the seroprevalence of fatteners was 0.3 and of breeding sows was 0.68 (Stukelj et al., 2004). In our Serology laboratory we tested annually 270 to 375 serum samples. Our tested farm could be classified into the level 1 according to revised Danish surveillance-and-control program for *Salmonella*. In our preliminary study we randomly selected 100 samples out of 375 tested in 2007 which would be the number of tested samples for that herd size according to Danish program. Seroprevalence to *Salmonellae* for year 2007 for mentioned farm was for all tested samples 12.8% for OD 40% and 24% for OD 20%. For randomly selected samples for the same year the prevalence was 7.5 % for OD 40% and 17% for OD 20%. We also compared results after testing with classification with weighted three months seroprevalence. Prevalence from all tested sera in the first three months in 2007 was 8% for OD 40% and 14% for OD 20%. In randomly selected samples for the same months prevalence was 7.5 % for OD 40% and 10% for OD 20%. Results from testing of all the samples and results for randomly selected samples show only differences in percentages but the classification level of the farm remains the same (Stukelj et al., 2009).

1.2 EU baseline studies of the prevalence of *Salmonella* in pigs

1.2.1 EU baseline study on the prevalence of *Salmonella* in slaughter pigs

To obtain an overview of the *Salmonella* prevalence in pigs in EU Member States (MSs) two baseline studies on the prevalence of *Salmonella* in slaughter and breeding pigs were conducted. The baseline study in slaughter pigs started on the 1st October 2006 and lasted till the 30th September 2007. Tested slaughter pigs were selected in slaughterhouses that together accounted for 80% of pigs slaughtered within each Member State (MS), which constituted the survey target population. Twenty-five EU MSs participated in the survey. Norway participated on a voluntary basis.

Slaughtered pigs with a live weight between 50 kg and 170 kg and their carcasses were randomly sampled in slaughterhouses representing at least 80% of MSs' total production of slaughtered pigs. The samples to take were stratified by the slaughterhouses' capacity (throughput) in the year 2005 and by the month. The day on which the samples were taken was also randomly chosen from all days of the month of sampling as was the slaughtered pig or its carcass from all scheduled pigs to slaughter on the selected slaughter day. From a selected slaughter pig at least 5 ileo-caecal lymph nodes weighing at least 15 grams were collected on a mandatory basis. The number of pigs to sample was 384 minimum and 2,400 maximum and was calculated for each MS. In addition, in order to assess the contamination of slaughter pig carcasses, 13 MSs (Austria, Belgium, Cyprus, Czech Republic, Denmark, France, Ireland, Latvia, Lithuania, Poland, Slovenia, Sweden and The United Kingdom) voluntarily sampled each at least 384 carcasses belonging to the slaughtered pigs of which lymph nodes were taken. This additional sampling was done by swabbing the surface of the carcass in a standardized way, after evisceration and before chilling. Moreover, 9 MSs (Cyprus, Denmark, France, Ireland, Lithuania, Slovenia, Sweden, The Netherlands and The United Kingdom) voluntarily collected a muscle sample (to extract meat juice) or a blood sample from all pigs selected for lymph node sampling for antibody detection examination. Samples were taken by the competent authority in each MS or under its supervision.

The EU live pig population totalled 160 million heads in 2005. The largest population was in Germany, 17% of the EU live pig population. Seven MSs (Germany, Spain, Poland, France, Denmark, The Netherlands and Italy) accounted for 74% of the total EU population. Conversely, several MSs had very small live pig populations. The EU slaughtered pig population totalled 240 million heads in 2005. The largest population was in Germany, 20% of the EU slaughtered pig population. Eight aforementioned MSs plus Belgium, accounted for 81% of the total EU slaughtered pig population. Conversely, several MSs had very small slaughtered pig populations.

The cleaned validated dataset comprised data on 19,159 slaughter pigs. On the sample-level the dataset contained 18,663 samples of lymph nodes, 5,736 carcass swabs and 5,972 serological samples originating from 25, 13 and 9 MSs, respectively. The dataset also included data on 408 lymph node samples from Norway. For slaughter pigs and of lymph node samples some invalid lymph node test results were excluded. A total of 934 slaughterhouses in the EU and nine in Norway were sampled, varying from three in Cyprus and Luxembourg to up to 400 in Poland (EFSA, 2008).

Observed prevalence of slaughter pigs infected with *Salmonella* spp. in lymph nodes

It is important to note that the absence of any *Salmonella* from the tested samples does not imply that a MS is *Salmonella* - free, as firstly the detection method has a sensitivity of less than 100%, so false negative results are plausible. Secondly, the prevalence within the MS may be too low for even one positive animal to be detected with the sample size that was used. *Salmonella* spp. was found in 24 out of the 25 MSs providing data on lymph node samples of slaughter pigs. No lymph node tested positive in Finland, whereas one pig tested positive in Norway. The observed EU-level prevalence was 10.3% (95% CI: 9.2; 11.5). The unweighted prevalence (10.8%) was included in the CI 95%. Within MSs, the prevalence varied between 0.0% and 29.0%. Serovar Typhimurium was isolated in all the 24 MSs reporting positive results for *Salmonella* in lymph nodes. One pig tested positive in Norway. The observed EU-level prevalence was 4.7% (95% CI: 4.1; 5.3). The unweighted prevalence (4.2%) was included in the CI 95% CI. At the MS-level, the observed prevalence was highest in Luxembourg (16.1%). Serovar Derby was isolated in 20 MSs. No lymph node tested positive for Derby in Cyprus, Estonia, Finland, Lithuania, Sweden and in Norway. The observed EU-level prevalence was 2.1% (95% CI: 1.8; 2.6). The unweighted prevalence (1.8%) was included in the CI 95% CI. At the MS-level, the observed prevalence was highest in France (6.5%). Serovars of *Salmonella* other than Typhimurium and Derby were found in lymph nodes of slaughter pigs from 24 MSs. The observed EU-level prevalence was 5.0% (95% CI: 4.4; 5.7). The unweighted prevalence (5.6%) was included in the CI 95%. At the MS-level, the observed prevalence was highest in Greece (17.2%).

The EU prevalence of 10.3% can be interpreted as showing that one in ten pigs slaughtered in the EU was infected with *Salmonella* when slaughtered. This infection may have arisen on the farm of origin or at any time during transport to slaughter or lairage. About half of the MSs had a *Salmonella* prevalence in lymph nodes above the EU average, while the other half had prevalence below the EU mean. This was also the case for serovar Typhimurium, but less true for Derby and for serovars other than these latter two, for which fewer MSs had figures above the EU mean. It is noteworthy that although there was a large variation in the slaughter pig *Salmonella* prevalence, the serovar distribution was not remarkably varying between the MSs, because two specific *Salmonella* serovars, Typhimurium and Derby, accounted for a major part of the positive findings at the EU-level and for most *Salmonella*-positive MSs. All 24 *Salmonella*-positive MSs isolated *Salmonella* Typhimurium and 20 detected *Salmonella* Derby. These two serovars are common serovars found in *Salmonella* infection cases in humans, and are both amongst the ten most frequently reported serovars in humans (EFSA, 2008).

Observed prevalence of carcasses contaminated with *Salmonella* spp.

Salmonella spp. was found in 11 out of the 13 MSs providing data on surface swabs-sampling of carcasses. No carcass swabs tested positive in Slovenia and Sweden. The observed 13 MS-group level prevalence was 8.3% (95% CI: 6.3; 11.0). At the MS-level, the observed prevalence was highest in Ireland (20.0%). For this 13 MS-group the observed prevalence of slaughter pigs infected with *Salmonella* spp. in lymph nodes was estimated as 9.6% (95% CI: 8.2%; 11.1%). Thus, one in 12 pig carcasses produced in this group of 13 MSs was contaminated with *Salmonella*. This estimation cannot as such be extrapolated to the level of the EU, because this group of MSs may not be representative for all MSs. One group of participating MSs had a prevalence above the weighted average (Belgium, France, Ireland and the United Kingdom), and the other one below the average (Austria, Cyprus, Czech Republic, Denmark, Latvia, Lithuania, Poland). This was the case for *Salmonella* spp., for serovar Typhimurium, and to a lesser extend for Derby. It was not the case for serovars other than the two latter ones.

Serovar Typhimurium was isolated in 10 MSs reporting positive results for *Salmonella* in carcass swabs. No carcass swabs tested positive in Latvia, Slovenia and Sweden. The observed 13 group-level prevalence was 3.9% (95% CI: 2.8; 5.5). At the MS-level, the observed prevalence was highest in Ireland (11.7%). Serovar Derby was isolated in 10 MSs. No carcass swabs tested positive in Cyprus, Slovenia and Sweden. The observed 13 MSs group-level prevalence was 2.6% (95% CI: 1.7; 3.9). At the MS level, the observed prevalence was highest in France (5.9%). Serovars of *Salmonella* other than Typhimurium and Derby were found on carcass swabs from 11 MSs. No carcass swabs tested positive in Slovenia and Sweden. The observed 13 group level prevalence was 2.3% (95% CI: 1.6; 2.5). At the MS-level, the observed prevalence was highest in France (4.8%).

It is again noteworthy that although there was a large variation in the prevalence of *Salmonella* contaminated carcasses, the serovar distribution was not remarkably varying between these MSs, because two specific *Salmonella* serovars, Typhimurium and Derby, accounted for a major part of the positive findings at the EU-level and for most *Salmonella*-positive MSs. The contamination of the carcasses occurred in the slaughterhouse and may have been due to infection within the pigs or from the slaughterhouse environment. For this 13-MS group the carcass swab *Salmonella* spp. prevalence appears to be similar to the lymph node prevalence. At the MS-level, the prevalence of contaminated carcass swabs tended to be similar or lower than the prevalence of slaughter pigs infected with *Salmonella* spp. in lymph nodes in 11 of the 13 MSs. Conversely, in two MSs (Belgium and Ireland) the prevalence of contaminated carcass swabs seemed higher than the prevalence of infected lymph nodes. However, sample size calculations have not been predicated for such comparison.

In this survey the carcass swab represents the closest sampled point to the exposure of the consumer, at the beginning of the food chain. Thus, since the imperative for control of *Salmonella* in pigs is the protection of public health, there is an argument that the carcass swab is the most appropriate measure of those utilised in this survey. Further, individual MSs might choose whether intervention at the farm, the slaughterhouse or some combined strategy afforded the best option for their particular circumstances (EFSA, 2008).

Observed prevalence of slaughter pigs with antibodies against *Salmonella*

Amongst the 9 participating MSs, two used the Salmotype Pig Screen® ELISA by Labor Diagnostik Leipzig, three MSs used the HerdCheck Swine *Salmonella*® ELISA by IDEXX, two MSs used an in house ELISA, one MS used the VetSign Porcine *Salmonella*® ELISA by Guildhay, and one MS used both the Salmotype Pig Screen® ELISA and the HerdCheck Swine *Salmonella*® ELISA. The NRLs used the cut-off of their choice. Eight MSs reported their results as relative optical densities (OD%) and one MS reported his results in S/P ratio (sample value related to positive control value). It was difficult to estimate the real seroprevalnece because of some inconclusive results, which could be counted as positive, negative or missing.

Seroprevalence (presence of *Salmonella* antibodies in meat juice or in sera) is a measure of the prior exposure of the pig to *Salmonella* infection. Due to the diversity of tests and cut-off points employed by the 9 MSs that chose to collect these samples, no group level prevalence can be estimated. The sensitivity and specificity of these tests is not precisely known and in most MSs, some inconclusive results were reported. The seroprevalence amongst these 9 MSs was estimated to have been as low as 2.2% (lower boundary of 95% CI, classifying inconclusive results as negative) in Sweden to as high as 41.6% (upper boundary of 95% CI, classifying inconclusive results as positive) in Cyprus.

The future value of testing of serological samples probably lies in their application within a MS for surveillance purposes and identification of positive herds, since these tests are relatively cheap, sample collection is straightforward and can be done by a slaughterhouse technician and in the case of meat samples, can be frozen for transport and batch testing. However, it should be recalled that these samples are poor predictors of the *Salmonella* status of the individual pig or carcass. This was further underpinned by the survey concordance-discordance results, at the MS-level, between the test for *Salmonella* spp. using lymph nodes and meat juice and sera samples. These analyses results revealed no to low agreement (EFSA, 2008).

Frequency distribution of *Salmonella* serovars in lymph nodes and carcass swabs

The serotyping of *Salmonella* isolates was mandatory according to the technical specifications of the survey. At least one isolate from each positive sample was to be typed according to the Kaufmann-White Scheme. Results from any sample where the serovar information was not available for any isolate were excluded from the final dataset. In total there were 2,600 *Salmonella*-positive lymph node samples. Two different *Salmonella* serovars were isolated from three *Salmonella*-positive lymph nodes. Eighty-seven different serovars were isolated from the lymph nodes of slaughter pigs across the EU. Serovars Typhimurium and Derby were highly predominant. Serovar Typhimurium was the most frequently reported serovar from the slaughter pigs' lymph nodes in EU and Norway, isolated in 40.0% of the *Salmonella* positive slaughter pigs, and reported by all (24) MSs having found *Salmonella* positive slaughter pigs and by Norway. The next common reported serovar was Derby, isolated from 14.6% of the positive slaughter pigs. Serovar Derby was also the second serovar most commonly isolated in terms of number of reporting MSs (20). Serovars Rissen and monophasic 4,[5],12:i:- were the third and the fourth most frequently recovered serovars, with an isolation rate in lymph nodes of 5.8% and 4.9%, respectively. Serovar Rissen was isolated in five MSs and *S.* 4,[5],12:i:- in eight MSs. Serovar Enteritidis was the fifth most common reported serovar and recovered in 19 MSs, in particular in Cyprus, Estonia, Poland and Slovenia where it was the most frequently isolated serovar in lymph nodes.

There were a total of 387 carcasses testing positive for *Salmonella* by surface swab-sampling in the 13 MSs. Thirty different serovars were isolated on the surface of the slaughter pig carcasses. Serovar Typhimurium was the most frequently recovered serovar from the surface of the slaughter pig carcasses in EU, representing 49.4% of the *Salmonella* positive carcasses. The second most frequent serovar was Derby (24.3% of the positive carcasses). The three next most frequent serovars were Infantis, Bredeney, and Brandenburg (3.4%, 2.1% and 1.8% of the positive carcasses, respectively). Serovar Typhimurium was the dominant serovar in 10 MSs. In Austria and in Poland, serovar Derby was isolated as frequently as Typhimurium.

A greater diversity of *Salmonella* serovars were isolated from lymph nodes than from carcass swabs, although there were five serovars that were only isolated from carcass swabs. Firstly, carcass swabs were collected from fewer MSs and secondly, the overall prevalence of *Salmonella* positive swabs was lower than that of lymph node samples within those MSs that tested both. The number of bacteria that may be collected from a carcass is also likely to be lower than the number found in the lymph node of an infected pig except in case of extreme contamination. Finally, the presence of *Salmonella* on a carcass swab may reflect post-slaughter contamination with serovars that exist in the slaughterhouse environment as well as infection originating from within the slaughtered pigs.

Serovar Typhimurium was isolated in all of the 24 MSs that found *Salmonella* in lymph node samples and in Norway. It was the most frequently isolated serovar in all MSs except Bulgaria (Derby), Cyprus (Enteritidis), Estonia (Enteritidis), Italy (Derby), Latvia (Brandenburg), Poland (Enteritidis), Slovenia (Enteritidis) and Slovakia (Derby). In six of these 8 MSs, serovar Typhimurium was the second most common serovar to be isolated whilst in Bulgaria, serovar Infantis was the second most prevalent serovar and in Latvia, where Derby came second. Serovar Typhimurium has long been recognised in many European countries as a common serovar amongst pigs although it has a wide host range and has also been isolated from domesticated mammals and poultry species. Overall, *S.* Typhimurium accounted for 40% of the serovars isolated in the survey.

In 18 of 24 MSs that isolated *Salmonella* from lymph nodes, serovar Derby was amongst the top three serovars to be isolated. In Spain and Portugal, serovar Derby was ranked fourth whilst it was not detected in Cyprus, Estonia, Lithuania or Sweden. It is widely recognised as a common serovar in pigs although it does occur in other livestock species. It accounted for 14.6% of the *Salmonella* isolated in this survey.

A wide range of other serovars were also detected, many in very low numbers. Serovar Enteritidis, which is usually associated with poultry, was found in 19 MSs and from 4.9% of all lymph node samples. It was as noted above, the most common isolate in Cyprus, Estonia, Poland, and Slovenia and the second most frequent isolate from Austria, Czech Republic, and Hungary. Serovar Enteritidis is the most frequent cause of human salmonellosis in the EU.

It can further be mentioned that *S.* Typhimurium and *S.* Derby were the most frequent serovars both in lymph nodes and on the surface of carcasses, suggesting that the serovars that exist in the slaughterhouse environment come mainly from the infected pigs that are slaughtered there. Overall, this survey demonstrates a wide variation in the distribution of *Salmonella* serovars in slaughter pigs and the presence of two dominant serovar in this species (EFSA, 2008).

Interpretation of the results from each of the three used survey tests

Salmonella infection results from ingestion or occasionally inhalation of viable bacteria. In pigs, infection within the intestinal tract may be followed by invasion of the cells of the gut and thence, infection is established in the intestinal lymph nodes. It is possible for pigs to ingest material containing *Salmonella* and for this to be in passive transit through the gut without actively establishing infection. Infected pigs may become carriers and excrete *Salmonella* in their faeces intermittently. Therefore, the presence of *Salmonella* within the lymph node is incontrovertible evidence that a pig is infected, as it is very unlikely that *Salmonella* can be isolated from lymph nodes of uninfected pigs and false positive results are rare. However, the test sensitivity is not 100% and there may therefore be false negative results. *Salmonella* excretion by carrier pigs is thought to be provoked by stress and may occur as the pigs are loaded and transported to the slaughterhouse. It is possible for pigs to become infected and for that infection to be transferred to the intestinal lymph nodes in a matter of hours. Therefore, a positive lymph node result may reflect infection on the farm of origin or during transport or lairage. The longer the duration of the transport and lairage phases, the more contaminated the environment during those phases, and the more stressful the conditions that are experienced, the greater the risk of infection occurring after departure from the farm.

Presence of *Salmonella* on carcass swabs reflects the surface contamination of the carcass. Although this may occur during transport or in the lairage, normal slaughterhouse practices including passing pigs through a scald tank and singeing to remove bristles act to reduce *Salmonella* contamination. Presence of *Salmonella* infection in the pig need not result in carcass contamination unless e.g. there is faecal leakage from the anus or the gut is accidentally nicked during processing. *Salmonella* may also survive in slaughterhouse environments, especially in equipment that is difficult to clean thoroughly. Poor hygiene in a slaughterhouse or amongst staff may also result in contamination of carcasses and one contaminated carcass may touch others, resulting in cross-contamination. Thus, the prevalence of positive carcass swabs is a product of the risk of infection within a pig, the risk that the infection is released to the exterior and the risk of cross-contamination from other carcasses or the slaughterhouse environment. It is predictable that presence of *Salmonella* in the gut is not completely associated with carcass contamination. It is also important to consider that the presence of *Salmonella* infection in the intestinal lymph nodes, which are removed from the carcass and are not consumed, may only represent a limited public health threat whilst a contaminated carcass is likely to be a greater risk to public health as the carcass is the start of the food chain.

Salmonella infection stimulates an immune response and circulating antibodies can be detected in blood, serum or meat juice. As antibodies persist beyond the time of infection, unsurprisingly a positive serological result is a poor indicator of current infection. Infection during transport to a slaughterhouse or in lairage does not result in a seropositive reaction, as there is insufficient time for a detectable immune response to occur before death. However, the prevalence of seropositive pigs does give a good estimate of the lifetime exposure to *Salmonella*. Therefore, it may be a valuable tool for surveillance of *Salmonella* infection on farms as part of a control programme (EFSA, 2008).

Conclusions

The main conclusions made by reporting team were:

- The survey provides valuable data for risk managers on the prevalence and distribution of *Salmonella* in EU MSs, and results are suitable to be used for setting targets for the reduction of the frequency of the *Salmonella* infection in slaughter pigs in the EU.
- Three tests were used in the survey: bacteriological tests of lymph nodes and of carcass swabs and a test for antibodies. *Salmonella* prevalence in lymph nodes reflects the infection of the pigs at the level of the primary production (i.e. on the farm and during subsequent transport and lairage). *Salmonella* contamination of the carcass may derive from the infection within the pig or from the slaughterhouse environment, whereas the presence of antibodies reflects past exposure of the pigs to *Salmonella*.
- The observed prevalence of slaughter pigs infected with *Salmonella* spp. varied widely amongst MSs.
- A large variety of serovars of *Salmonella* were isolated from ileo-caecal lymph nodes of slaughter pigs in the EU.
- A more limited range of serovars was identified on the surface of carcasses.
- With regard to seroprevalence, the observed estimates in slaughter pigs varied among the 9 participating MSs. However, these seroprevalence estimates are not directly comparable because of different tests and different thresholds used within participating MSs. No prevalence was therefore estimated at the MS-group level. Credible estimate of prevalence amongst these MSs varied from as low as 2% to as high as 42% (EFSA, 2008).

1.2.2 EU baseline study on the prevalence of *Salmonella* in holdings with breeding pigs

European Union Baseline survey on the prevalence of *Salmonella* in holdings with breeding pigs was carried out at farm level to determine the prevalence of *Salmonella* in pig breeding holdings. The herds were randomly selected from holdings constituting at least 80% of the breeding pig population in a Member State.

Sampling took place between January 2008 and December 2008. A total of 1,609 holdings housing and selling mainly breeding pigs (sows or boars of at least six months of age kept for breeding purposes) (breeding holdings) and 3,508 holdings housing breeding pigs and selling mainly pigs for fattening or slaughter (production holdings) from 24 European Union Member States, plus Norway and Switzerland were included in the survey. In each selected breeding and production holding, fresh voided pooled faecal samples were collected from 10 randomly chosen pens, yards or groups of breeding pigs over six months of age, representing the different stages of production of the breeding herd (maiden gilts, pregnant pigs, farrowing and lactating pigs, pigs in the service area, or mixed). The pooled samples from each holding were tested for the presence of *Salmonella* and the isolates were serotyped.

The overall European Union prevalence of *Salmonella*-positive holdings with breeding pigs was 31.8% and all but one participating Member State detected *Salmonella* in at least one holding. Twenty of the 24 Member States isolated *Salmonella* in breeding holdings and at European Union level 28.7% of the holdings were estimated to be positive for *Salmonella*. This prevalence varied from 0% to 64.0% among the Member States. The estimated European Union prevalence of breeding holdings positive to serovar Typhimurium and to serovar Derby was 7.8% and 8.9%, respectively. Twenty-one of the 24 Member States isolated *Salmonella* in production holdings and at the European Union level 33.3% of the production holdings were estimated to be positive for *Salmonella*. This prevalence varied from 0% to 55.7% among the Member States. The estimated European Union prevalence of production holdings positive for serovars Typhimurium and Derby was 6.6% and 9.0%, respectively. For the two non-Member States, Switzerland detected *Salmonella* in both breeding and production holdings while Norway did not detect any *Salmonella* in its surveyed holdings.

The number of different *Salmonella* serovars isolated in breeding holdings and production holdings across the European Union was 54 and 88, respectively. Serovar Derby was the most frequently isolated serovar in both breeding and production holdings, detected in 29.6% and 28.5% of the *Salmonella*-positive holdings, respectively. The next most commonly isolated serovar was serovar Typhimurium accounting for 25.4% and 20.1% of *Salmonella*-positive breeding holdings and production holdings, respectively. These serovars were also commonly found in the EU-wide baseline survey of fattening pigs at slaughter in 2006-2007. The next most frequently reported serovars were serovars London, Infantis and Rissen both in breeding and production holdings and each accounted for approximately 7% of the positive holdings, in each type of holding. Also *Salmonella* isolates with the incomplete antigenic formula 4,[5],12:i:-, which are likely to be related to the recent emergence of monophasic serovar Typhimurium, were reported by several Member States.

Salmonella infection in breeding pigs may be transmitted to slaughter pigs through trade and movement of live animals and contamination of holding, transport, lairage and slaughter facilities. This may lead to *Salmonella*-contamination of pig meat and consequently to human disease. Further studies in surveillance and control methods for *Salmonella* in breeding pigs as well as in the public health importance of consumption of meat from culled breeding pigs are recommended. Also investigations on the epidemiology of monophasic serovar Typhimurium would be welcome. The results of this survey provide valuable information for the assessment of the impact of *Salmonella* transmission originating from holdings with breeding pigs as a source of *Salmonella* in the food chain. These baseline prevalence figures may be used for the setting of targets for the reduction of *Salmonella* in breeding pigs, to follow trends and to evaluate the impact of control programmes (EFSA, 2009a).

1.3 Objectives of our investigation

The objectives of our investigations were to obtain an overview on *Salmonella* prevalence in pigs in Slovenia, which was part of EU Baseline study on the prevalence of *Salmonella* in slaughter pigs in 2008. Within this study Slovenia was one of the 9 countries that voluntarily included also detection of antibodies against *Salmonella* in meat juice. To assess the suitability of antibody detection for *Salmonella* we had previously monitored one of our big holdings already in 2007 (Stukelj et al., 2009).

2. Baseline study on the prevalence of *Salmonella* in slaughter pigs in Slovenia

2.1 Materials and methods

2.1.1 Pigs and holdings

In the EU baseline study on the prevalence of *Salmonella* in slaughter pigs 440 pigs from 178 holdings in Slovenia were tested. Almost a half of the pigs (212 or 48% of all tested) originated from small holdings (163 or 92% of all tested), which were represented in this study by only one to three pigs. For *Salmonella* isolation intestinal lymph nodes (minimum 15 grams) and carcass surface swabs were collected at slaughter and a piece of either diaphragm or neck muscles were collected for detection of antibodies in meat juice. Samples were sampled by official veterinarians and proceeded to Veterinary Faculty, National Reference Laboratory for *Salmonella*.

2.1.2 Detection of *Salmonella*

Isolation and identification were performed according to ISO/FDIS 6579, Annex D: 2007. We used Buffered peptone water (Biolife) for pre-enrichment, enrichment on Modified semisolid Rappaport-Vasilliadis agar (MSRV, Biocar) and plating on Xylose-lisine-desoxicholat agar (XLD, Biolife) and Rambach agar (Merck). *Salmonella* suspicious colonies were identified biochemically either by API 20 E (Biomérieux) or Crystal Enteric/nonfermenter ID kit (BBL). Serovars were identified by slide agglutination with STATENS SERUM INSTITUT *Salmonella* antisera according to White-Kauffman-Le Minor scheme (Grimont and Weill, 2007).

2.1.3 Antibody detection

The diaphragm samples were stored in plastic bags in freezer at -18° C. Before testing with ELISA, bags were taken from the freezer and the diaphragm samples were thawed, the angles of the plastic bags were cut and the meat juices from each bag were poured over to the micro tubes. The samples were prepared for further testing.

The Swine Salmonella Antibody IDEXX ELISA allows rapid screening for the presence of antibodies to three *Salmonella enterica* serogroups indicating swine herds' exposure to the bacteria. The assay is designed to detect antibodies to *Salmonella* in swine serum, plasma and meat juice. LPS antigen (serogroups O:4 (B), O:7 (C1) and O:9 (D1)) is coated on 96-well plates.

The presence or absence of antibody to *Salmonella* in the sample was determined by relating the absorbance value at 650 nm of the unknown to the positive control mean by calculating the sample to positive (S/P) ratio. In many countries and/or laboratories the results are calculated in OD% referring to a set of standard sera, defined according to the Danish Mix-ELISA system. To obtain a result comparable to this OD% scale, a correlation factor has been experimentally determined. The S/P value was divided by this factor to give an approximate OD% value. Samples with OD% equal or grater than 40% (S/P = 1. 0) were considered positive in general screening, and samples with OD% equal or grater than 20% (S/P = 0. 5) were considered positive in more stringent screening.

2.2 Results

2.2.1 Prevalence of *Salmonella* in pigs and holdings

All the carcass swabs tested negative for *Salmonella*. From lymph nodes of 28 pigs (6.36%) from 18 holdings (10.11%) we isolated *Salmonella enterica* subsp. *enterica*, belonging to 13 serovars, including four of five serovars of public health importance: Enteritidis (7 pigs from 6 holdings), Typhimurium (3 pigs from 3 holdings), Virchow (2 pigs from 2 holdings) and Infantis (1 pig from 1 holding). All the serovars belonged to sero-groups O:4 (B), O:7 (C1) and O:9 (D1), which are covered by IDEXX ELISA, used for antibody detection. From some bigger holdings, represented by 11 to 65 pigs, we isolated two to four different serovars.

For antibody detection we used two criteria: OD 40% and OD 20%. In IDEXX ELISA at OD 20% 91 (20.68%) pigs from 45 (25.28%) holdings tested positive. At OD 40% 48 (10.91%) pigs from 25 (14.04%) holdings, tested positive. This means that 52.75% of pigs positive at OD 20% from 55.56% serologically positive holdings reacted with high antibody titres.

Of 178 holdings in 102 (57.30%) holdings, represented by 165 (37.50%) pigs, all pigs tested negative in both tests. Another 165 pigs (37.50%) originating from 16 positive holdings (either by culture or ELISA) tested negative in both tests. All together 110 (25.00%) pigs from 50 (28.09%) holdings tested positive either by culture or ELISA and 9 (8.18%) of these 110 pigs from 6 holdings (12.00%) tested positive in both tests. Of 18 holdings positive by culture, 13 (72.22% of positives) were represented by 122 pigs (27.73% of all pigs). Of these 122 pigs 67 (54.92%) tested positive either by culture or ELISA or both and 55 (45.08%) tested negative both by culture and ELISA. Of 5 (27.78%) holdings with pigs testing positive only by culture and represented by 9 pigs (2.05% of all pigs), 6 pigs were positive by culture and 3 negative. Percentage of positive pigs (either by culture or ELISA) within holdings varied considerably. In the holdings represented by at least 10 pigs the range of ELISA positive pigs at OD 20% was from 5.00% to 53.33% with the average 22.91% and at OD 40% it was from 0.00% to 53.33% with the average of 15.46%. The results of culture and ELISA positive pigs and holdings are presented in Figure 1 and 2 respectively.

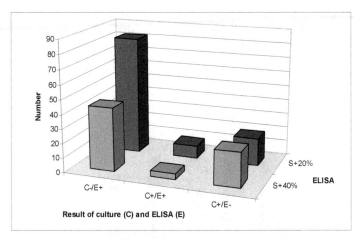

Fig. 1. Results of culture and ELISA positive pigs (N = 110).

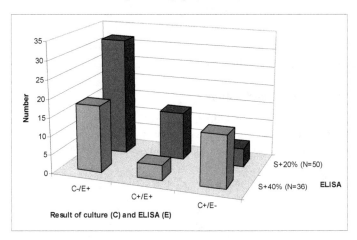

Fig. 2. Results of culture and ELISA positive holdings (N = 50).

2.2.2 Suitability of diaphragm and neck muscle meat juice

We also compared the results of ELISA of meat juice from diaphragm and neck muscles. For 9 pigs we did not have the data on the sampling site, so the results of altogether 439 pigs were processed. Meat juices of 304 (70.53%) pigs were from diaphragm and 127 (29.47%) from neck muscles. From diaphragm 71 (23.36%) meat juices were positive and from neck muscles 17 (13.39%) meat juices were positive. In the chi-square test the difference was statistically significant (t = 4.88, P< 0.05). Since different holdings were represented by different number of pigs, we compared also holdings from which only samples of diaphragm muscles, only neck muscles or both were tested. We had data for 175 holdings, of which 132 (75.43%) were represented only by diaphragm meat juices, 34 (19.43%) by neck muscle juices and 9 (5.14%) by both meat juices. Of the holdings with only diaphragm meat juices 35 (26.52%) had at least one positive pig. Of the holdings with only neck muscles juice 7 (20.59%) had at least one positive pig. Of the holdings with both juices 5 (55.56%) had at least one positive pig. Since the number of holdings with both juices was low and the proportion of pigs with either of juices varied greatly within holdings, we compared only holdings with one type of juice. In the chi-square test the difference was not significant (t = 0.24, P <0.05). The comparison of diaphragm and neck muscles' ELISA are presented also in Figure 3.

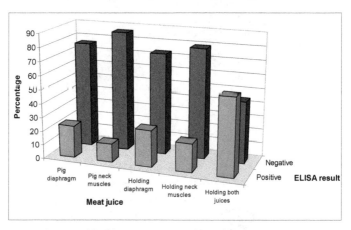

Fig. 3. Comparison of pig and holding meat juices' results.

2.2.3 Comparison of ELISA and culture

We found no correlation between culture and ELISA results. In the chi-square test the difference between them was statistically highly significant (t = 39.523, P< 0.005). We present the results of culture and ELISA in the holdings represented with at least 10 pigs in the Figure 4.

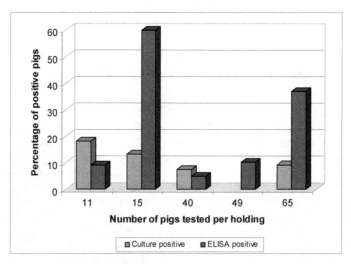

Fig. 4. Comparison of ELISA and culture.

2.2.4 Sampling for *Salmonella* reduction program

What happens if we do not have enough results over the whole year is presented in the figures 5, 6 and 7. In the figure 5 we present the results of holding A. The sampling covers all the twelve month, but the number of samples is too small, so the results are not reliable enough. We isolated three different serovars of *Salmonella*. The first was serovar Virchow in December and the next was Derby in January. In January and February high levels of antibodies were detected. Till August when serovar Coeln was isolated, there were no isolates. Its effect on seroconversion can not be estimated due to low number of samples.

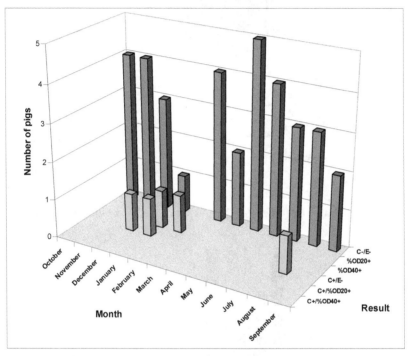

Fig. 5. Results of holding A over the year (N = 40 pigs tested; C = culture, E = ELISA).

In the holdings B and C there was no sampling in the second half of the study (spring and summer). In the holding B only 7 months were covered, and in the holding C only 6 months. In the holding B we found only some seroconversion, but no positive culture for *Salmonella* (Figure 6).

In the holding C we found seroconversion and culture positive pigs, but we didn't have an overview over the whole year. In October it seems that seroconversion remained from previous infection. In October we isolated serovar Choleraesuis var. Decatur, in November serovar Heidelberg and in December serovar Enteritidis. In February we isolated serovar Infantis and in March Enteritidis. From April to September, when the rate of *Salmonella* infection in humans is usually the highest, there were no samples. Some seroconversion was detected in October, which increased till January when the highest number of pigs positive at %OD 40 was detected. (Figure 7).

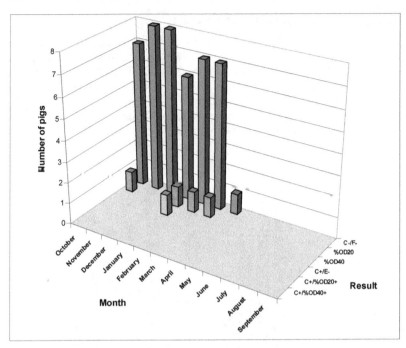

Fig. 6. Results of holding B over the year (N = 49 pigs tested; C = culture, E = ELISA).

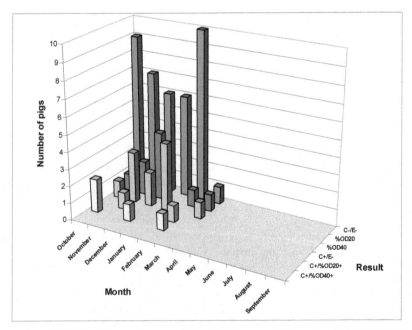

Fig. 7. Results of holding C over the year (N = 65 pigs tested; C = culture, E = ELISA).

2.3 Discussion

Current surveillance of *Salmonella* in pigs involves intensive and expensive scheme for herd classification. Isolation of *Salmonella* from lymph nodes is believed to reflect long-term exposure at herd level, but might indicate infection during transport and lairage. Bacteriological techniques for *Salmonella* detection are reported to have very high specificity (up to 100%) but sensitivity is low (Funk et al., 2000). In the EU Interlaboratory comparison study food II (2007), organized by CRL-*Salmonella*, the sensitivity of use of MSRV and XLD media depended on serovar and number of colony forming units (cfu) in samples. It ranged from 54.3% for 10 cfu of serovar Enteritidis to 100.0% for 50 cfu of serovar Typhimurium (Kuijpers et al., 2008). Besides, *Salmonella* are usually localized focally, so thorough homogenisation of samples is important. In faecal and dust samples competitive microflora might also lower the recovery of *Salmonella*.

ELISA detects specific antibodies against *Salmonella* and therefore it indicates past or resent exposure and different serological stages. According to producers manual the IDEXX Swine *Salmonella* Ab ELISA test specificity is 99.4%. Among *Salmonella* negative herds, some might be misclassified due to small number of pigs tested per holding - in 126 (70.79%) holdings in our Baseline study only one. More herds would be expected to be positive, if more samples had been collected. To improve herd test sensitivity, more samplings of herds would have been desirable (Baptista et al., 2010). CRL-*Salmonella* organized comparability of different ELISAs on the detection of *Salmonella* spp. antibodies in meat juice and serum. Ten national reference laboratories participated, using four different ELISA kits. The kits were designed to detect antibodies against *Salmonella* serogroups O:4 (B) and O:7 (C1) and one of them also against O:9 (D1). Laboratories used different OD% cut-off values from 15% to 40%. The comparison of results between laboratories was difficult. In nine of ten NRLs the results were significantly different from the results of CRL (Berk et al., 2008). Similar results were observed by Vico who compared three commercial enzyme-linked immunosorbent assays for meat juice samples. When these three kits were used in the same herd, the results deffered substantially. Thus caution is advised if it is decided to use these assays for herd health classivication in *Salmonella* control programs (Vico et al., 2010).

In our study, the seroprevalence at cut-off 20 OD% was in pigs 20.7% and in holdings 25.3%. The seroprevalence for cut-off 40 OD% in pigs was 10.9% and in the holdings 14%. Results from the bacteriological testing showed 6.4% positive pigs from 10.1% positive holdings. The results of statistical analysis showed the poor correlations between serology and bacteriology in pigs and in holdings.

In the instructions for sampling in the baseline study both meat juices from neck muscles and diaphragm were treated as equivalent. Our results also show that at the holding level there were no significant differences, although at pig level the difference seemed significant. We attribute this to the difference of seroprevalence between holdings and the differences in the numbers of pigs tested per single holding.

EU Member States approach the problem of reduction of *Salmonella* prevalence in pigs with different reduction programs. They categorize holdings in categories regarding seroprevalence. Danish *Salmonella* surveillance and control program in slaughter pigs was introduced in 1995 and started with cut-off for positive serology result 40 OD%. In August 2001 a new assignment was introduced which among others included reduction from cut-off 40 OD% to 20 OD% in the interpretation of the individual meat juice sample results (Alban et al., 2002). German *Salmonella* surveillance and control program in slaughter pigs classified herds in three categories: I (0-20%), II (20-40%), III (<40%) by their percentage of yearly positive samples, which was re-calculated quarterly. The number of participating herds increased continuously since the start of the monitoring program, with regional differences in the degree of participation. In the forth quarter of 2008, 81.9% of the herds were allocated to category I, 14.0% to category II and 4.0% to category III. However, the prevalence of *Salmonella* tended to decrease in herds that participated over of long period (Merle et al., 2011). In Slovenia only one holding sent samples monthly, so it was used in our preliminary study. Seroprevalence to *Salmonellae* for year 2007 for the mentioned holding was for all tested samples 12.8% for OD 40% and 24% for OD 20%. For randomly selected samples for the same year the prevalence was 7.5 % for OD 40% and 17% for OD20 % (Stukelj et al., 2009). In the baseline study in 2008 this holding was represented by 9 pigs, sampled in four months over summer and only one pig tested positive at OD 20%. The example of the three holdings (A, B, C) from the baseline study clearly shows the necessity of monthly testing of relevant number of pigs. In Danish *Salmonella* surveillance and control program herds with annual kill less than 100 pigs were excluded; they were considered insignificant, because of pigs from such herds only constituted around 1% of the total number of pigs slaughtered at the time of study. Also relatively more animals would need to be sampled to estimate the prevalence in these herds with an acceptable precision. The minimal number of tested pigs was 60 per year (Alban et al., 2001). In the baseline study 168 holdings (94.4% of all tested) in Slovenia were represented only by 1-5 pigs. Also in Slovenia such a program for monthly testing with relevant number of pigs would be appropriate but adapted to the high percentage of small herds. Sandberg et al. reported that the unit for testing should be the herd rather than the individual animal. The sampling should focus on the larger herds that supply most of the meat in the market and on the herds that distribute sows and piglets to other herds and can thus contribute to the spread of *Salmonella* among herds (Sandberg et al., 2002).

Sørensen et al. reported that they found no linear association between the proportion of positive lymph nodes and herd serology. In general, the highest proportion of positive pigs was observed for finishers originating from herds with seroprevalences varying from 61-70% (Sørensen et. al., 2004). CRL-*Salmonella* came to the same conclusion that there is no correlation between antibody levels and detection of *Salmonella* from lymph-nodes and carcass swabs (Berk et al., 2008). Österberg et al. reported that the seroconversion and excreting of *Salmonella* were serovar and dose-dependent. Pigs inoculated with levels of 10^6 and 10^9 cfu of serovar Derby produced specific antibodies, while pigs inoculated with 10^3 cfu of Derby or serovar Cubana produced no detectable antibody levels (Österberg et al., 2009). Within our study we compared individual pigs instead of serological statuses of the farm to *Salmonella*. The *Salmonella* prevalence in lymph nodes in individual pig is irrespective of herd serology. The presence of *Salmonella* in the lymph nodes may be caused by an infection so early in the pig's life that the serological response is no longer there, but bacteria has remained in the lymph nodes, or the pig has been infected very recently (Sørensen et. al., 2004). We, too, had cases, where pigs were culture positive and serologically negative. The *Salmonella* bacteria probably related to infection in the herd, but this infection may have occurred several weeks or months prior to slaughter. If the aim is to monitor the *Salmonella* prevalence in the herd, than herd serology is better indication than in caecal lymph nodes. Additionally, the presence of *Salmonella* in intestinal lymph nodes has a negligible impact on food safety as they are neither cut nor eaten. Usually leaking of intestinal content is more likely and more dangerous cause of carcass contamination with *Salmonella* and other enteric pathogens, so the technology and the way of handling pigs and their carcasses in slaughterhouse is very important. The results of De Busser et al. indicate that the lairage area is primary source of *Salmonella* in slaughter pigs and the carcass contamination originates from the environment rather than from the pig (inner contamination) itself (De Busser et al., 2011). Despite this, some countries used analyses of caecal lymph nodes to measure the *Salmonella* prevalence in pigs and herds.

The strong correlation between bacteriological findings and herd serology indicates that despite the fact that most *Salmonella* infections are silent in pigs, they nevertheless undergo an infectious process that results in immune response. The question is how *Salmonella* should be measured? Bacteriological measures as well as the measure of antibodies are strongly correlated. Therefore, four bacteriologically tested sites (carcass surface, pharix, lymph nodes, caecal content) and herd serology can in principle be used. The results of the study conducted by Sørensen et. al. demonstrated a strong association between herd serology measured by use of Danish mix - ELISA and the presence of *Salmonella* in caecal - contents, or carcass surface, and in pharynges, but not in caecal lymph nodes. This applies to Danish conditions where the transport time and duration of lairage is short (Sørensen et. al., 2004). The transport time and duration of lairage is short also in Slovenia, so similar measures would be indicated. In this study we tested by culture only lymph nodes and compared the results with serology. The results were not comparable, which was also expected from other studies (Sørensen et. al., 2004). This can also be expected in the case that *Salmonella* remains only in intestine and is occasionally excreted in environment.

3. Conclusion

For food safety assurance both approaches can be valuable. Antibody detection is an indicator of possible previous or on-going *Salmonella* infection in a herd, while *Salmonella* detection by culture in faeces, lymph nodes or animal environment indicates possible threat of food contamination, especially with serovars of public health importance. To obtain a reliable overview of the *Salmonella* prevalence in individual holdings regular monthly testing of relevant number of pigs is mandatory. Hygienic measures during pig production in holdings and in food production in slaughterhouses and food production plants are the key to reduction of *Salmonella* problem in humans.

4. Acknowledgment

Our research was financed by Ministry for Agriculture, Forestry and Food, Veterinary Administration of Republic of Slovenia and co-financed by EU Commission.

5. References

Alban, L., Stege, H. & Dahl, J. (2002). The new classification system for slaughter-pig herds in the Danish *Salmonella* surveillance-and-control program. *Preventive veterinary medicine*, 53 (1-2), pp. 133-46

Bager, F. & Petersen, J. (1991). Sensitivity and specificity of different methods for the isolation of *Salmonella* from pigs. *Acta Vet Scand*, 32, pp. 473-481

Baggsen, D.L. & Christensen, J. (1997). Distribution of *Salmonella enterica* serotypes and phage types in Danish pig herds, *Proceedings of the 2nd International Symposium on Epidemiology and Control of Salmonella in Pork*, ISBN 87-601-1838-5,Copenhagen, Denmark, Aug 1997,pp. 107-109

Baptista, F.M., Alban, L., Nielsen, L.R., Domingos, I., Pomba, C. & Almeida V. (2010). Use of Herd Information for Predicting *Salmonella* Status in Pig Herds. *Zoonoses and Public Health*, 57, 1,pp. 49-59

Benchop, J., Hazelton, M.L., Stevenson, M.A., Dahl, J., Morris, R.S. & French, N.P. (2008). Descriptive spatial epidemiology of subclinical *Salmonella* infection in finisher pig herds: application of a novel method of spatially adaptive smoothing. *Vet Res;* 39 pp. 2

Berk, P.A., van de Heijden, H.M.J.F., Mooijman, K.A. (2008) Comparability of different ELISA's on the detection of *Salmonella* spp. antibodies in meat juice and serum. RIVM Report 330604007/2008 Bilthoven, the Netherlands. pp. 23-48

Davies, R.H., McLaren, I.M. & Bedford, S. (1999). Distribution of *Salmonella* contamination in two pig abattoirs. In *Proceedings of the Third International Symposium on the Epidemiology and Control of Salmonella in Pork*, Washington, DC, 5–7 August. pp. 286–288

De Busser, E.V., Maes, D., Houf, K., Dewulf, J., Imberechts, H., Bertrand, S. & De Zutter, L. (2011). Detection and characterization of *Salmonella* in lairage, on pig carcasses and intestines in five slaughterhouses. *International Journal of food Microbiology*, 145, pp. 279-286

EFSA, European Food Safety Authority (2008). Report of the Task Force on Zoonoses Data Collection on the analysis of the baseline survey on the prevalence of *Salmonella* in slaughter pigs, part A. *EFSA Journal*, 135, pp. 1-111

EFSA, European Food Safety Authority (2009a). Analysis of the baseline survey on the prevalence of *Salmonella* in holdings with breeding pigs, in the EU, 2008, Part A: *Salmonella* prevalence estimates, EFSA Journal 2009; 7(12): [93 pp.].

EFSA, European Food Safety Authority (2009b). The community summary report on trends and sources of zoonoses, zoonotic agents, antimicrobial resistance and foodborne outbrakes in the European Union in 2007. *EFSA Journal*, 223, pp. 1-313

EFSA, European Food Safety Authority (2010). The Community Summary Report on Trends and Sources of Zoonoses, Zoonotic Agents and Food-Borne Outbreaks in the European Union in 2008. EFSA Journal; 2010(1):1496, pp. 23-25

Fedorka-Cray, P.J., Whipp, S.C., Isaacson, R.E., Nord, N. & Langer, K. (1994). Transmission of *Salmonella* Tyhimurium to swine. *Veterinary Microbiology*, 41, 333–344

Fedorka-Cray, P.J., Gray, J.T. & Wray, C. (2000). *Salmonella* infections in pigs. In : *Salmonella in Domestic animals*, Wray, C., Wray, A., pp. 191-207, CAB International, ISBN 0-85199–261–7, Wallingford

Funk, J.A., Davies, P.R. & Nicholas, M.A. (2000). The effect of fecal sample weight on detection of *Salmonella* enteritica in swine feces. *J. Vet. Diagn. Invest.*, 12, *pp. 412-418*

Griffith, R.W., Schwartz, K.J. & Meyerholz, D.K. (2006). *Salmonella*, In: *Diseases of swine*, Straw, B.E., Zimmerman, J.J., D´Allaire, S. & Taylor, D.J., pp. 739–754, Blackwell Publishing, ISBN-13 : 978-0-8138-1703-3, Ames, Iowa, USA

Grimont P.A.D., Weill F.X. (2007). Taxonomy and nomenclature of the genus *Salmonella*. In: Antigenic formulae of the *Salmonella* serovars, pp. 6-12, WHO Collaborating Centre for Reference and Reasearch on *Salmonella*, 9th ed.ition, Institut Pasteur, Paris, France

Hopp, P., Wahlstrom, H. & Hirn, J. (1999). A common *Salmonella* control programme in Finland, Norway and Sweden. *Acta Vet. Scand.*, Suppl. 91, 45– 49

Hurd, H.S., McKean, J.D., Wesley, I.V. & Karriker, L.A. (2001). The effect of lairage on *Salmonella* isolation from market swine. *Journal of Food Protection*, 64, 939–944

Kuijpers, A.F.A., Veenman, C., Kassteele van de, J. & Mooijman, K.A. (2008). *EU interlaboratory comparison study food II (2007), Bacteriological detection of Salmonella in minced beef.*, RIVM Report 330604010/2008, Bilthoven, the Netherlands, pp. 49

Merle, R., Kösters, S., May, T., Portsch, U., Blaha, T. & Kreienbrock, L. (2011). Serological *Salmonella* monitoring in German pig herds: Results of the years 2003-2008. Preventive Veterinary medicine, 99 (2-4), pp. 229-233

Nielsen, B., Baggesen, D., Bager, F., Haugegaard, J. & Lind, P. (1995). The serological response to *Salmonella* serovars Typhimurium and Infantis in experimentally infected pigs: The time course followed with an indirect anti-LPS ELISA and bacteriological examinations. *Vet Microbiol*; 47, pp. 205-218.

Nielsen, B. (2002). Pork safety - A world overview. *Proceedings of the 17th International pig veterinary society congress* , Ames, Iowa, USA, July 2002;pp. 121-135

Österberg, J., Lewerin, S. & Wallgren, P. (2009). Patterns of excretion and antibody responses of pigs inoculated with *Salmonella* Derby and *Salmonella* Cubana. *Veterinary Record*, 165 (14), pp. 404-408

Rostagno, M.H., Eicher, S.D., & Lay Jr., D.C. (2010). Does pre-slaughter stress affect pork safety risk? *Proceedings of the 21st IPVS Congress*, Vancouver, Canada, ISBN 90-5864-086-8, July2010, pp. 176

Sandberg, M., Hopp, P., Jarp, J. & Skjerve, E. (2002). An evaluation of the Norwegian *Salmonella* surveillance and control program in live pig and pork. *International Journal og Food Microbiology*, 72, pp. 1-11

Seidler, T., Alter, T., Krüger, M. & Fehlhaber, K. (2001). Transport stress-consequences for bacterial translocation, endogenouscontamination and bacterialactivity of serum of slaughter pigs. Berliner und Münchener Tierärztliche Wochenschrift 114, pp. 375-377

Sørensen , L.L., Alban, L , Nielsen, B & Dahl, J. (2004). The correlation between *Salmonella* serology and isolation of *Salmonella* in Danish pigs at slaughter. *Veterinary Microbiology* 101, pp. 131-141

Statistical office of the Republic of Slovenia, 18.7.2011, Available from http://www.stat.si/prikaziPDF.aspx?ID=2931

Stukelj, M. & Valencak Z. (2004). Control of salmonellosis in Slovenia-preliminary results, *Proceedings of the 18th International pig veterinary society congress*, Hamburg, Germany, ISBN 13: 9780521760591, July 2004, pp. 688

Stukelj, M., Golinar Oven, I. & Valencak, Z. (2009). Assessment of *Salmonella* prevalence on large pig farm with serological testing in finishing pigs. *Medicine veterinarie*, 62, 8 Juny 2009, pp.1031-1038

Swanenburg, M., Urlings, H.Λ.Р., Keuzenkamp, D.A. & Snijders, J.M.A. (2001). Salmonella in the lairage of pig slaughterhouses. *Journal of Food Protection*, 64, 12–16

Taylor, D.J. (2006). Salmonellosis, In: *Pig diseases*, Taylor, D.J., pp 150-155, St Edmundsbury Press Ltd, ISBN 0 9506932 7 8, Cambridge, Great Britain

Turney, I. (2003). Serologic basis for assessment of subclinical *Salmonella* infection in swine: Part 1. *Journal of Swine health and production*, 11, 5, pp. 247-251

Van der Heijden, H.M.J.F., Boleij, P.H.M., Loeffen, W.L.A., Bongers, J.H., Van der Wolf, P.J. & Tielen, M.J.M. (1998). Development and validation of an inderect ELISA for the detection of antibodies against *Salmonella* in swine, *Proceedings of the 15th IPVS congress*, Birmingham, England, ISBN 1555442749, July1989, pp. 69

Van der Wolf, P.J., Bongers, J.H., Elbers, A.R., Franssen, F.M., Hunneman, W.A., van Exsel, A.C. & Tielen, M.J. (1999). Salmonella infections in finishing pigs in The Netherlands: bacteriological herd prevalence, serogroup and antibiotic resistance of isolates and risk factors for infection. *Veterinary Microbiology* 67, 263–275.

Van Winsen, R.L., van Nes, A., Keuzenkamp, D., Urlings, H.A.P., Lipman, L.J.A, Biesterveld, S., Snijders, J.M.A., Verheijden, J.H. M. & van Knapen, F. (2001). Monitoring of transmission of *Salmonella enterica* serovars in pigs using bacteriological and serological detection methods. *Veterinary Microbiology*, 80, pp. 267-274

Vico, J.P., Engel, B., Buist, W.G., Mainar-Jamie, R.C. (2010). Evaluation of three commercial enzyme-linked immunosorbent assays for detection of antibodies agains *Salmonella* spp. in meat juice from finishing pigs in Spain. *Zoonoses and Public Health*, 57 (Suppl. 1), pp. 107-114

Wood, R.L., Pospischil, A. & Rose, R. (1989). Distribution of persistant *Salmonella* Typhimurium infection in internal organs in swine. *American Journal of Veterinary Research*, 50, 7, July 1989, pp. 1015-1021

4

Neutrophil Cellular Responses to Various *Salmonella typhimurium* LPS Chemotypes

Anna N. Zagryazhskaya[1], Svetlana I. Galkina[1], Zoryana V. Grishina[1],
Galina M. Viryasova[1], Julia M. Romanova[2], Michail I. Lazarenko[3],
Dieter Steinhilber[4] and Galina F. Sud'ina[1]

[1]*A.N.Belozersky Institute of Physico-Chemical Biology*
Moscow State University, Moscow
[2]*The Gamaleya Research Institute of Epidemiology and Microbiology, Moscow*
[3]*National Research Center for Hematology, Moscow*
[4]*Institute of Pharmaceutical Chemistry*
Johann Wolfgang Goethe University Frankfurt, Frankfurt am Main
[1,2,3]*Russia*
[4]*Germany*

1. Introduction

The first line of defense against invading bacteria is provided by the innate immune system, and polymorphonuclear leukocytes (PMNL) contribute to bacterial clearance by uptake and intracellular killing of microbes. Lipopolysaccharides (LPS, endotoxin), a major component of the outer membranes of Gram-negative bacteria, is shed into the environment and acts as a highly potent proinflammatory substance. About 15 – 25% of the bacterial surface in *Salmonella typhimurium* was found to be covered by LPS (Mühlradt et al., 1974). LPS initiates the cascade of pathophysiological reactions called endotoxin shock. LPS released from Gram-negative bacteria induces a strong priming of superoxide production (Guthrie et al., 1984) and facilitates the rapid elimination of the bacteria. However, an excessive activation of neutrophils could be self-destructive in septic shock. A number of mediators, such as cytokines, nitric oxide and eicosanoids, are responsible for most of the manifestations caused by LPS. The toxic and other biological properties of LPS are due to the action of endogenous mediators, which are formed following interaction of LPS with cellular targets (Galanos & Freudenberg, 1993). Biological activities of LPS have been well established, but some uncertainty remains regarding to the responses to various LPS chemotypes.

LPS are phosphorylated glycolipids that possess complex chemical structures (Müller-Loennies et al., 2007). LPS are composed of covalently linked structural domains: lipid A, an oligosaccharide core, and O- polysaccharide (or O- antigen) (Raetz & Whitfield, 2002). Lipid A is the minimal biologically active unit of LPS and is thus called the 'endotoxic principle' of LPS. The full chemical structures of lipid A from *E. coli* and *Salmonella enterica* serovar Typhimurium (S. Typhimurium) were identified in 1983, and the similarity of their structures was proved (Takayama et al., 1983; Alexander & Rietschel, 2001, review). Lipid A is the hydrophobic portion of the molecule. The hydrophilic polysaccharide portion may be further subdivided into the O-specific and the core oligosaccharide. Bacteria which contain an O- polysaccharide have a smooth colony appearance when grown on agar plates and therefore this type of LPS is referred to as smooth(S)-type LPS. The outer parts of LPS (O-polysaccharide) interact with the host immune system. Westphal and al. established that the O-polysaccharide component contained the serologically active determinants (the species-specific bacterial O-antigen) (Westphal, 1978; Westphal & Luederitz, 1961). Currently, based on O-antigens (O-polysaccharides), *Salmonella* strains have been classified into over 50 serogroups (Fitzgerald et al., 2007).

The presence of O-antigen in LPS is irrelevant for bacterial invasion of epithelial cells; in contrast, a core structure is necessary for adhesion and subsequent entry of S. Typhimurium into epithelial cells (Bravo et al., 2011). Mutant bacteria (rough mutants) produce LPS with short oligosaccharide chains but not O- polysaccharide. Chemical analysis of LPS from such *Salmonella* mutants distinguished Ra from Re chemotypes: Ra describes the largest core structure and Re was assigned to the smallest core structure. LPS from rough mutants, so-called Ra, Rb, Rc, Rd and Re LPS, mainly differ in the length of the core oligosaccharide, while the lipid-A portion is assumed to be identical. The chemical structures of *Salmonella* LPS have been investigated in many details (Olsthoorn et al., 1998; Perepelov et al., 2010).

Neutrophil-mediated innate host defense mechanisms include phagocytosis of bacteria. Upon activation, polymorphonuclear leukocytes (PMNL, neutrophil), produce signicant amounts of leukotriene B4 (LTB4) in addition to several cytokines and inammatory mediators, and thus recruit other neutrophils to the site of inammation. LTB4 is one of the most potent chemotactic compounds produced in macrophages and neutrophils (Toda et al., 2002). Stimulation of leukotriene B4 synthesis in PMNLs plays a role in stimulation of phagocytosis and bacterial killing (Mancuso et al., 2001). The key enzyme of LT synthesis in neutrophils is 5-lipoxygenase (5-LO), which metabolizes arachidonic acid (AA), first to 5S-hydroperoxyeicosatetraenoic acid (5-HPETE), and then to leukotriene A4 (LTA 4) (Samuelsson, 1983). Unstable LTA4 intermediate is converted to 5S,12R-dihydroxy-6,14-cis-8,10-trans-eicosatetraenoic acid (leukotriene B4, LTB4) and (non-enzymatically) to its isomers. The 5-LO metabolite LTB4 is a proinflammatory mediator that activates neutrophils, thus changing their shape and promoting their binding to endothelium by inducing the expression of cell-adhesion molecules. The localization of leukocytes to the site of inflammation results in endothelial and other tissue damage, i.e. metabolites of 5-LO contribute to the multiple organ injury and dysfunction during inflammatory process (Collin et al., 2004; Cuzzocrea et al., 2003; 2004). Any modulation of the activity of PMNL is a potential cause of the altered immune response to infection. The phagocytosis of microorganisms by PMNL is enhanced by LPS. And though *Salmonella*-LPS related complications have been successfully blunted with 5-LO inhibitors (Matera et al., 1988; Altavilla et al., 2009), little is known about phagocytosis and 5-LO products regulation by LPS chemotypes.

Effects of structurally different LPS types upon neutrophil functions were examined. Ruchaud-Sparagano et al. (Ruchaud-Sparagano et al., 1998) investigated the mechanisms of LPS action by examining the effect of smooth and rough chemotypes of LPS in stimulating neutrophil beta2 integrin activity and fMLP-induced respiratory burst. They reported just kinetic differences in the action of rough LPS and smooth LPS: rough LPS acts more rapidly than S-LPS to cause functional alterations in neutrophils. Similar results were obtained on neutrophils in whole blood: again just kinetic difference was observed between R- and S-LPS in the expression of cell surface receptors CD11b and CD11c on neutrophils (Gomes et al., 2010). Nevertheless, the rough mutant as well as S LPS differ in some distinct physico-chemical properties. Due to these differences, it was found a lower fluidity of S LPS chemotype than Ra and Re mutants (Luhm et al., 1998). It was established that the bioactivity of LPS was dependent on the length of their core oligosaccharides, and endotoxin-induced cytokine secretion decreased with decreasing sugar moiety (and increasing fluidity) in the order S ≥ Ra>Rc>Re LPS (Luhm et al., 1998). Comparative evaluation of the endotoxic properties of LPS preparations by using the LAL assay showed that endotoxic activity of the rough Re mutant SL1102, the rough Ra mutant TV119, and the smooth strain SH4809 of *Salmonella* Typhimurium increased in the order S < Ra < Re (Shnyra et al., 1993).

When neutrophils were challenged with *Salmonella minnesota* smooth-strain and rough-strain mutants (Ra, Rb2, RcP-, Rd1P- and Re) as well as with lipid A, in the case of luminol-dependent chemiluminescence (respiratory burst), lipid A was the most potent stimulus, with the response decreasing as molecular complexity increased, with S- LPS equally potent as Ra LPS (Pugliese et al., 1988). An oxygen-independent system in the antimicrobial effects of neutrophils is also sensitive to LPS chemotype. As the carbohydrate content of the mutant LPS decreased, the bacteria became less resistant to the oxygen-independent bactericidal activity of neutrophils (Okamura & Spitznagel, 1982). Based on these data, one can conclude that there are qualitative as well as quantitative effects of the carbohydrate moieties of LPS. We report here that various LPS forms from *Salmonella typhimurium* bacteria significantly differ in their ability to influence adhesion, phagocytosis as well as formation of 5-LO products, and reactive oxygen and nitrogen species in human neutrophils.

2. Materials and methods

Zymosan A from *Saccharomyces cerevisiae*, lipopolysaccharides from *Salmonella* enterica serovar Typhimurium (the source strain for smooth form is ATCC 7823, rough strains from *Salmonella typhimurium* TV119 (Ra mutant) and SL1181 (Re mutant)), N^{ω}-Nitro-L-arginine methyl ester hydrochloride (L-NAME), staurosporine from *Streptomyces sp.* were from Sigma (St. Louis, MO, USA and Steinheim, Germany). S. Typhimurium virulent strain C53 was a kind gift of Prof. F. Norel (Pasteur Institute, France) (Kowarz et al., 1994). Bacteria were grown in Luria–Bertani broth and washed twice using physiological salt solution with centrifugation at 2000 g. The concentration of the stock suspension was 1×10^9 CFU/mL. The bacteria were opsonized with 5% fresh normal human serum (NS) from the same donor whose blood was used for preparation of neutrophils. NS was prepared by clotting and centrifugation of fresh whole blood at room temperature. In some experiments, the NS was decomplemented by heat inactivation for 30 min at 56°C (heat inactivated serum, HIS). Nitrate/Nitrite fluorometric assay kit was from Cayman Chemical (Ann Arbor, MI, USA). Ficoll-Paque was purchased from Pharmacia (Uppsala, Sweden). Human serum albumin, fraction V (HSA) was from Calbiochem (La Jolla, CA, USA). Hepes and o-phenylenediamine were from Fluka (Deisenhofen, Germany). Phosphate buffered saline (PBS) was purchased from Gibco (Paisley, UK, Scotland, UK). Dextran T-500 was from Pharmacosmos (Holbaek, Denmark). High-pressure liquid chromatography (HPLC) solvents were purchased from Chimmed (Moscow,

Russia). Prostaglandin B2 was from Cayman Chemical Company (Ann Arbor, USA). Hank's balanced salt solution (with calcium and magnesium but without phenol red and sodium hydrogen carbonate, HBSS), HBSS modified (without calcium, magnesium, phenol red and sodium hydrogen carbonate), Dulbecco's PBS (with magnesium, but without calcium), cytochrome c from horse heart were purchased from Sigma (Steinheim, Germany).

2.1 Human neutrophil and red blood cell (RBC) isolation

PMNLs were isolated from freshly drawn EDTA-anticoagulated donor blood by standard techniques, as previously described (Sud'ina et al., 2001). Leukocyte-rich plasma was prepared by sedimentation of RBCs with 3% dextran T-500 at room temperature. Granulocytes were purified by centrifugation of leukocyte rich plasma through Ficoll-Paque (density 1.077 g/mL) followed by hypotonic lysis of the remaining RBCs. PMNLs were washed twice with PBS, resuspended at 10^7/mL (purity 96–97%, viability 98–99%) in Dulbecco's PBS containing 1 mg/mL glucose (without $CaCl_2$), and stored at room temperature. RBCs were isolated from EDTA-anticoagulated donor blood by sequential centrifugation (at 1100 rpm) and washing with PBS. After three washes, the cells were resuspended at 2.7×10^9/mL in PBS and stored at room temperature.

2.2 Preparation of collagen-, fibronectin- or HUVEC-coated surfaces

Plastic tissue-culture 24-well plates (Corning Incorporated, Corning, NY, USA) were coated with 75 µg/ml type I collagen or 15 µg/ml fibronectin for 24h. Prior to use, the protein coated surfaces were washed, incubated for 1 h in PBS with 0.1% human serum albumin, and then thoroughly washed with PBS. Human umbilical vein endothelial cells (HUVEC), passages 1–3, were maintained in medium 199 containing 10% fetal calf serum (FCS), 3.5 units/ml heparin (Fluka, Deisenhofen, Germany), 50 µg/ml endothelial cell growth factor (ICN, Ohio, USA), 10 U/ml penicillin and 10 mg/ml streptomycin. The cells were passaged using trypsin-EDTA solution (500 BAEE units trypsin and 180 mg EDTA/ml in PBS), and seeded on 24-well plates (Galkina et al., 2004). One day before the experiments, the monolayers were washed and medium was replaced with the same medium containing 2% FCS, rather than 10 %.

2.3 Preparation of lipopolysaccharides (LPS) solutions and opsonized zymosan (OZ)

Lipopolysaccharides from *Salmonella enterica* serovar *typhimurium* were solubilized in PBS (1 mg/ml) by vortexing, heated in a water bath to 60°C for 30 min, cooled to room temperature, and subjected to one more cycle of heating to 60°C and cooling to room temperature. Zymosan A particles from *Saccharomyces cerevisiae* were suspended in PBS and boiled for 5 min. After cooling to room temperature, the prepared suspension was washed with PBS and opsonized by adding 20-30 % freshly prepared autologous human normal serum for 30 min at 37°C, washed 3 times with PBS and resuspended in the Hank's balanced salts medium containing 10 mM Hepes (HBSS/Hepes).

2.4 PMNL adhesion assay

Myeloperoxidase activity was used to measure PMNL attachment under static conditions to collagen or HUVEC adsorbed on to plastic surfaces. For measuring PMNL adhesion, HUVECs grown in 24-well plates were washed once with HBSS. PMNLs (10^6/well) were added to a coated 24-well culture plate in 500 µl of HBSS/Hepes medium. After 30 min of incubation with or without the additives in a CO_2 incubator at 37°C to allow neutrophil adherence, wells were washed twice with 500 µl of PBS solution for removal of non-adherent PMNLs. The extent of adherence was measured after the addition of detergent and a myeloperoxidase substrate, as described (Schierwagen et al., 1990; Sud'ina et al., 1998). A solution (300µl) of 5.5mM o-phenylenediamine and 4mM H_2O_2 in buffer (67mM Na_2HPO_4, 35mM citric acid and 0.1% Triton X-100, pH5) was added to each well, and after 5 min the reaction was stopped by the addition of an equal volume of 1M H_2SO_4. Standard dilutions of PMNLs with or without tested compounds were used for calibration.

2.5 Phagocytosis experiments

PMNLs (5×10^6/ml) were placed into 6-well plates (2 ml/well) containing collagen- of fibronectin-coated coverslips for 30 min of incubation with tested compounds. Then 0.25 mg/ml of opsonized zymosan (OZ) was added for another 5 min. The cells were gently washed with PBS, and then fixed for 30 min in HBSS medium modified, with 10 mM HEPES and 2.5% glutaraldehyde. After gentle washing with PBS, the samples were examined by phase contrast microscopy. The number of OZ particles ingested was counted and the data were expressed as a phagocytic index, which was derived by multiplying the portion of PMNLs containing at least one ingested target by the mean number of phagocytosed targets per positive PMNL. Data were obtained from ~ 100 cells per coverslip.

2.6 Scanning electron microscopy

Cells were fixed for 30 min in 2.5% glutaraldehyde, postfixed for 15 min with 1% osmium tetroxide in 0.1 M cacodylate (pH 7.3), dehydrated in an acetone series, critical-point dried with liquid CO2 as the transitional fluid in a Balzers apparatus, sputter-coated with gold–palladium, and observed at 15 kV with a Camscan S-2 (Tescan, USA) or JSM-6380 (JEOL, Germany) scanning electron microscope.

2.7 Nitrite measurement

Nitric oxide, derived from the conversion of L-arginine to L-citrulline, reacts with molecular oxygen to form nitrite and nitrate (Moncada & Higgs, 1993). NO production was measured as total nitrite concentration in the sample after enzymatic conversion of nitrate to nitrite by nitrate reductase. A highly sensitive fluorometric assay for nitrite measurements, which is based on the acid-catalyzed ring closure of 2,3-diaminonaphtalene (DAN) with formation of highly fluorescent product 2,3-aminonapthotriasole in the presence of nitrite, was used to probe PMNLs for NO production (Nath & Powledge, 1997). For this purpose, PMNLs (2×10^7/ml) were incubated with compounds tested for 30 min, then OZ was added for the next 30 min, reaction was stopped by centrifugation (400g, 10 min) and supernatant was filtered though 10 000 Mr cutoff microcentrifuge filters (Millipore corporation, USA) at 14 000g for 30 min at room temperature. The ultrafiltration step was necessary to remove any trace amounts of zymosan particles and hemoglobin which may be present in PMNL samples due to red cells contamination, which strongly interferes with the fluorescent measurements (Misko et al., 1993). Nitrite measurements in the prepared supernatants were performed in triplicate using Nitrate/Nitrite fluorometric assay kit (Cayman Chemical, Ann Arbor, MI,

USA) according to manufacturer's protocol at excitation and emission wavelengths of 360 and 430 nm, respectively, by plate reader Infinite 200 (Tecan Group Ltd., Mainz, Germany). All compounds added to PMNLs were tested for their autofluorescence within the spectrum region in the assay buffer.

2.8 Superoxide measurement

PMNL incubations on collagen- and fibronectin-coated surfaces were performed as described for PMNL adhesion assay. 50 µM cytochrome c, tested compounds and 300 u/ml superoxide dismutase (SOD), were added (as indicated) to the medium prior to the cells. The plates were incubated at 37 °C for 30 min, then OZ was added or not for another 30 min. The incubation was stopped by cooling to 4°C, and cytochrome c reduction was measured as the increase in $\Delta 550/535$ (the change in the ratio of absorbances at 550 and 535nm). Reduction of 10 µM cytochrome c produced an increase in $\Delta 550/535$ of 0.18 absorbance unit.

2.9 Assay of reactive oxygen species

The formation of active oxygen by neutrophils stimulated with phorbol 12-myristate 13-acetate (PMA), LPS chemotypes and OZ was monitored by measuring luminol-enhanced luminescence as described earlier (Sud'ina et al., 1991). Chemiluminescence was monitored in a 1251 LKB luminometer, using 1 µM luminol. Measurements were made every 5 min over a 30 min period at 37°C.

2.10 Incubations for studies of arachidonic acid (AA) metabolism and leukotriene (LT) synthesis

PMNLs suspension (2×10^7 cells) was incubated in 6 ml HBSS/Hepes medium at 37 °C with or without agents tested for 30 min, and then stimulated by the addition of opsonized Salmonella (OS) or OZ for 20 min. The incubations were stopped by addition of an equal volume of methanol at -20°C with prostaglandin B2 (PGB2) as an internal standard. The samples were stored at -20°C. The denatured cell suspension was centrifuged (at 2000 rpm), which yielded supernatants designated as water/methanol extracts.

2.11 Lipoxygenase product analysis

The water/methanol extracts were purified by solid-phase extraction using C18 Sep-Paks (500mg), which was conditioned first with methanol, then with water. 5-LO metabolites were extracted with 1.4 ml methanol, the samples were evaporated, redissolved in 35 µl methanol/water (2:1) and chromatographed by reversed-phase HPLC. The purified samples were injected into a 5 µm Nucleudur C18 column (250 mm×4.6 mm; Macherey-Nagel, Dueren, Germany). The products were eluted at 0.7 ml/min in a linear gradient from 30 to 100% solvent B: the eluents consisted of methanol/acetonitrile/water/acetic acid/triethylamine in the ratios (solvent A) 25/25/50/0.05/0.08 and (solvent B) 50/50/0/0.05/0.04, and elution was monitored using a UV detector at 280 nm and 238 nm. Products of the 5-LO pathway that were measured included leukotriene B4 (LTB4), 5-hydroxyeicosatetraenoic acid (5-HETE), 20-hydroxy-LTB4 (ω-OH-LTB4) and iso-LTB4 [5(S),12(S,R)-dihydroxy-all-trans-eicosatetraenoic acids], identified by their co-elution with authentic standards. The respective extinction coefficients and their ratios to that of the internal standard were used to quantify products.

2.12 Statistics

Statistical analysis was performed using the Student's t-test. Statistical significance was assumed, where probability values of less than 0.05 were obtained. Results are reported as mean ± SD of the data of at least three independent experiments.

3. Results and discussion

Neutrophils are professional phagocytes and the first line of defense of innate immune system at bacterial challenge (Borregaard, 2010). Circulating lipopolysaccharides released from bacteria may activate neutrophils. LPS elicit wide spectra of biological responses in human body. When activating an immune response, they may produce pathologically imbalanced immune response, - that is why they are called "endotoxins". Depending on a dose, they may stimulate antigen-specific immune response, and they are added as accompanying agents in vaccination (Baldridge et al., 1999; Gereda et al., 2000). High levels of endotoxins cause endotoxic shock (Rietschel et al., 1996). LPS induce numerous cellular signals. In the focus of the current work, we should stress on NO formation and leukotriene synthesis. Superproduction of NO in sepsis results in the disturbed blood flow (Li & Forstermann, 2000), which originates from contractile disfunction of smooth muscle cells and impaired PMNL chemotaxis (Lopez-Bojorquez et al., 2004). LPS interaction with leukocytes signal to phosphorylation of phospholipase A2 and arachidonic acid (AA) release from cell membranes (Doerfler et al., 1994). AA and its methabolites are the compounds with high biological activity.

In this work, we investigated effects of *Salmonella* enterica serovar typhimurium LPS species of various chemotypes (from deep rough Re mutant consisting of the lipid A and the KDO (3-deoxy-D-manno-oct-2-ulosonic acid) residues, rough Ra mutant with complete core, and S form) on cellular responses of human PMNLs. LPS chemotype structures are schematically presented in Fig.1.

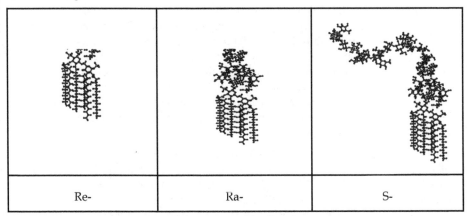

| Re- | Ra- | S- |

Fig. 1. Schematic presentation of LPS molecules: Re-, Ra- and S-LPS.

3.1 Neutrophil adhesion

Regulation of neutrophil adhesiveness is generally considered to be a key element in the development of inflammatory reactions. Neutrophils are known to spread on a protein-coated surface, a process that has been interpreted as "frustrated" phagocytosis. To elucidate whether Salmonella LPS of various chemotypes selectively influenced the number of adherent neutrophils, an adhesion study was performed. PMNL adhesion to collagen-coated surface was crucially increased by Ra LPS (Salmonella LPS from Ra mutant TV119). The effect was slightly lower on the surfaces coated by endothelial monolayer (Fig.2).

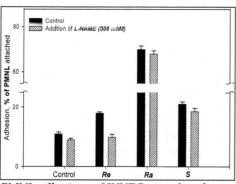
PMNL adhesion to HUVEC-coated surfaces

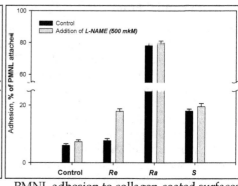
PMNL adhesion to collagen-coated surfaces

Fig. 2. PMNL attachment to HUVEC- and collagen-coated plastic surface determined as described in Methods after 30 min incubation without (control) or with various LPS forms, and 500 µM L-NAME. Cell attachment has been expressed as a percentage of PMNLs adhered in relation to the total number of PMNLs added.

Priming of PMNLs with the LPS chemotypes induced cell activation including NO and superoxide release, as well as an increase in intracellular calcium concentration. The experiments with nonselective NOS inhibitor L-NAME at 500 µM demonstrated that only Re LPS (Salmonella LPS from Re mutant SL1181) was sensitive to NOS inhibition (Fig.2). The antioxidant agent diphenileleiodonium (DPI) that inhibits NADPH oxidase-mediated ROS formation, and also inhibits other flavo-enzymes such as NO synthase and xanthine oxidase (Wind et al., 2010), did not affect LPS-induced PMNL attachment (data not shown). LPS- induced intracellular calcium concentration varied in the order Ra ≥ S > Re (Zagryazhskaya et al., 2010). Taking into account the slight sensitivity of PMNL attachment to NO synthesis inhibitors, we can propose that the divalent cation requirements for the Mac-1 and LFA-1-dependent processes of adhesion (Graham & Brown, 1991; Wright & Jong, 1986) may limit the role for NO and superoxide in the specificity of these LPS chemotypes in PMNL adhesion, in the serum-free medium.

The addition of heat inactivated normal serum (HIS) markedly decreased neutrophil adhesion, but the selective prominent increase of the neutrophil attachment induced by Ra LPS chemotype was evident (Fig.3). It has been published that serum enhanced LPS-induced production of nitric oxide in J774.1 and BAM3 macrophage-like cell line (Ohki et al., 1999). Human serum albumin is known to increase iNOS expression in the lung of rats (Jakubowski et al., 2009). In our assay 500 µM L-NAME partially reversed the effect of serum (Fig.3), which supported the hypothesis that nitric oxide mediates smoothing out the specificity of various LPS chemotypes in the presence of serum. NO is known to reduce adhesion molecules expression on neutrophils (Kubes et al., 1991; 1994; Banick et al., 1997; Kosonen et al., 1999), and we observed decreased attachment in the presence of serum (Fig.3). We propose that in the serum-containing medium, along with other factors, NO also plays a role in the specificity of LPS chemotypes in PMNL adhesion.

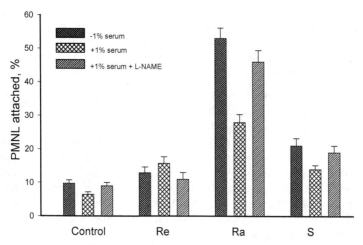

Fig. 3. Effect of serum on PMNL attachment, induced by LPS chemotypes. PMNLs (5×10^5/well) were added to a collagen-coated 24-well culture plate in 500 µl of HBSS/Hepes medium, without (control) or with 1% serum, 5 µg/ml LPS and 500 µM L-NAME. After 30 min of incubation in a CO_2 incubator at 37 °C to allow leukocyte adherence, wells were washed twice with 500 µl of PBS solution, for removal of non-adherent PMNLs. The extent of adherence was measured after the addition of detergent and a myeloperoxidase substrate, and has been expressed as a percentage of PMNLs adhered in relation to the total number of PMNLs added.

3.2 NO and ROS formation in neutrophils

NO release is an important endogenous regulatory mechanism of inflammatory response. Human neutrophils were evaluated for their ability to generate nitric oxide more than 30 years ago (Schmidt et al., 1989; Wright et al., 1989). Nitric oxide synthase (NOS) enzymes in neutrophils were characterised, and the data on the expression of NOS isoforms are contradictory (Amin et al., 1995; Carreras et al., 1996; Greenberg et al., 1998; Cedergren et al., 2003; de Frutos et al., 2001; Molero et al., 2002; Saini et al., 2006; Chatterjee et al., 2007; 2008). The level of NO synthesis in PMNLs is comparable to one in endothelial cells, and therefore contributes significantly to the amount of NO in circulation (Miles et al., 1995; Wright et al., 1989). It is proposed that NO synthesis in neutrophils is of great physiological significance, as it modulates neutrophil function at sites of inflammation. NO participates in activation of a newly described mechanism of immune defense as formation of neutrophil extensions (cytonemes), when neutrophils do not phagocyte, but bind bacteria extracellularly (Galkina et al., 2009).

Lipopolysaccharides are well known for their ability to elicit the release of NO from eukaryotic cells including macrophages, neutrophils, and endothelial cells (Jean-Baptiste, 2007; Titheradge, 1999; Tsutsui et al., 2009). Endotoxemia is often associated with increased NO (Evans et al., 1993; Szabo et al., 1993; Gomez-Jimenez et al., 1995). NO is a unique "messenger". The biological half-life of NO is rather long - several seconds (Lancaster & Ignarro, 2002), and this molecule easily passes through cell membranes, and can interact with transition metals forming nitrosyl complexes and influencing activity of many enzymes (Korhonen et al., 2005). A reaction of particular biological relevance is the reaction of NO with superoxide with the formation of OONO- (peroxynitrite, PN) (Beckman et al., 1990). Concentration of superoxide increases up to 0.1 µM during inflammatory responses (Zweier et al., 1989), but the spectrum of reactive oxygen/nitrogen species depends on the balance of NO and superoxide within the local chemical environment (Jourd'heuil et al., 1999; 2001).

Conditions for production and release of NO in human PMNLs are still largely unknown. We recently published a paper on the influence of various LPS differing in their chain length on NOS activity in opsonized zymosan stimulated human PMNLs. We observed significant difference between Re and Ra forms of S. Typhimurium LPS in the capacity to induce NO release: Re LPS was twice more potent than Ra LPS (Zagryazhskaya et al., 2010). It is known that LPS activates protein kinase C (PKC) in macrophages and PMNLs. Increased PKC activity may inhibit NOS activity and staurosporine was shown to reverse this inhibition (Muniyappa et al., 1998). Therefore, we tested if staurosporine (St), the nonselective PKC inhibitor, could influence the specific LPS effects on NO synthesis, observed in our studies. The results are presented in Fig. 4.

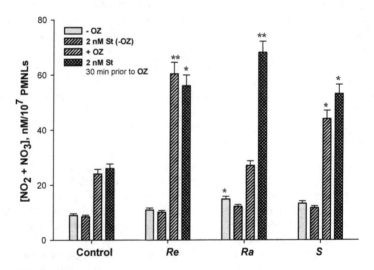

Fig. 4. Modulation of nitrite production in PMNLs by staurosporine. PMNLs (2×10^7/ml) were incubated for 30 min at 37 °C in the presence or absence of additives, (as specified): 5 µg/ml of different LPS forms, 2 nM staurosporine (St) and then stimulated for 30 min with 2 mg/ml OZ. * P < 0.05 vs corresponding control. ** P < 0.01 vs corresponding control.

Staurosporine (1-2 nM) inhibited a small increase in nitrate/nitrite level, produced by Ra and S LPS in the absence of OZ, slightly increased NO production caused by OZ alone, and partially reversed NO synthesis in OZ-stimulated PMNLs, primed by various LPS. Ra LPS form, which produced the minimal increase in NO synthesis (and even decrease in some experiments), caused the maximal NO production in the presence of staurosporine, Re and S LPS were less active. L-NAME, NOS inhibitor, significantly decreased NO production in the presence of St (data not shown) indicating staurosporine influence on NOS activity in human PMNLs, primed with LPS. These experiments demonstrated significant difference in NO production between LPS species and confirmed the role for NO in the specificity of LPS chemotypes.

LPS-priming of phagocytic leukocytes leads to nicotinamide adenine dinucleotide (NADPH) oxidase activation and potent generation of reactive oxygen species (ROS) upon stimulation (Curnutte & Babior, 1974; Drath & Karnovsky, 1975), and this process is often referred to as the respiratory burst. We determined the capacity of various LPS species to modulate superoxide anion (O_2^-) production measured as cytochrome c reduction, as well as ROS production measured as luminol-dependent chemiluminescence, in PMLNs prior to or without their activation by OZ. The most potent O_2^- production was detected in Ra-primed cells in which we observed approximately 5-fold increase in the production level detected in control cells (Zagryazhskaya et al., 2010). It is noteworthy that activation of the cells with OZ dramatically increased O_2^- generation in both LPS-primed and control cells, while relative values of the LPS effects were diminished. In luminol-dependent chemiluminescence, again, the efficacy pattern Ra > S > Re was found (Fig. 5). Ra LPS was the most potent chemotype in ROS and O_2^- release from LPS-treated PMNL.

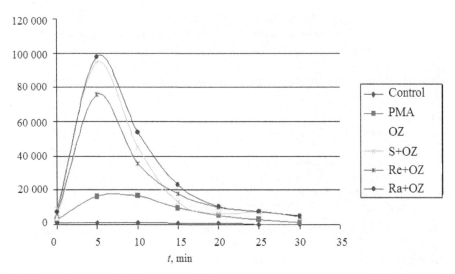

Fig. 5. Effect of various LPS forms on ROS formation in PMNLs, as recorded by luminol-enhanced luminescence (y-axis, arbitrary units). PMNLs were incubated at 37 °C in the presence or absence of additives, as specified: 100 nM PMN, 5 µg/ml of different LPS forms, 0.2 mg/ml OZ.

The addition of heat inactivated normal serum (HIS) decreased superoxide release, but the effect of Ra LPS chemotype was still maximal (Fig.6). Supposedly, serum increased LPS-induced NO release, which neutralized superoxide and resulted in smoothing out the specificity of various LPS chemotypes (Fig.6.).

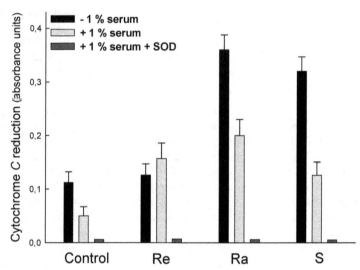

Fig. 6. Effect of serum on LPS-induced superoxide production in human PMNLs. 10^6 PMNLs/ml were incubated for 30 min at 37 °C without (control) or with 5 µg/ml LPS on a collagen-coated surface, in the medium without (- 1% serum) or with 1 % serum (heat inactivated human serum). When indicated, 300 u/ml superoxide dismutase (SOD) were added. Superoxide production was measured as cytochrome c reduction, as described in Methods.

3.3 NO levels dictate the signaling pathway to phagocytosis and LT synthesis in PMNL

We addressed the role for NO in phagocytosis of opsonized zymosan (OZ) influenced by various LPS chemotypes. OZ was prepared by incubating zymosan particles (dried cell walls of Saccharomyces cerevisiae) with autologous serum. Using phase-contrast microscopy we determined the phagocytic index in the cells exposed to 1 µg/ml of LPS species (Re, Ra, S) from *Salmonella* enterica serovar typhimurium for 30 min prior to OZ addition (for additional 5 min). The role for NO in distinct effects of various LPS chemotypes is clearly evident in phagocytosis of OZ by PMNL (Fig. 7).

Ra LPS mutant caused maximal increase in the index as compared to the control measurement. The S- and Re- forms were less effective than the Ra-form and the resulting pattern of their efficacy can be presented as Ra >S >Re.

In Fig. 8 scanning electron microscopy photos illustrate phagocytosis of OZ by untreated PMNLs (Control + OZ) and cells preincubated for 30 min with 1 µg/ml Ra LPS (Ra LPS + OZ). Scanning electron microscopy studies revealed that LPS-treated neutrophils engulfed simultaneously more particles of opsonized zymosan (OZ) than control cells (Fig.8).

We investigated how the LPS forms augment neutrophil phagocytosis in the presence of NOS inhibitor (500 µM L-NAME). L-NAME produced an increase in the effects of Re and S LPS chemotypes and furthermore attenuated the difference between effects of various LPS chemotypes (Fig.7). These data pointed out that the potency of different LPS chemotypes to activate neutrophil phagocytosis is largely due to their ability to induce NO synthesis.

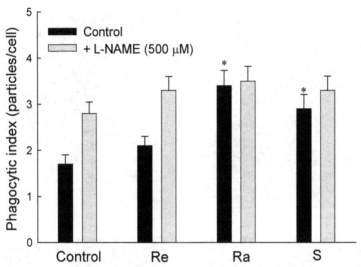

Fig. 7. Effect of LPS chemotypes on OZ uptake by PMNL. Phagocytic index was assessed by light microscopy 5 min after OZ addition to PMNLs, pretreated for 30 min at 37 °C without (control) or with 1 µg/ml of different LPS species and 500 µM L-NAME. * P < 0.05 vs corresponding control.

The phagocytosis of zymosan is a good experimental model to study leukotriene synthesis in PMNLs. Leukotrienes constitutes a family of inflammatory mediators, being formed in PMNL during phagocytosis of bacteria or zymosan particles. The opsonization of zymosan with normal serum resulted in enhanced activation of LT synthesis in PMNLs (Fig.9). LPS-priming of neutrophils further increased LT synthesis, with the effect increasing in the order Re < S < Ra LPS, as we published recently (Zagryazhskaya, et al., 2010). 100 µM and 500 µM L-NAME decreased the specificity of LPS chemotypes (Zagryazhskaya, et al., 2010).

In contrast, 10 µM diphenileleiodonium chloride (DPI) emphasized the effects of Re and S LPS (Fig 10). The antioxidant agent diphenileleiodonium inhibits NADPH oxidase-mediated ROS formation, and also inhibits other flavo-enzymes such as NO synthase and xanthine oxidase (Wind et al., 2010). In our experiments, inhibition of NO release attenuated the specificity of LPS chemotypes, but when we simultaneously inhibited NO and ROS formation, the chemotypes demonstrated the most prominent variation of their effects on LT synthesis (Fig.10). We suggest that LT synthesis is regulated by various LPS chemotypes via multiple mechanisms, and peroxynitrite is involved in this regulation. The specificity of LPS species is mainly dependent on nitric oxide generation induced by LPS. The published data concerning NO interaction with 5-LO admit various interpretations however, the most recent findings support the inhibitory effect of NO on 5-LO synthetic capacity (Coffey et al., 2000). Furthermore, we revealed that minimal NO synthesis facilitated OZ uptake, adhesion, LT and O2- production, as it was observed in the cells primed with Ra LPS. As soon as we inhibited NOS with L-NAME, the other LPS forms, Re and S chemotypes, exhibited comparable capacity to stimulate OZ uptake, LT and O2- production.

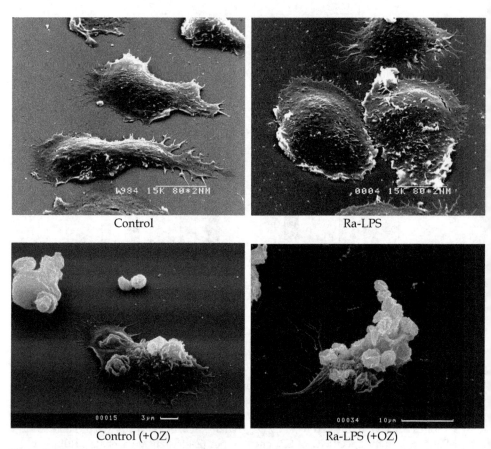

Fig. 8. Scanning electron microscopy of PMNLs untreated (control) or treated with 1 µg/ml Ra-LPS for 30 min (Ra-LPS). OZ uptake in untreated PMNLs (Control + OZ) and PMNLs exposed to Ra-LPS chemotype (Ra-LPS + OZ).

NO can inhibit 5-lipoxygenase directly (Coffey, et al., 2000) and via activation of soluble guanilate cyclase (Coffey, et al., 2008). Peroxynitrite (PN), formed by NO and superoxide, can cause inhibition of 5-LO (Coffey, et al., 2001), as well as 5-LO activation by increasing 'peroxide tone' of the cell (Goodwin et al., 1999; Ullrich & Kissner, 2006). Complex interplay between NO, superoxide and PN is obviously involved in fine regulation of LT synthesis. When we inhibited both NO and PN in incubations with DPI, we revealed huge activation of LT synthesis in Ra- and S- LPS primed cells (Fig 10).

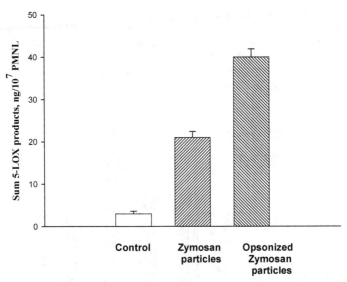

Fig. 9. Opsonized zymosan (OZ) activates LT synthesis in PMNLs. PMNLs suspension (2 × 10^7 cells) was incubated at 37 °C for 30 min, then zymosan particles and opsonized zymosan particles were added where indicated for 20 min. Products of the 5-LO pathway were determined by HPLC, as described in Materials in Methods.

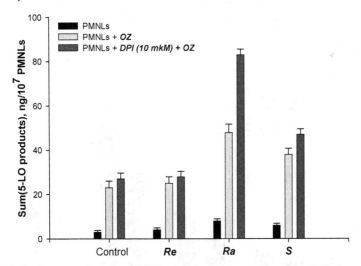

Fig. 10. 5-LO products synthesis in PMNLs is selectively regulated by LPS chemotypes. PMNLs (2 × 10^7) were treated for 30 min at 37 °C without (control) or with 5 μg/ml of various LPS forms and 10 μM diphenileleiodonium chloride (DPI), and then stimulated for 30 min with 2 mg/ml OZ.

Red blood cells are known to consume NO (Romero et al., 2006). In the vascular space, where phagocytes are relatively rare, particles that have been opsonized by complement are immobilized to the surface of red blood cells for further clearance by phagocytes (Pilsczek et al., 2005). When we added RBC in incubations with neutrophils, we found higher effects of LPS on LT synthesis in the presence of red blood cells. This effect was observed in PMNL interaction with OZ (Fig.11) and with opsonized bacteria (Fig.12). Establishing which mechanisms of NOS and NADPH-oxidase activation and signaling are essential for phagocytosis and 5-LO activation is the next objective.

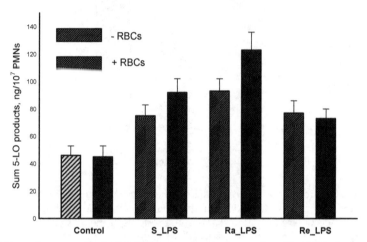

Fig. 11. 5-LO products synthesis in PMNLs is selectively regulated by LPS chemotypes in the presence of red blood cells (RBC). PMNLs (2×10^7) without or with RBC (5×10^7) were preincubated for 30 min without (control) or with 5µg/ml LPS, then stimulated for 20 min with 2mg/ml OZ.

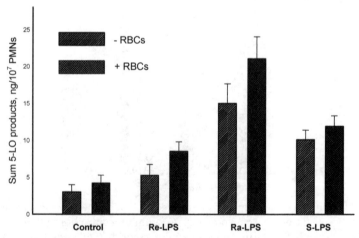

Fig. 12. 5-LO products synthesis in PMNLs is selectively regulated by LPS chemotypes in the presence of red blood cells (RBC). PMNLs (2×10^7) without or with RBC (5×10^7) were preincubated for 30 min without (control) or with 5µg/ml LPS, then stimulated for 20 min with 3×10^8 OS (opsonized *S. Typhimurium* cells).

4. Conclusion

The regulation of neutrophil adherence, phagocytosis and leukotriene synthesis by various LPS chemotypes from S. Typhimurium has received little attention in scientific literature, which prompted us to this study. We presented data on the regulation of neutrophil cellular responses by three LPS species from *Salmonella enterica* serovar *typhimurium* with different increasing chain lengths, namely Re mutant SL 1181 (lipid A + 2 KDO residues), Ra mutant TV 119 (comprising lipid A and complete core) and S - form smooth LPS which possesses all three main components of endotoxin structure (lipid A, core and O-antigen). Our investigation supports the hypothesis that NO plays a crucial role in regulation of LPS-induced phagocytosis and leukotriene synthesis in neutrophils. High levels of endogenous NO inhibit 5-LO activity and leukotriene synthesis, and erythrocytes constitute an important 'sink' for NO and its product peroxynitrite. When excess NO is consumed by red blood cells, we found distinct and significant priming of neutrophils by LPS chemotypes. We conclude that LPS and red blood cells mediate activation of leukotriene synthesis in PMNL using NO release as intra- and intercellular regulatory mechanism. These data with LPS chemotypes contribute to the understanding of the basic factors involved in the regulation of neutrophil responses to LPS.

Author for correspondence: Galina F. Sud'ina, A.N.Belozersky Institute of Physico-Chemical Biology, Moscow State University, Moscow 119991, Russia; tel. +7-495-9393174; fax +7-495-9393181; e-mail: sudina@genebee.msu.ru

5. Acknowledgments

This work was supported by the Russian Foundation for Basic Research grants 10-04-01479 and 09-04-00367.

6. References

Alexander, C. & Rietschel, E.T. (2001). Bacterial lipopolysaccharides and innate immunity. *Journal of Endotoxin Research*, Vol.7. No.3, pp. 167-202, ISSN 1743-2839.

Altavilla, D.; Squadrito, F.; Bitto, A.; Polito, F.; Burnett, B.P.; Di Stefano, V. & Minutoli L. (2009). Flavocoxid, a dual inhibitor of cyclooxygenase and 5-lipoxygenase, blunts pro-inflammatory phenotype activation in endotoxin-stimulated macrophages. *British Journal of Pharmacology*, Vol.157, No.8, pp.1410-1418, ISSN 1476-5381.

Amin, A.R.; Attur, M.; Vyas, P.; Leszczynska-Piziak, J.; Levartovsky, D.; Rediske, J.; Clancy, R.M.; Vora, K.A. & Abramson, S.B. (1995). Expression of nitric oxide synthase in human peripheral blood mononuclear cells and neutrophils. *Journal of inflammation*, Vol.47, No.4, pp.190–205, ISSN 1078-7852.

Baldridge, J.R. & Crane, R.T. (1999). Monophosphoryl lipid A (MPL) formulations for the next generation of vaccines. *Methods*, Vol.19, No.1, pp.103-107, ISSN 1095-9130.

Banick, P.D.; Chen, Q.; Xu, Y.A. & Thom, S.R. (1997). Nitric oxide inhibits neutrophil beta 2 integrin function by inhibiting membrane-associated cyclic GMP synthesis. *Journal of cellular physiology*, Vol.172, No.1, pp.12-24, ISSN 1097-4652.

Beckman, J. S.; Beckman, T. W.; Chen, J.; Marshall, P. A. & Freeman, B. A. (1990). Apparent hydroxyl radical production by peroxynitrite: implications for endothelial injury fromnitric oxide and superoxide. *Proceedings of the National Academy of Sciences of the United States of America*, Vol.87, No.4, pp.1620–1624, ISSN 1091-6490.

Borregaard, N. (2010). Neutrophils, from marrow to microbes. *Immunity*, Vol. 33, No.5, pp. 657–670, ISSN 1097-4180.

Bravo, D.; Hoare, A.; Silipo, A.; Valenzuela, C.; Salinas, C.; Alvarez, S.A.; Molinaro, A.; Valvano, M.A. & Contreras, I. (2011). Different sugar residues of the lipopolysaccharide outer core are required for early interactions of *Salmonella* enterica serovars Typhi and Typhimurium with epithelial cells. *Microbial pathogenesis*, Vol.50, No.2, pp.70–80, ISSN 1096-1208.

Carreras, M.C.; Poderoso, J.J.; Cadenas, E. & Boveris, A. (1996). Measurement of nitric oxide and hydrogen peroxide production from human neutrophils. Methods in enzymology, Vol.269, pp.65–75, ISSN 1557-7988.

Cedergren, J.; Follin, P.; Forslund, T.; Lindmark, M.; Sundqvist, T.; & Skogh T. (2003). Inducible nitric oxide synthase (NOS II) is constitutive in human neutrophils. *APMIS*, Vol.111, No.10, pp.963-968, ISSN 1600-0463.

Chatterjee, M.; Saluja, R.; Kanneganti, S.; Chinta, S. & Dikshit, M. (2007). Biochemical and molecular evaluation of neutrophil NOS in spontaneously hypertensive rats. *Cellular and molecular biology (Noisy-le-Grand, France)*, Vol. 53, No.1, pp.84–93, ISSN 1165-158X.

Chatterjee, M.; Saluja, R.; Kumar, V.; Jyoti, A.; Kumar, Jain G.; Kumar, Barthwal M. & Dikshit, M. (2008). Ascorbate sustains neutrophil NOS expression, catalysis, and oxidative burst. *Free radical biology & medicine*, Vol.45, No.8, pp.1084–1093, ISSN 1873-4596.

Coffey, M.J.; Phare, S.M. & Peters-Golden, M. (2000) Prolonged exposure to lipopolysaccharide inhibits macrophage 5-lipoxygenase metabolism via induction of nitric oxide synthesis. *Journal of immunology*, Vol.165, No.7, pp.3592-3598, ISSN 1550-6606.

Coffey, M.J.; Phare, S.M. & Peters-Golden, M. (2001). Peroxynitrite-induced nitrotyrosination of proteins is blocked by direct 5-lipoxygenase inhibitor zileuton. *The Journal of pharmacology and experimental therapeutics*, Vol.299, No.1, pp.198-203, ISSN 1521-0103.

Coffey, M.J.; Phare, S.M.; Luo, M. & Peters-Golden, M. (2008). Guanylyl cyclase and protein kinase G mediate nitric oxide suppression of 5-lipoxygenase metabolism in rat alveolar macrophages. *Biochimica et biophysica acta*, Vol. 1781, No.6-7, pp.299-305, ISSN 0006-3002.

Collin, M.; Rossi, A.; Cuzzocrea, S.; Patel, N.S.; Di Paola, R.; Hadley, J.; Collino, M.; Sautebin, L. & Thiemermann, C. (2004). Reduction of the multiple organ injury and dysfunction caused by endotoxemia in 5-lipoxygenase knockout mice and by the 5-lipoxygenase inhibitor zileuton. *Journal of leukocyte biology*, Vol.76, No.5, pp.961-970, ISSN 1938-3673.

Curnutte, J.T. & Babior, B.M. (1974). Biological defense mechanisms. The effect of bacteria and serum on superoxide production by granulocytes. *The Journal of clinical investigation*, Vol.53, No.6, pp.1662-1672, ISSN 1558-8238.

Cuzzocrea, S.; Rossi, A.; Serraino, I.; Di Paola, R.; Dugo, L.; Genovese, T.; Britti, D.; Sciarra, G.; De Sarro, A.; Caputi, A.P. & Sautebin, L. (2003). 5-lipoxygenase knockout mice exhibit a resistance to acute pancreatitis induced by cerulein. *Immunology*, Vol.110, No.1, pp.120-130, ISSN 1365-2567.

Cuzzocrea, S.; Rossi, A.; Serraino, I.; Di Paola, R.; Dugo, L.; Genovese, T.; Britti, D.; Sciarra, G.; De Sarro, A.; Caputi, A.P. & Sautebin, L. (2004). Role of 5-lipoxygenase in the multiple organ failure induced by zymosan. *Intensive care medicine*, Vol.30, No.10, pp.1935-1943, ISSN 1432-1238.

Doerfler, M.E.; Weiss, J.; Clark, J.D. & Elsbach P. (1994). Bacterial lipopolysaccharide primes human neutrophils for enhanced release of arachidonic acid and causes phosphorylation of an 85-kD cytosolic phospholipase A2. *The Journal of clinical investigation*, Vol.93, No.4, pp.1583-1591, ISSN 1558-8238.

Drath, D.B. & Karnovsky, M.L. (1975). Superoxide production by phagocytic leukocytes. *The Journal of experimental medicine*, Vol.141, No.1, pp.257-262, ISSN 1540-9538.

Evans, T.; Carpenter, A.; Kinderman, H. & Cohen, J. (1993). Evidence of increased nitric oxide production in patients with the sepsis syndrome. *Circulatory shock*, Vol.41, No.2, pp.77-81, ISSN 0092-6213.

Fitzgerald, C.; Collins, M.; van Duyne, S.; Mikoleit, M.; Brown, T. & Fields, P. (2007). Multiplex, bead-based suspension array for molecular determination of common Salmonella serogroups. *Journal of Clinical Microbiology*, Vol.45, No.10, pp.3323–3334, ISSN 1098-660X.

de Frutos; Sánchez de Miguel, L.; Farré, .;, Gómez, J.; Romero, J.; Marcos-Alberca, P.; Nuñez, A.; Rico, L. & López-Farré, A. (2001). Expression of an endothelial-type nitric oxide synthase isoform in human neutrophils: modification by tumor necrosis factor-alpha and during acute myocardial infarction. *Journal of the American College of Cardiology*, Vol.37, No.3, pp.800-807, ISSN 1558-3597.

Galanos, C. & Freudenberg, M.A. (1993). Mechanisms of endotoxin shock and endotoxin hypersensitivity. *Immunobiology*, Vol.187, No.3-5, pp.346-356, ISSN 1878-3279.

Galkina, S.I.; Dormeneva, E.V.; Bachschmid, M.; Pushkareva, M.A.; Sud'ina, G.F. & Ullrich V. (2004). Endothelium-leukocyte interactions under the influence of the superoxide-nitrogen monoxide system. *Medical science monitor*, Vol.10, No.9, pp.BR307-BR316, ISSN 1643-3750.

Galkina, S.I.; Romanova, J.M.; Stadnichuk, V.I.; Molotkovsky, J.G.; Sud'ina, G.F. & Klein, T. (2009). Nitric oxide-induced membrane tubulovesicular extensions (cytonemes) of human neutrophils catch and hold *Salmonella* enterica serovar Typhimurium at a distance from the cell surface. *FEMS immunology and medical microbiology*, Vol.56, No.2, pp.162-171, ISSN 1574-695X.

Gereda, J.E.; Leung, D.Y.; Thatayatikom, A.; Streib, J.E.; Price, M.R.; Klinnert, M.D. & Liu, A.H. (2000). Relation between house-dust endotoxin exposure, type 1 T-cell development, and allergen sensitisation in infants at high risk of asthma. *Lancet*, Vol.355, No.9216, pp.1680-1683, ISSN 1474-547X.

Gomes, N.E.; Brunialt, M.K.; Mendes, M.E.; Freudenberg, M.; Galanos, C. & Salomão R. (2010). Lipopolysaccharide-induced expression of cell surface receptors and cell activation of neutrophils and monocytes in whole human blood. *Brazilian journal of medical and biological research*, Vol.43, No.9, pp.853-858, ISSN 1414-431X.

Gomez-Jimenez, J.; Salgado, A.; Mourelle, M.; Martin, M.C.; Segura, R.M.; Peracaula, R. & Moncada, S. (1995). L-arginine: nitric oxide pathway in endotoxemia and human septic shock. *Critical care medicine*, Vol.23, No.2, pp.253-258, ISSN 1530-0293.

Goodwin, D.C.; Landino, L.M. & Marnett, L.J. (1999). Reactions of prostaglandin endoperoxide synthase with nitric oxide and peroxynitrite. *Drug metabolism reviews*, Vol.31, No.1, pp.273-294, ISSN 1097-9883.

Graham, I.L. & Brown, E.J. (1991). Extracellular calcium results in a conformational change in Mac-1 (CD11b/CD18) on neutrophils. Differentiation of adhesion and phagocytosis functions of Mac-1. *Journal of immunology*, Vol.146, No.2, pp.685-691, ISSN 1550-6606.

Greenberg, S.S.; Ouyang, J.; Zhao, X. & Giles, T.D. (1998). Human and rat neutrophils constitutively express neural nitric oxide synthase mRNA. *Nitric Oxide*, Vol.2, No.3, pp.203-212, ISSN 1089-8611.

Guthrie, L.A., McPhail,L.C , Henson, P.M. & Johnston, R.B. Jr. (1984). Priming of neutrophils for enhanced release of oxygen metabolites by bacterial lipopolysaccharide. Evidence for increased activity of the superoxide-producing enzyme. *The Journal of experimental medicine*, Vol.160, No.6, pp.1656-1671, ISSN 1540-9538.

Jakubowski, A.; Maksimovich, N.; Olszanecki, R.; Gebska, A.; Gasser, H.; Podesser, B,K.; Hallström, S.; & Chlopicki, S. (2009). S-nitroso human serum albumin given after LPS challenge reduces acute lung injury and prolongs survival in a rat model of endotoxemia. *Naunyn-Schmiedeberg's archives of pharmacology*, Vol.379, No.3, pp.281-290, ISSN 1432-1912.

Jean-Baptiste, E. (2007). Cellular mechanisms in sepsis. *Journal of intensive care medicine*, Vol.22, No.2, pp.63-72, ISSN 1525-1489.

Jourd'heuil, D.; Miranda, K. M.; Kim, S. M.; Espey, M. G.; Vodovotz, Y.; Laroux, S.; Mai, C. T.; Miles, A. M.; Grisham, M. B. & Wink, D. A. (1999). The oxidative and nitrosative chemistry of the nitric oxide/superoxide reaction in the presence of bicarbonate. *Archives of biochemistry and biophysics*, Vol.365, No.1, pp.92-100, ISSN 1096-0384.

Jourd'heuil, D.; Jourd'heuil, F.L.; Kutchukian, P.S.; Musah, R.A.; Wink, D.A. & Grisham MB. (2001). Reaction of superoxide and nitric oxide with peroxynitrite. Implications for peroxynitrite-mediated oxidation reactions in vivo. *The Journal of biological chemistry*, Vol.276, No.31, pp.28799-28805, ISSN 1083-351X.

Korhonen, R.; Lahti, A.; Kankaanranta, H. & Moilanen E. (2005). Nitric oxide production and signaling in inflammation. *Current drug targets. Inflammation and allergy*, Vol.4, No.4, pp.471-479, ISSN 1568-010X.

Kosonen, O.; Kankaanranta, H.; Malo-Ranta, U. & Moilanen E. (1999). Nitric oxide-releasing compounds inhibit neutrophil adhesion to endothelial cells. *European journal of pharmacology*, Vol.382, No.2, pp.111-117, ISSN 1879-0712.

Kowarz, L.; Coynault, C.; Robbe-Saule, V. & Norel F. (1994). The *Salmonella* typhimurium katF (rpoS) gene: cloning, nucleotide sequence, and regulation of spvR and spvABCD virulence plasmid genes. *Journal of bacteriology*, Vol.176, No.22, pp.6852-6860, ISSN 1098-5530.

Kubes, P.; Suzuki, M. & Granger DN. (1991). Nitric oxide: an endogenous modulator of leukocyte adhesion. *Proceedings of the National Academy of Sciences of the United States of America*, Vol.88, No.11, pp.4651-4655, ISSN 1091-6490.

Kubes, P.; Kurose, I. & Granger, D.N. (1994). NO donors prevent integrin-induced leukocyte adhesion but not P-selectin-dependent rolling in postischemic venules. *The American journal of physiology*, Vol.267, No.3, Pt.2, pp.H931-H937, ISSN 0002-9513.

Lancaster Jr., J. R. (2002). The physical properties of nitric oxide, In: *Nitric oxide – biology and pathobiology*. L.J.Ignarro, (Ed.), 209-224, Academic Press, San Diego.

Li, H. & Forstermann, U. (2000). Nitric oxide in the pathogenesis of vascular disease. *The Journal of pathology*, Vol.190, No.3, pp.244-254, 1096-9896, ISSN 1096-9896.

Lopez-Bojorquez, L.N.; Dehesa, A.Z. & Reyes-Teran G. (2004). Molecular mechanisms involved in the pathogenesis of septic shock. *Archives of medical research*, Vol.35, No.6, pp.465-479, ISSN 1873-5487.

Luhm, J.; Schromm, A.B.; Seydel, U.; Brandenburg, K.; Wellinghausen, N.; Riedel, E.; Schumann, R.R. & Rink L. (1998). Hypothermia enhances the biological activity of lipopolysaccharide by altering its fluidity state. *European journal of biochemistry*, Vol.256, No.2, pp.325-333, ISSN 1432-1033.

Mancuso, P.; Nana-Sinkam, P. & Peters-Golden, M. (2001). Leukotriene B4 augments neutrophil phagocytosis of Klebsiella pneumoniae. *Infection and immunity*, Vol.69, No.4, pp.2011-2016, ISSN 1098-5522.

Matera, G.; Cook, J.A.; Hennigar, R.A.; Tempel, G.E.; Wise, W.C.; Oglesby, T.D. & Halushka, P.V. (1988). Beneficial effects of a 5-lipoxygenase inhibitor in endotoxic shock in the rat. *The Journal of pharmacology and experimental therapeutics*, Vol.247, No.1, pp.363-371, ISSN 1521-0103.

Miles, A.M.; Owens, M.W.; Milligan, S.; Johnson, G.G.; Fields, J.Z.; Ing, T.S.; Kottapalli, V.; Keshavarzian, A. & Grisham, M.B. Nitric oxide synthase in circulating vs. extravasated polymorphonuclear leukocytes. *Journal of leukocyte biology*, Vol.58, No.5, pp.616–622, ISSN 1938-3673.

Misko, T.P.; Schilling, R.J.; Salvemini, D.; Moore, W.M. & Currie, M.G. (1993) A fluorometric assay for the measurement of nitrite in biological samples. *Analytical biochemistry*, Vol.214, No.1, pp.11-16, ISSN 1096-0309.

Molero, L.; Garcia-Duran, M.; Diaz-Recasens, J.; Rico, L.; Casado, S. & Lopez-Farre, A. (2002). Expression of estrogen receptor subtypes and neuronal nitric oxide synthase in neutrophils from women and men: regulation by estrogen. *Cardiovascular research*, Vol.56, No.1, pp.43-51, ISSN 1755-3245.

Moncada, S. & Higgs, A. (1993) The L-arginine-nitric oxide pathway. *The New England journal of medicine*, Vol.329, No.27, pp.2002-2012, ISSN 1533-4406.

Mühlradt, P.F.; Menzel, J.; Golecki, J.R. & Speth V. (1974). Lateral mobility and surface density of lipopolysaccharide in the outer membrane of *Salmonella* typhimurium. *European journal of biochemistry*, Vol.43, No.3, pp.533-539, ISSN 1432-1033.

Müller-Loennies, S.; Brade, L. & Brade H. (2007). Neutralizing and cross-reactive antibodies against enterobacterial lipopolysaccharide. International journal of medical microbiology, Vol.297, No.5, pp.321-340, ISSN 1618-0607.

Muniyappa, R.; Srinivas, P.R.; Ram, J.L.; Walsh, M.F. & Sowers, J.R. (1998). Calcium and protein kinase C mediate high-glucose-induced inhibition of inducible nitric oxide synthase in vascular smooth muscle cells. *Hypertension*, Vol.31, No.1, Pt.2, pp.289-295, ISSN 1524-4563.

Nath, J. & Powledge, A. (1997) Modulation of human neutrophil inflammatory responses by nitric oxide: studies in unprimed and LPS-primed cells. *Journal of leukocyte biology*, Vol.62, No.6, pp.805-816, ISSN 1938-3673.

Okamura, N. & Spitznagel, J.K. (1982). Outer membrane mutants of *Salmonella* typhimurium LT2 have lipopolysaccharide-dependent resistance to the bactericidal activity of anaerobic human neutrophils. *Infection and immunity*, Vol.36, No.3, pp.1086-1095, ISSN 1098-5522.

Ohki, K.; Amano, F.; Yamamoto, S. & Kohashi O. (1999). Suppressive effects of serum on the LPS-induced production of nitric oxide and TNF-alpha by a macrophage-like cell line, WEHI-3, are dependent on the structure of polysaccharide chains in LPS. *Immunology and cell biology*, Vol.77, No.2, pp.143-152, ISSN 1440-1711.

Olsthoorn, M.M.; Petersen, B.O.; Schlecht, S.; Haverkamp, J.; Bock, K.; Thomas-Oates, J.E. & Holst O. (1998). Identification of a novel core type in Salmonella lipopolysaccharide. Complete structural analysis of the core region of the lipopolysaccharide from *Salmonella* enterica sv. Arizonae O62. *The Journal of biological chemistry*, Vol.273, No.7, pp.3817-3829, ISSN 1083-351X.

Perepelov, A.V.; Liu, B.; Shevelev, S.D.; Senchenkova, S.N.; Hu, B.; Shashkov, A.S.; Feng, L.; Knirel, Y.A. & Wang L. (2010). Structural and genetic characterization of the O-antigen of *Salmonella* enterica O56 containing a novel derivative of 4-amino-4,6-dideoxy-D-glucose. *Carbohydrate research*, Vol.345, No.13, pp.1891-1895, ISSN 1873-426X.

Pilsczek, F.H.; Nicholson-Weller, A. & Ghiran I. (2005). Phagocytosis of Salmonella montevideo by human neutrophils: immune adherence increases phagocytosis, whereas the bacterial surface determines the route of intracellular processing. The Journal of infectious diseases, Vol.192, No.2, pp.200–209, ISSN 1537-6613.

Pugliese, C.; LaSalle, M.D. & DeBari, V.A. (1988). Relationships between the structure and function of lipopolysaccharide chemotypes with regard to their effects on the human polymorphonuclear neutrophil. *Molecular immunology*, Vol.25, No.7, pp.631-637, ISSN 1872-9142.

Raetz, C.R. & Whitfield, C. (2002). Lipopolysaccharide endotoxins. Annual review of biochemistry, Vol.71, pp.635-700, ISSN 1545-4509.

Rietschel, E.T.; Brade, H.; Hols,t O.; Brade, L.; Muller-Loennies, S.; Mamat, U.; Zahringer, U.; Beckmann, F.; Seydel, U.; Brandenburg, K.; Ulmer, A.J.; Mattern, T.; Heine, H.; Schletter, J.; Loppnow, H.; Schonbeck, U.; Flad, H.D.; Hauschildt, S.; Schade, U.F.; Padova, F.D.; Kusumoto, S. & Schumann R.R. (1996). Bacterial endotoxin: Chemical constitution, biological recognition, host response, and immunological detoxification. *Current topics in microbiology and immunology*, Vol.216, pp.39-81, ISSN 0070-217X.

Romero, N.; Denicola, A. & Radi, R. (2006). Red Blood Cells in the Metabolism of Nitric Oxide-derived Peroxynitrite. *IUBMB Life*, Vol.58, No.10, pp.572-580, ISSN 1521-6551.

Ruchaud-Sparagano, M.H.; Ruivenkamp, C.A.; Riches, P.L.; Poxton, I.R. & Dransfield I. (1998). Differential effects of bacterial lipopolysaccharides upon neutrophil function. *FEBS Letters*, Vol.430, No.3, pp.363-369, ISSN 1873-3468.

Samuelsson, B. (1983). Leukotrienes: mediators of immediate hypersensitivity reactions and inflammation. Science, Vol.220, No.4597, pp.568–575, ISSN 1095-9203.

Saini, R.; Patel, S.; Saluja, R.; Sahasrabuddhe, A.A.; Singh, M.P.; Habib, S.; Bajpai, V.K. & Dikshit, M. (2006). Nitric oxide synthase localization in the rat neutrophils: Immunocytochemical, molecular, and biochemical studies. *Journal of leukocyte biology*, Vol.79, No.3, pp.519–528, ISSN 1938-3673.

Schierwagen, C.; Bylund-Fellenius, A.C. & Lundberg, C. (1990) Improved method for quantification of tissue PMN accumulation measured by myeloperoxidase activity. *Journal of pharmacological methods*, Vol.23, No.3, pp.179-186, ISSN 0160-5402.

Schmidt, H.H.; Seifert, R. & Böhme, E. (1989). Formation and release of nitric oxide from human neutrophils and HL-60 cells induced by a chemotactic peptide, platelet activating factor and leukotriene B4. *FEBS Letters*, Vol.244, No.2, pp.357-360, ISSN 1873-3468.

Shnyra, A.; Hultenby, K. & Lindberg, A.A. (1993). Role of the physical state of Salmonella lipopolysaccharide in expression of biological and endotoxic properties. *Infection and immunity*, Vol.61, No.12, pp.5351-5360, ISSN 1098-5522.

Sud'ina, G.F.; Tatarintsev, A.V.; Koshkin, A.A.; Zaitsev, S.V.; Fedorov, N.A. & Varfolomeev, S.D. (1991). The role of adhesive interactions and extracellular matrix fibronectin from human polymorphonuclear leukocytes in the respiratory burst. *Biochimica et biophysica acta*, Vol.1091, No.3, pp.257-260, ISSN 0006-3002.

Sud'ina, G.F.; Mirzoeva, O.K.; Galkina, S.I.; Pushkareva, M.A. & Ullrich V. (1998). Involvement of ecto-ATPase and extracellular ATP in polymorphonuclear granulocyte-endothelial interactions. *FEBS letters*, Vol.423, No.2, pp.243-248, ISSN 1873-3468.

Sud'ina, G.F.; Brock, T.G.; Pushkareva, M.A.; Galkina, S.I.; Turutin, D.V.; Peters-Golden, M. & Ullrich, V. (2001) Sulphatides trigger polymorphonuclear granulocyte spreading on collagen-coated surfaces and inhibit subsequent activation of 5-lipoxygenase. *The Biochemical journal*, Vol.359, Pt.3, pp.621-629, ISSN 1470-8728.

Szabó, C.; Mitchell, J.A.; Thiemermann, C. & Vane, J.R. (1993). Nitric oxide-mediated hyporeactivity to noradrenaline precedes the induction of nitric oxide synthase in endotoxin shock. *British journal of pharmacology*, Vol.108, No.3, pp.786-792, ISSN 1476-5381.

Takayama, K.; Qureshi, N. & Mascagni P. (1983). Complete structure of lipid A obtained from the lipopolysaccharides of the heptoseless mutant of *Salmonella* typhimurium. *The Journal of biological chemistry*, Vol.258, No.21, pp.12801-12803, ISSN 1083-351X.

Titheradge, M.A. (1999). Nitric oxide in septic shock. *Biochimica et biophysica acta*, Vol.1411, No.2-3, pp.437-455, ISSN 0006-3002.

Toda, A.; Yokomizo, T. & Shimizu, T. (2002). Leukotriene B4 receptors. *Prostaglandins & other lipid mediators*, Vol.68–69, pp.575–585, ISSN 1098-8823.

Tsutsui, M.; Shimokawa, H.; Otsuji, Y.; Ueta, Y.; Sasaguri, Y. & Yanagihara, N. (2009). Nitric oxide synthases and cardiovascular diseases: insights from genetically modified mice. *Circulation journal*, Vol.73, No.6, pp.986-993, ISSN 1347-4820.

Ullrich, V. & Kissner, R. (2006). Redox signaling: bioinorganic chemistry at its best. *Journal of inorganic biochemistry*, Vol.100, No.12, pp.2079-2086, ISSN 1873-3344.

Westphal, O. (1978). Bacterial polysaccharides, In: *Complex carbohydrates*, E.F. Neufeld & V. Ginsburg, (Ed.), Volume 50, Methods in enzymology. University of Virginia. Pp. 1–6.

Westphal, O.; Luederitz, O. (1961). Chemistry of bacterial O-antigens. *Pathologia et microbiologia*, Vol.24, pp.870-889, ISSN 0031-2959.

Wind, S.; Beuerlein, K.; Eucker, T.; Müller, H.; Scheurer, P.; Armitage, M.E.; Ho, H.; Schmidt, H.H. & Wingler, K. (2010). Comparative pharmacology of chemically distinct NADPH oxidase inhibitors. *British journal of pharmacology*, Vol.161, No.4, pp.885-898, ISSN 1476-5381.

Wright, S.D. & Jong, M.T. (1986). Adhesion-promoting receptors on human macrophages recognize Escherichia coli by binding to lipopolysaccharide. *The Journal of experimental medicine*, Vol.164, No.6, pp.1876-1888, ISSN 1540-9538.

Wright, C.D.; Mülsch, A.; Busse, R. & Osswald, H. (1989). Generation of nitric oxide by human neutrophils. *Biochemical and biophysical research communications*, Vol.160, No.2, pp.813-819, ISSN 1090-2104.

Zagryazhskaya, A.N.; Lindner, S.C.; Grishina, Z.V.; Galkina, S.I.; Steinhilber, D. & Sud'ina, G.F. (2010). Nitric oxide mediates distinct effects of various LPS chemotypes on phagocytosis and leukotriene synthesis in human neutrophils. *The international journal of biochemistry & cell biology*, Vol.42, No.6, pp.921-931, ISSN 1878-5875.

Zweier, J. L.; Kuppusamy, P.; Williams, R.; Rayburn, B. K.; Smith, D.; Weisfeldt, M. L. & Flaherty, J. T. (1989). Measurement and characterization of postischemic free radical generation in the isolated perfused heart. *The Journal of biological chemistry*, Vol.264, No.32, pp.18890–18895, ISSN 1083-351X.

5

The Central Nervous System Modulates the Immune Response to *Salmonella*

Rafael Campos-Rodríguez[1], Andres Quintanar Stephano[2],
Maria Elisa Drago-Serrano[3], Edgar Abarca-Rojano[1], Istvan Berczi[2,4],
Javier Ventura-Juárez[5] and Alexandre Kormanovski[1]
[1]*Sección de Estudios de Posgrado e Investigación, Escuela Superior de Medicina*
Instituto Politécnico Nacional
[2]*Departamento de Fisiología y Farmacología, Centro de Ciencias Básicas*
Universidad Autónoma de Aguascalientes, Aguascalientes
[3]*Departamento de Sistemas Biológicos*
Universidad Autónoma Metropolitana Unidad Xochimilco
[4]*Department of Immunology, Faculty of Medicine, University of Manitoba, Winnipeg*
[5]*Departamento de Morfología, Centro de Ciencias Básicas*
Universidad Autónoma de Aguascalientes, Aguascalientes
[1,2,3,5]*México*
[4]*Canada*

1. Introduction

Salmonella infection induces an immune response, the first and principal element of which is a local activation in the intestine. This intestinal response and the systemic response of the immune system have multidirectional interactions with the nervous and endocrine systems (Berczi and Szentivanyi 2003). The central nervous system (CNS) signals the immune system via hormonal and neural pathways, and the immune system signals the CNS through various cytokines. Whereas most information regarding these interactions is related to functions of the systemic immune response (Berczi, Nagy et al. 1981; Chrousos 1995; Madden and Felten 1995; Elenkov, Wilder et al. 2000; Webster, Tonelli et al. 2002; Berczi and Szentivanyi 2003), much less is known about the interactions between the hypothalamus, the pituitary, and local gastrointestinal immune reactions (Berczi, Nagy et al. 1981; Ottaway 1991; Bienenstock 1992; Chrousos 1995; Madden and Felten 1995; Elenkov, Wilder et al. 2000; Webster, Tonelli et al. 2002; Berczi and Szentivanyi 2003; Campos-Rodriguez, Quintanar-Stephano et al. 2006).

The CNS regulates the intestinal immune system through the three divisions of the autonomic nervous system: sympathetic, parasympathetic, and enteric. By signals sent along sympathetic and parasympathetic fibers, the CNS controls the enteric nervous system (ENS),which in turn regulates gastrointestinal functions, including immune functions (Cooke 1986; Ottaway 1991; Gonzalez-Ariki and Husband 1998; Bueno 2000; Spiller 2002). Moreover, the CNS regulates the mucosal immune system through the hypothalamic–pituitary–adrenal (HPA) axis, an essential part of whichis glucocorticoid production (Sternberg 2001; Webster, Tonelli et al. 2002; Jarillo-Luna, Rivera-Aguilar et al. 2008).

Ongoing research to clarify the bidirectional communication between the immune and central nervous systemshas in part been carried out byproducing electrolytic or pharmacologic lesions in several areas of the brain, such basal ganglia, striatum, hypothalamus, hippocampus and thalamus, and then observing the resulting immune response. This approach has been used in our recent studies (Campos-Rodriguez, Quintanar-Stephano et al. 2006; Rivera-Aguilar, Querejeta et al. 2008; Quintanar-Stephano, Abarca-Rojano et al. 2010) to observe the effect of brain lesions on the immune response to *Salmonella* and one of its main components, lipopolysaccharide (LPS). It has been found that brain lesions modify the number and functions of lymphocytes in the spleen, thymus and blood (Jankovic and Isakovic 1973; Payan, McGillis et al. 1986).

The aim of this chapter is to describe the effects of CNS lesions on the immune response to *Salmonella*. The mechanisms are explored by which these lesions affect the systemic and intestinal immune responses. Since the production of intestinal IgAis fundamental in the protection against *Salmonella* invasion, an evaluation is made of the role of neurotransmitters, glucocorticoids and neuroendocrine molecules in the regulation of such production.

2. Hypophysectomy and neurointermediatepituitary lobectomy reduce the humoral immune response to *Salmonella enterica* serovar Typhimurium

The hypothalamus induces the secretion of anterior pituitary hormones, and in this way the CNS can have both an immunostimulatory and immunosuppressor effect. In this sense, the immune response is stimulated mainly by the release of growth hormone (GH) and prolactin (PRL) (Berczi, Nagy et al. 1981; Block, Locher et al. 1981; Nagy and Berczi 1981; Berczi, Nagy et al. 1984; Edwards, Yunger et al. 1991; Nagy and Berczi 1991; Edwards, Arkins et al. 1992; Nagy and Bercz 1994; Madden and Felten 1995; Berczi and Szentivanyi 2003), and inhibited by the hypothalamic-pituitary-adrenocortical (HPA) axis, which causes the release of adrenocorticotropin (ACTH), which in turn stimulates the secretion of adrenocortical glucocorticoids (Chrousos 1995; Sternberg 2001; Webster, Tonelli et al. 2002). This increase of circulating glucocorticoids (GCs) is caused when the HPA axis is activated during many bacterial and viral infections.

In vivo, we have demonstrated that arginine vasopressin (AVP) released from the posterior pituitary affects humoral and cell mediated immune responses (Organista-Esparza, Tinajero-Ruelas et al. 2003; Quintanar-Stephano, Kovacs et al. 2004; Quintanar-Stephano, Organista-Esparza et al. 2004; Quintanar-Stephano, Chavira-Ramirez et al. 2005; Quintanar-Stephano, Organista-Esparza et al. 2005; Quintanar-Stephano, Abarca-Rojano et al. 2010). Regarding *Salmonella enterica* serovar Typhimurium (*Salmonella typhimurium*) infection, there is experimental evidence that pituitary hormones have a protective effect (Edwards, Yunger et al. 1991; Edwards, Ghiasuddin et al. 1992). For instance, the increased susceptibility to intraperitoneal *Salmonella typhimurium* infection found in hypophysectomized (HYPOX) rats is countered by GH treatment, which restores normal resistance. In intact rats and mice, GH and PRL enhance resistance to *Salmonella typhimurium* infection through an increase in phagocytosis and intracellular destruction of bacteria by peritoneal macrophages. *Salmonella* or other challenges to the immune system, such as immobilization stress and burn injury, increase the levels of GCs, which in turn increase bacterial translocation from the gastrointestinal tract to the mesenteric lymph nodes (Jones, Minei et al. 1990; Fukuzuka, Edwards et al. 2000; Dunn, Ando et al. 2003).

All of the aforementioned suggests that anterior and posterior pituitary hormones participate as stimulating factors in the control of systemic and intestinal immune responses to *Salmonella*. To further explore this idea, we investigated the systemic and intestinal immune responses in HYPOX and neurointermediate pituitary lobectomy (NIL) rats orally infected with nonlethal doses of *Salmonella typhimurium* (Campos-Rodriguez, Quintanar-Stephano et al. 2006).The most relevant results are that the kinetics of intestinal *Salmonella* elimination in sham-operated (SHAM), HYPOX and NIL groupswas similar with no clinical signs of salmonellosis and no mortality.However, nine days after inoculation, the number of *Salmonella typhimurium* cells in Peyer's patches and spleens of HYPOX and NIL groups was higher than in the sham-operated group (P <0.001) (Fig. 1), andthere were a greater number of bacteria in HYPOX than NIL animals (P <0.01). The fact that the total or partial ablation of the hypophysis increased susceptibility to infection after oral inoculation with *Salmonella typhimurium* means that the pituitary gland is required for protection against infection by intraperitoneal *Salmonella* inoculation (Edwards, Yunger et al. 1991).

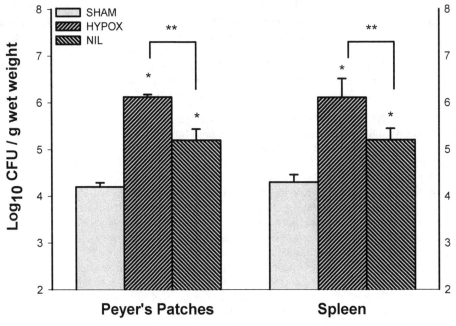

Fig. 1. Persistence of serovar Typhimurium infection in Peyer's patches and spleen. Sham-operated (SHAM), HYPOX, and NIL groups were orally infected and sacrificed 9 days postinoculation. Tissues were aseptically removed and processed for bacterial quantification. Data are expressed as means SD of results from four to six rats per group. In Peyer's patches and spleens, bacterial counts were significantly higher in HYPOX and NIL groups than in the sham-operated group (*, P < 0.001) and significantly higher in the HYPOX group than in the NIL group (**, P < 0.01). (From Campos-Rodriguez et al. Hypophysectomy and Neurointermediate Pituitary Lobectomy Reduce Serum Immunoglobulin M (IgM) and IgG and Intestinal IgA Responses to *Salmonella enterica* Serovar Typhimurium Infection in Rats. Infect Immun. 2006; 74(3):1883-1889).

Most pituitary hormones directly or indirectly modulate inflammatory/immune responses. For example, adrenocorticotropin increases the secretion of GCs, which in turn stimulate the immune function at physiological doses (Munck and Naray-Fejes-Toth 1992; Reichlin 1993; Chrousos 1995; Wiegers and Reul 1998; Sapolsky, Romero et al. 2000). GH, PRL, TSH and - endorphin produced in the anterior pituitary as well as the AVP released from the posterior pituitary have immunopotentiating and proinflammatory properties (Heijnen, Kavelaars et al. 1991; Navolotskaya, Malkova et al. 2002; Klein 2003). Therefore, the differences between NIL and HYPOX rats may be related to the amount of hormones that regulate the immune response located in the anterior and posterior pituitary. Another possible factor is that the partial or total removal of the pituitary may affect the activity of phagocytes, the principal cells of the innate immunity involved in killing Salmonella typhimurium (Mittrucker and Kaufmann 2000; Kirby, Yrlid et al. 2002). It has been demonstrated that peritoneal macrophages from HYPOX rats have an impaired tumor necrosis factor alpha response to in vitro lipopolysaccharide stimulation and are less effective in killing Salmonella typhimurium than those derived from rats with intact pituitaries (Edwards, Lorence et al. 1991). GH injections enhanced resistance of both intact and HYPOX rats following a challenge with Salmonella typhimurium (Edwards, Lorence et al. 1991; Edwards, Ghiasuddin et al. 1992). The enhanced resistance is correlated with the ability of peritoneal macrophages from these animals to generate toxic oxygen metabolites, such as superoxide anion and hydrogen peroxide (Edwards, Ghiasuddin et al. 1992). In addition, GH activates human monocytes for enhanced reactive oxygen intermediate production in vitro (Warwick-Davies, Lowrie et al. 1995; Warwick-Davies, Lowrie et al. 1995; Navolotskaya, Malkova et al. 2002).

An analysis of the secretion of intestinal IgA specific to outer membrane proteins of Salmonella shows that the titers of the specific intestinal IgA response was significantly lower in HYPOX and NIL animals than in the sham-operated group (P< 0.001, Fig. 2), and was also lower in the HYPOX than NIL rats (P <0.001)(Campos-Rodriguez, Quintanar-Stephano et al. 2006).The fact that HYPOX induced a more marked decrease in the humoral immune responses to outer membrane proteins of Salmonella typhimurium than NIL suggests that the hormones melanocyte stimulating hormone (MSH), AVP, and oxytocin from the neurointermediate pituitary lobe may affect adaptive immune responses. The direct anti-inflammatory effects of MSH on immunocytes have been described previously (Catania and Lipton 1993; Blalock 1999; Luger, Scholzen et al. 2003; Taylor 2003). Since NIL eliminates the intermediate lobe— the main source of pituitary -MSH— an increased inflammatory response to Salmonella typhimurium infection may be expected. However, our results show that -MSH from the intermediate pituitary lobe is not involved in the immune response to Salmonella typhimurium infection. Further experiments are required to test this possibility.

Furthermore, in the aforementioned study levels of IgG and IgM were also significantly lower in the HYPOX and NIL animals than in the sham-operated group (Fig. 2), and in HYPOX rats than in the NIL group (Campos-Rodriguez, Quintanar-Stephano et al. 2006). The cause of these reduced humoral immune responses may be the decreased secretion of the neurointermediate pituitary hormones. In previous experiments, we found that in NIL rats there are decreased humoral and cell-mediated immune responses, including: (i) decreased hemagglutination, IgG and IgM responses to sheep red blood cells (Organista-Esparza, Tinajero-Ruelas et al. 2003; Quintanar-Stephano, Kovacs et al. 2004), (ii) decreased contact hypersensitivity to dinitrochlorobenzene (Quintanar-Stephano, Kovacs et al. 2004), and (iii) protection against EAE (Quintanar-Stephano, Chavira-Ramirez et al. 2005). In agreement with these previous findings, our results suggest that the higher colonization of the Peyer's patches and spleens and the decreased IgG, IgM, and IgA responses to *Salmonella typhimurium* may be due to AVP deficiency in the NIL animals. In another study we found that in both HYPOX and NIL rats, there was a decrease in the IgM response to the LPS of *Salmonella typhimurium* (Quintanar-Stephano, Abarca-Rojano et al. 2010). These results support the view that hormones from both pituitary lobes play an important stimulatory/modulatory role in both humoral and cell-mediated immune responses.

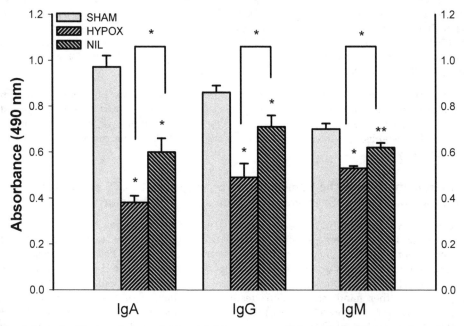

Fig. 2. Intestinal IgA and serum IgG and IgM response to *Salmonella typhimurium*. Intestinal IgA or serum IgG and IgM antibodies were quantified by ELISA using *Salmonella typhimurium* surface antigens. Serum and gut samples were obtained 9 days postinoculation. The samples were assayed in triplicate, and the titers were expressed as the absorbance at 490 nm. Data are expressed as means SD of results from four to six rats per group. The immunoglobulin levels were significantly lower in HYPOX and NIL groups than in the sham operated group (*, P < 0.001; **, P < 0.01) and significantly lower in the HYPOX group than in the NIL group (*, P < 0.001). (From Campos-Rodriguez et al. Hypophysectomy and Neurointermediate Pituitary Lobectomy Reduce Serum Immunoglobulin M (IgM) and IgG and Intestinal IgA Responses to *Salmonella enterica* Serovar Typhimurium Infection in Rats. Infect Immun. 2006; 74(3):1883-1889).

Finally, intestinal elimination of *Salmonella typhimurium* HYPOX and NIL rats was similar to that seen in sham-operated animals. However, it is known that HYPOX animals develop an increased susceptibility to intraperitoneal *Salmonella typhimurium* infection, and that GH and PRL treatments protect the rats against the disease (Edwards, Yunger et al. 1991; Edwards, Ghiasuddin et al. 1992). Similarly, PRL increases resistance to infection in normal mice after intraperitoneal inoculation of *Salmonella typhimurium* (Di Carlo, Meli et al. 1993; Meli, Raso et al. 1996).

Since the immune responses are PRL and GH dependent and no pituitary hormones are produced in the HYPOX animals, how can the formation of anti-*Salmonella typhimurium* IgG, IgM, and IgA immunoglobulins be explained? Perhaps part of the answer lies in an unpublished study with HYPOX animals. After surgery a gradual increase was observed in the plasma PRL levels, which after 7 to 9 weeks post-operation reached 50% of the levels of this hormone found in intact animals (Nagy and Berczi 1991; Quintanar-Stephano and A. Organista- Esparza, unpublished). Although the source of this non-pituitary PRL is not known, one possibility is from T lymphocytes (Draca 1995; Stevens, Ray et al. 2001). The fact that HYPOX rats had a higher number of *Salmonella typhimurium* cells in Peyer's patches and spleen than sham operated and NIL rats suggests that the low serum IgG and IgM and intestinal IgA immunoglobulin levels in HYPOX rats may be due to the insufficient immune-stimulating effect of the non-pituitary PRL (Nagy and Berczi 1991). However, further studies are needed to evaluate this suggestion.

In summary, it can be concluded that through different mechanisms, hormones from both the anterior and neurointermediate pituitary lobes play an important role in the control of systemic and gastrointestinal immune responses to *Salmonella*. However, more experiments are needed to establish the interactions between the hypothalamo-neurohypophysial (AVP) and immune systems.

3. Striatum modulates the humoral immune response to LPS and outer membrane proteins of *Salmonella enterica* serovar Typhimurium

The striatum is implicated in movement and learning (Costall, Naylor et al. 1972; Pycock 1980; Graybiel 1995), and there is increasing evidence that it is involved in the modulation of immune responses, although such evidence is contradictory. Bilateral electrolytic lesions of the caudate nucleus of rats do not reduce the intensity of cell-mediated immune responses or the production of antibodies to bovine serum albumin (BSA) (Jankovic and Isakovic 1973). On the other hand, such lesions result in a reduction of the antibody immune response to sheep red blood cells (SRBC) (Devoino, Alperina et al. 1997; Devoino, Cheido et al. 2001; Nanda, Pal et al. 2005; Rivera-Aguilar, Querejeta et al. 2008). In addition, the destruction of dopaminergic neurons in the substantia nigra or dopaminergic terminals in the caudate nucleus by in situ injection of 6-hydroxydopamine decreases the antibody response and impairs cell-mediated immunity (Deleplanque, Vitiello et al. 1994; Devoino, Alperina et al. 1997; Devoino, Cheido et al. 2001; Filipov, Cao et al. 2002).Furthermore, bilateral lesions of nigrostriatal pathways induced by systemic injections of the neurotoxin 1-methyl-4-phenyl-1,2,3,6- tetrahydropymidine reduce the number of leukocytes, alter lymphocyte populations,decrease proliferation of T lymphocytes induced bymitogens or alloantigens, and modify the synthesis of cytokines (Renoux, Biziere et al. 1989; Bieganowska, Czlonkowska et al. 1993; Shen, Hebert et al. 2005; Engler, Doenlen et al. 2009).

These findings suggest that the nigrostriatal dopaminergic system has an immunostimulatory effect on the humoral and cell-mediated immune response. To test the hypothesis that GABAergic medium-sized spiny neurons in the striatum modulate the humoral immune response, in rats with a bilateral lesion of the striatum provoked by the injection of quinolinic acid we analyzed this response to several antigens (both T-independent and T-dependent antigens), including LPS and outer membrane proteins of *Salmonella typhimurium*. Quinolinic acid produces axon-sparing lesions that result in a loss of GABAergic medium-sized spiny neurons (MSP) in the striatum, while the dopaminergic terminal network originating from cell bodies in the substantia nigra remains unchanged (McGeer and McGeer 1976; Schwarcz, Whetsell et al. 1983; Beal, Kowall et al. 1986).

3.1 Bilateral lesion of the striatum decreases the humoral immune response to TNP-LPS

The serum levels of IgG and IgM antibodies anti-trinitrophenol-lipopolysaccharide (TNP-LPS)(Fig. 3, panel A and B), and the IgA antibodies anti-TNP-LPS in intestinal fluid (Fig. 3, panel C) were significantly lower in rats with a bilateral lesion of the striatum compared with the control group (P < 0.01). These results show that the lesions of the striatum had a prolonged effect on the immune response to this T-independent antigen, indicating that the striatum modulates this type of humoral immune response. On the contrary, a bilateral lesion of the striatum increased the humoral immune response to T-dependent antigens (ovoalbumin, lysozyme and bovine serum albumin).

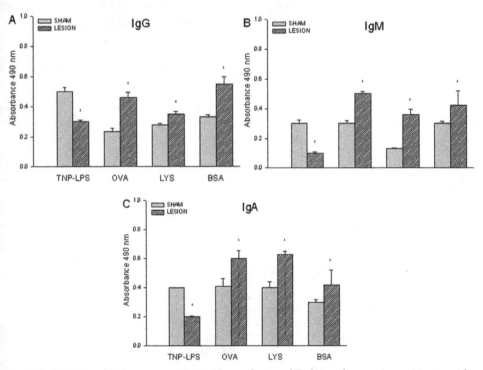

Fig. 3. IgG, IgM and IgA response to T-independent and T-dependent antigens in rats with bilateral lesion of striatum. The antibody response to TI and TD antigens was analyzed in rats that had been lesioned 25 days before immunization. The serum IgM and IgG levels as well as the intestinal IgA levels to the T-independent antigen (TNP-LPS) were significantly lower in lesioned rats than in the sham-operated rats (*P < 0.01). On the contrary, the antibody levels to all the T-dependent antigens (OVA, lysozyme, and BSA) were significantly higher in the lesioned group than in sham operated group (*P < 0.01). (From Rivera-Aguilar et al. Role of the striatum in the humoral immune response to thymus-independent and thymus-dependent antigens in rats. ImmunolLett 2008;120:20-28).

Although the mechanisms by which the lesion of the striatal MSP neurons leads to a decrease in the immune response to TNP-LPS (TI type 1 antigen) are not known, it is likely that they are related to defects in B lymphocyte activation. In fact, the number of IgM+ B cells in the marginal zone of the spleen was significantly lower in lesioned rats than in the control group.However, the mechanisms by which striatal lesions reduce the population of B cells in the spleen are at the present unknown, as are the mechanisms involved in the maturation, selection and long-term survival of immature peripheral B cells (Thomas, Srivastava et al. 2006).

We also found that striatal lesions caused a reduction in the expression of the gene for caveolin-1 and in the number of lymphocytes caveolin-1+ in the spleen (Fig. 4). Caveolin-1, expressed on B-lymphocytes, down-regulates tyrosine phosphorylation of Btk, a molecule that participates in B-cell activation and signaling (Vargas, Nore et al. 2002; Medina, Williams et al. 2006). Caveolin-1 deficient mice have a reduced response to both type 1 and type 2 thymus-independent antigens, but have a normal response to thymus-dependent antigens (Medina, Williams et al. 2006). Therefore, it is possible that the reduced response to TNP-LPS is caused by the decreased expression of caveolin-1 in B cells that respond to TI antigens.

3.2 Bilateral lesion of striatum increased the humoral immune response to outer membrane proteins (OMP) of *Salmonella enterica* serovar Typhimurium

To evaluate whether the increase in the humoral immune response to protein antigens was a general effect in rats with striatal lesions, we analyzed the IgG immune response to outer membrane proteins of *Salmonella typhimurium*. The levels of IgG in serum were significantly higher in rats with a bilateral lesion than in sham-operated animals (Table 1, P < 0.001). These results support the idea that a bilateral lesion of striatum increases the humoral immune response to T-dependent antigens.

Group	SHAM	Lesion of CN	P
Saline	0.100 ± 0.050	0.091 ± 0.060	
10^7*	0.282 ± 0.008	0.519 ± 0.023	< 0.001 ‡
10^8*	0.524 ± 0.020	0.650 ± 0.050	0.004 ‡

* 10^7 or 10^8 CFU of *Salmonella enterica* Serovar Typhimurium were administered i.p 7 days before the serum was collected and the titers were determined by ELISA. The data are presented as mean ± standard deviation of the absorbance at 490 nm (n = 4-6 rats per group).
‡ Difference in IgG levels between sham-operated and rats with lesion of striatum were significant as determined by the non-paired Student *t* test. Representative results from two independent experiments are shown.

Table 1. IgG antibody response to proteins of *Salmonella enterica* Serovar Typhimurium rats with bilateral lesion of striatum

The mechanisms involved in the increase of the immune response to OMP and other TD antigens in rats with a bilateral lesion in striatum are not clear. One possibility is that cytokines produced by the inflammatory cells in the injured area of the brain increase the antibody production. However, we did not find inflammatory cells in these areas and the expression of mRNA for cytokines did not increase (Fig. 4, panel A). Another possibility is that alterations in the HPA axis contribute to the observed changes in the humoral immune response. Nevertheless, we did not find any increase in the expression of mRNA for prolactin in the hypophysis of rats with a bilateral lesion.

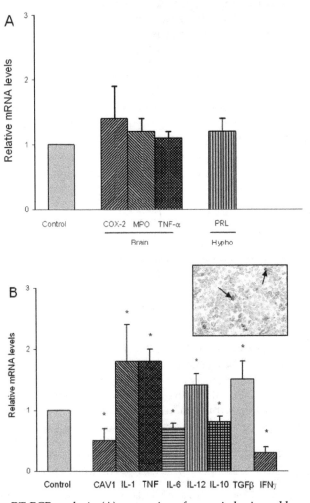

Fig. 4. Real-time RT-PCR analysis: (A) expression of genes in brain and hypophysis. Samples were collected 15 days after bilateral lesion of striatum and the mRNA expression of cyclooxygenase (COX)-2,myeloperoxidase (MPO), tumor necrosis factor-alpha (TNF-α), and prolactin (PRL) was measured by real-time RT-PCR. (B) expression of genes in spleen. The expression of caveolin-1 (CAV1), TNF-α, Interleukin (IL)-6, IL-12, IL-10, Transforming growth factor-beta (TGF-β), Interferon-gamma (IFN-γ) was measured by real-time RT-PCR, as detailed in materials and methods. Data represent the mean ± S.D (n = 6). *P < 0.05 compared with sham rats. Insert: immunolocalization of caveolin-1 positive cells in the splenic marginal zone, 400×. (From Rivera-Aguilar et al. Role of the striatum in the humoral immune response to thymus-independent and thymus-dependent antigens in rats. ImmunolLett 2008;120:20-28).

Because the response to OMP requires CD4+ T cells and cytokines,we analyzed that population aswell as the expression of genes for cytokines in the spleen. The number of CD4+ T cells in the spleen was significantly higher in lesioned rats than in the control group, and that increase could explain the augmented immune response to TD antigens (Table 2). Although the mechanism by which striatal lesions increase the number of naïve CD4+ T cells in the spleen is unknown, one possibility is that high corticosterone levels promote the migration of lymphocytes from the blood to the spleen, as occurs from the blood to other tissues (Dhabhar 2001). The other possibility, that the population of CD4+ T cells was activated by antigens and costimulators, is ruled out by the fact that CD4+ T cells did not express the gene for interleukin (IL)-2 , since this cytokine is produced by activated CD4+ T cells (Jenkins, Khoruts et al. 2001),On the other hand, the increase in their number can be explained by a greater migration of CD+ T cells from the blood into the spleen, although the mechanism of this possible migration remains unclear.

Spleniccell	SHAM	Lesion of Striatum	P
IgM +	34 ± 3	25 ± 3	< 0.001‡
IgG +	5 ± 0.4	5 ± 0.6	1.0
CD4+	5 ± 0.4	8 ± 1.2	< 0.05‡

The data are presented as mean ± standard deviation of the number of positive cells for IgM, IgG, Caveolin-1, and CD4+ (n = 4-6 rats per group).
‡Differences in number of cells between sham-operated and rats with lesion of striatum were significant as determined by the Student t test. Representative results from two independent experiments are shown.

Table 2. Lymphocytes and Caveolin-1+ cells in the spleen of rats with bilateral lesion of striatum

Whereas in lesioned rats the expression of genes for IL-1, tumor necrosis factor (TNF), IL-12 and transforming growth factor-beta (TGF-β) increased, the expression of genes for IL-6, IL-10 and interferon-gamma (IFN-γ) decreased (Fig. 4, panel B). Although this pattern of cytokine production could contribute to the activation of the immune system (Trinchieri 1998; Pestka, Krause et al. 2004), further studies are needed to elucidate the role of cytokines in these changes. However, the fact that CD4+ T cells did not express the gene for IL-2, and that an increased expression of the gene for TGF-β was foundprobably explains the higher synthesis of IgA antibodies observed in lesioned rats, since TGF-β stimulates the production of IgA antibodies (Li, Wan et al. 2006).

Finally, the higher corticosterone levels found in lesioned rats compared with the control group (221. 8±53 ng/ml versus 24.6±12 ng/ml; mean ± S.D.; P < 0.001) could contribute to the changes observed in the immune response. Since glucocorticoids have opposite effects on the TI and TD antibody responses (Addison and Babbage 1981; Garvy and Fraker 1991), high corticosterone concentrations may depress TI responses and stimulate TD responses. In addition, given that physiological glucocorticoid concentrations enhance immunoglobulin production in vitro and in vivo (Ambrose 1964; Halliday and Garvey 1964; Fauci, Pratt et al. 1977; Gonzalez-Ariki and Husband 1998), the rise in corticosterone levels that we found might explain the increase in the immune response to TD antigens. However, pharmacological studies are required to elucidate the role of glucocorticoids in mediating the effects of striatal lesions on immune function.

In summary, our results indicate that striatal GABAergic medium-sized spiny neurons probably modulate the humoral immune response to *Salmonella* outer membrane proteins (OMP) through mechanisms related to the function of B and T cells, the expression of caveolin-1, and andchanges in serum levels of corticosterone.

4. Pathways for the CNS regulation ofthe immune response to *Salmonella* in the intestinal mucosa

In the intestinal mucosa, main site of entry of *Salmonella*, the CNS may regulate the immune response to *Salmonella* by modulating the activity of the HPA axis and the activity of the autonomic nervous system.

4.1 Role of the HPA axis in the immune response to *Salmonella*

The activity of the hypothalamus-pituary-adrenal axis results in the release of the corticotropin-releasing factor (CRF), the adrenocorticotropin hormone (ACTH) and glucocorticoidsinto the circulatory system (Wilder 1995; Webster, Tonelli et al. 2002; Charmandari, Tsigos et al. 2005; Gunnar and Quevedo 2007). Glucocorticoids released from the adrenal gland are delivered to the intestinal mucosa through blood circulation. Glucocorticoids inhibit mucosal inflammation through activation of glucocorticoid receptors present on epithelial cells and intestinal lymphocytes (Boivin, Ye et al. 2007; Jarillo-Luna, Rivera-Aguilar et al. 2008; Fujishima, Takeda et al. 2009; Resendiz-Albor, Reina-Garfias et al. 2010). Also, GCs increase bacterial translocation from the gastrointestinal tract to the mesenteric lymph nodes (Jones, Minei et al. 1990; Fukuzuka, Edwards et al. 2000; Dunn, Ando et al. 2003).

4.2 Role of the autonomic nervous system in the immune response to *Salmonella*

The CNS can modulate the activity of the autonomic nervous system (the adrenergic and cholinergic nervous system) and evoke the neuronal release of norepinephrine (NE), acetylcholine (ACh) and other neurotransmitters in peripheral tissues, including the intestinal mucosa (Felten, Felten et al. 1987; Kulkarni-Narla, Beitz et al. 1999; Kohm and Sanders 2001; Tracey 2002; Green, Lyte et al. 2003; Pavlov, Wang et al. 2003; Sternberg 2006; Sanders and Kavelaars 2007; Schmidt, Xie et al. 2007; Chrousos 2009; Kvetnansky, Sabban et al. 2009). These mediators may influence the function of the intestinal mucosa and its associated surface bacterial populations.

4.2.1 Sympathetic nervous system and the immune response to *Salmonella*

The sympathetic or adrenergic division of the autonomic nervous system is associated with a dual mode of regulation of inflammatory responses (Hasko and Szabo 1998; Elenkov, Wilder et al. 2000). Epinephrine (adrenaline), secreted from the adrenal medulla, and norepinephrine (noradrenaline), which is both secreted from the adrenal medulla and released from sympathetic nerve axons, modulate the release of cytokines and inflammation through adrenoceptors on immune cells (Hasko and Szabo 1998; Elenkov, Wilder et al. 2000).

There is strong immunohistochemical evidence for catecholaminergic innervation of Peyer's patches, the inductive sites for mucosal immunity and the main entry site for *Salmonella*. In addition, adrenergic receptors are expressed on neurons, epithelial cells and other cellular components of the intestinal mucosa (Kulkarni-Narla et al. 1999) (Nijhuis, Olivier et al. ; Kulkarni-Narla, Beitz et al. 1999; Green, Lyte et al. 2003; Chiocchetti, Mazzuoli et al. 2008; Lyte, Vulchanova et al. 2011).

Norepinephrine (NE), released within the intestinal wall during activation of the sympathetic nervous system, has a wide variety of actions at the intestinal mucosa. Norepinephrine participates in the host-*Salmonella* interaction by enhancing the growth of Salmonella *enterica* and other enteropathogens, such as enterohemorrhagic *Escherichia coli* O157:H7 (EHEC) and *Yersinia enterocolitica* (Freestone, Haigh et al. 2007; Green and Brown 2010; Lyte, Vulchanova et al. 2011). This same neurotransmitter substancealters mucosal attachment, and therefore the invasiveness, of serovars of *Salmonella enterica* by acting on cells of the intestinal mucosa that express adrenoroceptors (Green and Brown 2010). In this same study, the electrical stimulation of enteric nerves increased *Salmonella typhimurium* internalization in ileal mucosa explants from swine (Schreiber, Price et al. 2007). These results suggest that enteric catecholaminergic nerves modulate *Salmonella* colonization of Peyer's patches at the earliest stages of infection, in part by altering epithelial uptake of bacteria(Brown and Price 2008).Furthermore, NE apparently activates the expression of virulence-associated factors in *Salmonella typhimurium*, including flagella-mediated motility (Bearson and Bearson 2008; Moreira, Weinshenker et al. 2010), and Type III protein secretion (Rasko, Moreira et al. 2008; Moreira, Weinshenker et al. 2010).Currently, the cellular mechanisms underlying these neurally mediated effects on *Salmonella* internalization in the intestinal mucosa are undefined.It has been proposed that catecholamines may regulate the sampling function of Peyer's patches in the control of the entry of pathogenic microbes or immune processing of the same at these intestinal sites (Green and Brown 2010).

4.2.2 The parasympathetic nervous system and the immune response to *Salmonella*

Efferent vagus nerve fibers innervate the small intestine and proximal colon of the gastro intestinal tract(Altschuler, Ferenci et al. 1991; Altschuler, Escardo et al. 1993),suggesting the possibility that cholinergic activity may modulate immune cells residing in, or recruited to, the densely innervated bowel wall(Van Der Zanden, Boeckxstaens et al. 2009). In fact, current knowledge indicates that the vagus nerve provides an important bi-directional communication circuit by which the brain modulates inflammation (Tracey 2002; Pavlov, Wang et al. 2003).

The presence of bacterial infection and inflammation can be detected by the sensory (afferent) vagus nerve and communicated to the nucleus tractus solitarus in the brainstem medulla oblongata. Neural communication between this other brainstem nuclei and "higher" brain structures, including the hypothalamus, are associated with the generation of brain-derived anti-inflammatory output through the efferent vagus nerve, which inhibits pro-inflammatory cytokine release and protects against systemic and mucosal inflammation. As acetylcholine is the principle parasympathetic neurotransmitter,this vagal function has been termed "the cholinergic anti-inflammatory pathway" (Borovikova, Ivanova et al. 2000; Tracey, Czura et al. 2001; Tracey 2002; Pavlov, Wang et al. 2003; Pavlov and Tracey 2005; Pavlov and Tracey 2006; Bonaz 2007; Gallowitsch-Puerta and Pavlov 2007; Tracey 2007; Tracey 2010).

Information about the role of the parasympathetic system and the immune response to *Salmonella* is scarce. In one study, in *Salmonella typhimurium*-stimulated groups, inflammatory pathological changes were seen in ileum and the mesenteric lymph node. Whereas *Salmonella* induced a decrease in the level of CD4+ T cells in peripheral blood, such levels were restored to normal by a subdiaphragmatic vagotomy. The vagus nerve is involved in the transmission of abdominal immune information to the brain during *Salmonella typhimurium* infection, and it plays an important role in the maintenance of the immune balance of the organism (Wang, Wang et al. 2002). In another study, the specific inhibition of acetylcholinesterase (AChE), the enzyme that degrades ACh, rendered animals more resistant to infection by a virulent strain of *Salmonella typhimurium*, which correlated with the efficient control of bacterial proliferation in spleen. Immunologically, inhibition of AChE enabled the animals to mount a more effective systemic (inflammatory and anti-microbial) response, and to secrete higher levels of interleukin-12, a key T helper type 1-promoting cytokine. Thus, in one model of Gram-negative bacterial infection, cholinergic stimulation was shown to enhance the anti-microbial immune response leading to effective control of bacterial proliferation and enhanced animal survival (Fernandez-Cabezudo, Lorke et al.).

Currently, there is no evidence that the cholinergic anti-inflammatory pathway inhibits or enhances the immune response to *Salmonella* in the intestinal mucosa. However, taking into account that the anti-inflammatory activity of the cholinergic nervous system is based on cholinergic signals that are linked to macrophages and other innate immune cells, whichare central to the control of *Salmonella* infection, it is likely that the cholinergic nervous system attenuates the inflammatory response to *Salmonella* (Jones and Falkow 1996; Mittrucker and Kaufmann 2000; Wick 2004).

5. Neuroendocrine regulation of intestinal IgA and protection against *Salmonella*

Glucocorticoids, catecholamines and acetylcholine regulate the secretion of Intestinal IgA, which in turn plays a key role in protecting against *Salmonella* infection.Therefore; this molecule may mediate the effects of the CNS on the immune response to this bacterium.

Secretory immunoglobulin A (S-IgA) is the most abundant intestinal immunoglobulin. By binding to antigens, such as microbes and toxins, S-IgA prevents them from attaching to or penetrating the mucosal surface (Mowat 2003; Fagarasan and Honjo 2004; Kaetzel 2005; Cerutti and Rescigno 2008; Macpherson, McCoy et al. 2008; Brandtzaeg 2009). IgA is secreted into the intestinal lumen due to the cooperation of local plasma cells with epithelial cells. The polymeric IgA (pIgA) secreted by plasma cells diffuses through the stroma and binds to the polymeric immunoglobulin receptor (pIgR) on the basolateral surface of the epithelial cells to form the pIgA–pIgR complex, which in turn is translocated to the apical surface of epithelial cells, where it is cleaved and secreted into lumen as S-IgA (Norderhaug, Johansen et al. 1999; Kaetzel 2005).

In the intestinal lumen, S-IgA protects against infection by inhibiting *Salmonella* adhesion to epithelial cells and M cells and the penetration of this bacterium into deeper tissues (Michetti, Mahan et al. 1992; Michetti, Porta et al. 1994; Mittrucker and Kaufmann 2000; Matsui, Suzuki et al. 2003). However, little is known about the neuroendocrine regulation of intestinal IgA (Schmidt, Eriksen et al. 1999; Schmidt, Xie et al. 2007; Reyna-Garfias, Miliar et al. 2010).

5.1 Glucocorticoids and IgA

Glucocorticoids have several diverse effects on the production and secretion of IgA in the intestine. They have been shown to increase or decrease intestinal IgA levels, effects which may be species-dependent (Alverdy and Aoys 1991; Spitz, Ghandi et al. 1996; Reyna-Garfias, Miliar et al. 2010; Lyte, Vulchanova et al. 2011). Other studies have demonstrated that GCsreduce the number of IgA-producing cells in Peyer's patches of mice (Martinez-Carrillo, Godinez-Victoria et al. 2011), decrease the number of intraepithelial lymphocytes (IEL) in the proximal small intestine of mice (Jarillo-Luna, Rivera-Aguilar et al. 2007; Jarillo-Luna, Rivera-Aguilar et al. 2008; Reyna-Garfias, Miliar et al. 2010),andincrease the levels of mRNA for pIgR in the proximal duodenum of suckling rats (Li, Wang et al. 1999).Thus,it may be through the liberation of GCs that the CNS regulates the production and secretion of intestinal IgA specific to *Salmonella*.

5.2 Noradrenaline and IgA

Although the intestinal tract is a major site for mucosal immunity and is extensively innervated, little is known about the adrenergic regulation of enteric S-IgA secretion.Norepinephrine stimulates S-IgA secretion by acting through alpha-adrenergic receptors in the colonic mucosa, and in this way may enhance mucosal defense *in vivo* (Schmidt, Xie et al. 2007)]. This neurotransmitter also significantly increases pIgR mRNA expression and intestinal IgA concentration (Reyna-Garfias, Miliar et al. 2010). The increased expression of pIgR might contribute to an increasedsecretion of S-IgA in the gut, andthus a greater protectionagainst pathogens including *Salmonella*. A sympathectomy decreases the number of IgA-positive lamina propria cells in the weanling rat (Gonzalez-Ariki and Husband 2000). Furthermore, NE has been found to slightly increase the number of IgA-immunoreactive cells in the intestinal wall of marathon runners (Nilssen, Oktedalen et al. 1998).Finally, we have found that catecholamines reduce the number of IgA-producing cells in Peyer's patches of mice(Martinez-Carrillo, Godinez-Victoria et al. 2011), decrease the number of IEL in the proximal small intestine of mice (Jarillo-Luna, Rivera-Aguilar et al. 2008), increase the IgA concentration in rat small intestine(Reyna-Garfias, Miliar et al. 2010), and reduce the intestinal IgA concentration in mice (Jarillo-Luna, Rivera-Aguilar et al. 2007).

Although the effect of noradrenaline or adrenaline (catecholamines) on the production of IgA antibodies specific to *Salmonella* has not been studied, it is possible that the release of these molecules by the activation of the sympathetic-adrenal medullary axis may modify the production and secretion of intestinal IgA specific to *Salmonella*.

5.3 Acetylcholine and IgA

Some data indicate that intestinal secretion of immunoglobulin A is stimulated by the muscarinic effect of cholinergic agonists, which suggest that the basal secretion of immunoglobulin A may be influenced by the parasympathetic nervous system (Wilson, Soltis et al. 1982; Freier, Eran et al. 1987; Freier, Eran et al. 1989; Schmidt, Xie et al. 2007). However, there is no information about the role of the parasympathetic nervous system in the secretion of IgA during infections by *Salmonella*.

6. Neurotransmitters and neuroendocrine molecules: Substance P, cholecystokinin, Somatostatin and the Macrophage migration inhibitory factor (MIF)

Apart from the immune regulatory role of the classic neurotransmitters, acetylcholine and norepinephrine, both the sympathetic and parasympathetic subdivisions of the autonomic nervous system include several subpopulations of neurons that express several neuropeptides related to the modulation of the immune response. In this sense,corticotropin-releasing hormone (CRH), neuropeptide Y (NPY), somatostatin, and galanin are found in postganglionic noradrenergic vasoconstrictive neurons, whereas vasoactive intestinal peptide (VIP), Substance P (SP), and calcitonin gene-related peptide are found in cholinergic neurons (Charmandari, Tsigos et al. 2005; Kvetnansky, Sabban et al. 2009).

There are even some gut neuropeptides, including SP, neuropeptide Y and neurotensin, that possess inherent antimicrobial activity (Brogden, Guthmiller et al. 2005). The role of neuropeptides and their receptors in the inflammatory response to *Salmonella* and other invasive pathogens has scarcely been analyzed.

6.1 Substance P

Substance P participates in the intestinal immune response to *Salmonella* in several ways. Oral infection with *Salmonella* increases SP and neurokinin A mRNA precursors, and the expression of substance P receptors in Peyer's patches, lymph nodes and spleen (Bost 1995; Kincy-Cain and Bost 1996; Pothoulakis and Castagliuolo 2003). Substance P increases resistance to *Salmonella* by improving the activity of macrophages, and increases the production of IFN-γ and IL-12, which are part of the initial response to *Salmonella* that helps limit bacterial growth and dissemination (Kincy-Cain, Clements et al. 1996; Kincy-Cain and Bost 1997; Weinstock 2003). Thus, it is postulated that SP and its receptor may contribute to the mounting of a coordinated early immune response against *Salmonella* infection (Pothoulakis and Castagliuolo 2003; Weinstock 2003).

6.2 Somatostatin

Somatostatin (SOM) exerts an active role in the regulation of mucosal inflammatory responses (Pothoulakis and Castagliuolo 2003). SOM released from neuronal and non-neuronal cells distributed throughout the gastrointestinal tract may modulate la inflammatory response to *Salmonella* infection by inhibiting the release of pro-inflammatory cytokines such as IL-10 and IL-8 from intestinal epithelial cells (Chowers, Cahalon et al. 2000; Pothoulakis and Castagliuolo 2003).

6.3 Macrophage migration inhibitory factor (MIF)

The cytokine macrophage migration inhibitory factor (MIF) exerts a multitude of biological functions. Notably, it induces inflammation at the interface between the immune system and the HPA axis (Flaster, Bernhagen et al. 2007). The role of MIF in infectious diseases has scarcely been studied. MIF-deficient (MIF(-/-) knockout mice do not control an infection with wild-type *Salmonella typhimurium*. Increased susceptibility is accompanied by decreased levels of IL-12, IFN-γ, and tumor necrosis factor alpha, and markedly increases of IL-1β levels. Additionally, compared with control animals, infected MIF (-/-) mice show elevated serum levels of nitric oxide and corticosterone. These results suggest that MIF is a key mediator in the host response to *Salmonella typhimurium*. Not only does MIF promote development of a protective Th1 response, but it also ameliorates disease by altering levels of reactive nitrogen intermediates and corticosteroid hormones, which both exert immunosuppressive functions (Koebernick, Grode et al. 2002). Epithelial MIF from cultured cells was found to be released predominantly from the apical side after *Salmonella* infection (Maaser, Eckmann et al. 2002).

6.4 Effect of these molecules on the production and secretion of IgA

The aforementioned molecules, in addition to their functions in the innate and cellular immune responses, affect the production and secretion of intestinal IgA. For example, the intravenous or intra-arterial injection of gut neuropeptides cholecystokinin, substance P and somatostatin increase S-IgA secretion in isolated loops of the rat small intestine and vascularly-perfused segments of the swine ileum (Wilson, Soltis et al. 1982; Freier, Eran et al. 1987; Freier, Eran et al. 1989; Schmidt, Xie et al. 2007).

7. Conclusion

CNS can regulate the immune response to *Salmonella* by the activation of both the HPA axis and the autonomic nervous system (including the sympathetic, parasympathetic and enteric divisions). Hormones, neurotransmitters, neuropeptides and neuroendocrine molecules mediate the effects of the CNS on the systemic and intestinal immune responses. In the intestinal mucosa, the CNS may modify the synthesis and secretion of IgA, which protects against the invasion by *Salmonella*.

8. Acknowledgment

We thank Bruce Allan Larsen for reviewing the use of English in this manuscript. This work was supported in part by grants from SEPI-IPN and from COFAA-IPN.

9. References

Addison, I. E. and J. W. Babbage (1981). "Thymus-dependent and thymus-independent antibody responses: contrasting patterns of immunosuppression." *Br J Exp Pathol* 62(1): 74-8.

Altschuler, S. M., J. Escardo, et al. (1993). "The central organization of the vagus nerve innervating the colon of the rat." *Gastroenterology* 104(2): 502-9.

Altschuler, S. M., D. A. Ferenci, et al. (1991). "Representation of the cecum in the lateral dorsal motor nucleus of the vagus nerve and commissural subnucleus of the nucleus tractus solitarii in rat." *J Comp Neurol* 304(2): 261-74.

Alverdy, J. and E. Aoys (1991). "The effect of glucocorticoid administration on bacterial translocation. Evidence for an acquired mucosal immunodeficient state." *Ann Surg* 214(6): 719-23.

Ambrose, C. T. (1964). "The Requirement for Hydrocortisone in Antibody-Forming Tissue Cultivated in Serum-Free Medium." *J Exp Med* 119: 1027-49.

Beal, M. F., N. W. Kowall, et al. (1986). "Replication of the neurochemical characteristics of Huntington's disease by quinolinic acid." *Nature* 321(6066): 168-71.

Bearson, B. L. and S. M. Bearson (2008). "The role of the QseC quorum-sensing sensor kinase in colonization and norepinephrine-enhanced motility of *Salmonella enterica* serovar Typhimurium." *Microb Pathog* 44(4): 271-8.

Berczi, I., E. Nagy, et al. (1984). "The influence of pituitary hormones on adjuvant arthritis." *Arthritis Rheum* 27(6): 682-8.

Berczi, I., E. Nagy, et al. (1981). "Regulation of humoral immunity in rats by pituitary hormones." *Acta Endocrinol (Copenh)* 98(4): 506-13.

Berczi, I. and A. Szentivanyi (2003). The immune.neuroendocrine circuitry. History and progress I. *Neuroimmmune Biology*. I. Berczi and A. Szentivanyi. Amsterdam, The Netherlands, Elsevier. 3: 495-536.

Bieganowska, K., A. Czlonkowska, et al. (1993). "Immunological changes in the MPTP-induced Parkinson's disease mouse model." *J Neuroimmunol* 42(1): 33-7.

Bienenstock, J. (1992). "Cellular communication networks. Implications for our understanding of gastrointestinal physiology." *Ann N Y Acad Sci* 664: 1-9.

Blalock, J. E. (1999). "Proopiomelanocortin and the immune-neuroendocrine connection." *Ann N Y Acad Sci* 885: 161-72.

Block, L. H., R. Locher, et al. (1981). "125I-8-L-arginine vasopressin binding to human mononuclear phagocytes." *J Clin Invest* 68(2): 374-81.

Boivin, M. A., D. Ye, et al. (2007). "Mechanism of glucocorticoid regulation of the intestinal tight junction barrier." *Am J Physiol Gastrointest Liver Physiol* 292(2): G590-8.

Bonaz, B. (2007). "The cholinergic anti-inflammatory pathway and the gastrointestinal tract." *Gastroenterology* 133(4): 1370-3.

Borovikova, L. V., S. Ivanova, et al. (2000). "Vagus nerve stimulation attenuates the systemic inflammatory response to endotoxin." *Nature* 405(6785): 458-62.

Bost, K. L. (1995). "Inducible preprotachykinin mRNA expression in mucosal lymphoid organs following oral immunization with Salmonella." *J Neuroimmunol* 62(1): 59-67.

Brandtzaeg, P. (2009). "Mucosal immunity: induction, dissemination, and effector functions." *Scand J Immunol* 70(6): 505-15.

Brogden, K. A., J. M. Guthmiller, et al. (2005). "The nervous system and innate immunity: the neuropeptide connection." *Nat Immunol* 6(6): 558-64.

Brown, D. R. and L. D. Price (2008). "Catecholamines and sympathomimetic drugs decrease early Salmonella Typhimurium uptake into porcine Peyer's patches." *FEMS Immunol Med Microbiol* 52(1): 29-35.

Bueno, L. (2000). "Neuroimmune alterations of ENS functioning." *Gut* 47 Suppl 4: iv63-5; discussion iv76.

Campos-Rodriguez, R., A. Quintanar-Stephano, et al. (2006). "Hypophysectomy and neurointermediate pituitary lobectomy reduce serum immunoglobulin M (IgM) and IgG and intestinal IgA responses to Salmonella enterica serovar Typhimurium infection in rats." *Infect Immun* 74(3): 1883-9.

Catania, A. and J. M. Lipton (1993). "alpha-Melanocyte stimulating hormone in the modulation of host reactions." *Endocr Rev* 14(5): 564-76.

Cerutti, A. and M. Rescigno (2008). "The biology of intestinal immunoglobulin A responses." *Immunity* 28(6): 740-50.

Cooke, H. J. (1986). "Neurobiology of the intestinal mucosa." *Gastroenterology* 90(4): 1057-81.

Costall, B., R. J. Naylor, et al. (1972). "On the involvement of the caudate-putamen, globus pallidus and substantia nigra with neuroleptic and cholinergic modification of locomotor activity." *Neuropharmacology* 11(3): 317-30.

Charmandari, E., C. Tsigos, et al. (2005). "Endocrinology of the stress response." *Annu Rev Physiol* 67: 259-84.

Chiocchetti, R., G. Mazzuoli, et al. (2008). "Anatomical evidence for ileal Peyer's patches innervation by enteric nervous system: a potential route for prion neuroinvasion?" *Cell Tissue Res* 332(2): 185-94.

Chowers, Y., L. Cahalon, et al. (2000). "Somatostatin through its specific receptor inhibits spontaneous and TNF-alpha- and bacteria-induced IL-8 and IL-1 beta secretion from intestinal epithelial cells." *J Immunol* 165(6): 2955-61.

Chrousos, G. P. (1995). "The hypothalamic-pituitary-adrenal axis and immune mediated inflammation." *N Engl J Med* 332(20): 1351-62.

Chrousos, G. P. (2009). "Stress and disorders of the stress system." *Nat Rev Endocrinol* 5(7): 374-81.

Deleplanque, B., S. Vitiello, et al. (1994). "Modulation of immune reactivity by unilateral striatal and mesolimbic dopaminergic lesions." *Neurosci Lett* 166(2): 216-20.

Devoino, L., E. Alperina, et al. (1997). "Involvement of brain dopaminergic structures in neuroimmunomodulation." *Int J Neurosci* 91(3-4): 213-28.

Devoino, L. V., M. A. Cheido, et al. (2001). "Involvement of the rat caudate nucleus in the immunostimulatory effect of DAGO." *Neurosci Behav Physiol* 31(3): 323-6.

Dhabhar, F. S., and McEwen, B. S (2001). Bidirectional effects of stress and glucocorticoid hormones on immune function: Possible explanations for paradoxical observations. *Psychoneuroimmunology* F. D. Ader R, Cohen N San Diego, Academic Press. I: 301- 338.

Di Carlo, R., R. Meli, et al. (1993). "Prolactin protection against lethal effects of Salmonella typhimurium." *Life Sci* 53(12): 981-9.

Draca, S. (1995). "Prolactin as an immunoreactive agent." *Immunol Cell Biol* 73(6): 481-3.

Dunn, A. J., T. Ando, et al. (2003). "HPA axis activation and neurochemical responses to bacterial translocation from the gastrointestinal tract." *Ann N Y Acad Sci* 992: 21-9.

Edwards, C. K., 3rd, S. Arkins, et al. (1992). "The macrophage-activating properties of growth hormone." *Cell Mol Neurobiol* 12(5): 499-510.

Edwards, C. K., 3rd, S. M. Ghiasuddin, et al. (1992). "In vivo administration of recombinant growth hormone or gamma interferon activities macrophages: enhanced resistance to experimental Salmonella typhimurium infection is correlated with generation of reactive oxygen intermediates." *Infect Immun* 60(6): 2514-21.

Edwards, C. K., 3rd, R. M. Lorence, et al. (1991). "Hypophysectomy inhibits the synthesis of tumor necrosis factor alpha by rat macrophages: partial restoration by exogenous growth hormone or interferon gamma." *Endocrinology* 128(2): 989-86.

Edwards, C. K., 3rd, L. M. Yunger, et al. (1991). "The pituitary gland is required for protection against lethal effects of Salmonella typhimurium." *Proc Natl Acad Sci U S A* 88(6): 2274-7.

Elenkov, I. J., R. L. Wilder, et al. (2000). "The sympathetic nerve--an integrative interface between two supersystems: the brain and the immune system." *Pharmacol Rev* 52(4): 595-638.

Engler, H., R. Doenlen, et al. (2009). "Time-dependent alterations of peripheral immune parameters after nigrostriatal dopamine depletion in a rat model of Parkinson's disease." *Brain Behav Immun* 23(4): 518-26.

Fagarasan, S. and T. Honjo (2004). "Regulation of IgA synthesis at mucosal surfaces." *Curr Opin Immunol* 16(3): 277-83.

Fauci, A. S., K. R. Pratt, et al. (1977). "Activation of human B lymphocytes. IV. Regulatory effects of corticosteroids on the triggering signal in the plaque-forming cell response of human peripheral blood B lymphocytes to polyclonal activation." *J Immunol* 119(2): 598-603.

Felten, D. L., S. Y. Felten, et al. (1987). "Noradrenergic sympathetic neural interactions with the immune system: structure and function." *Immunol Rev* 100: 225-60.

Fernandez-Cabezudo, M. J., D. E. Lorke, et al. "Cholinergic stimulation of the immune system protects against lethal infection by Salmonella enterica serovar Typhimurium." *Immunology* 130(3): 388-98.

Filipov, N. M., L. Cao, et al. (2002). "Compromised peripheral immunity of mice injected intrastriatally with six-hydroxydopamine." *J Neuroimmunol* 132(1-2): 129-39.

Flaster, H., J. Bernhagen, et al. (2007). "The macrophage migration inhibitory factor-glucocorticoid dyad: regulation of inflammation and immunity." *Mol Endocrinol* 21(6): 1267-80.

Freestone, P. P., R. D. Haigh, et al. (2007). "Blockade of catecholamine-induced growth by adrenergic and dopaminergic receptor antagonists in Escherichia coli O157:H7, Salmonella enterica and Yersinia enterocolitica." *BMC Microbiol* 7: 8.

Freier, S., M. Eran, et al. (1989). "A study of stimuli operative in the release of antibodies in the rat intestine." *Immunol Invest* 18(1-4): 431-47.

Freier, S., M. Eran, et al. (1987). "Effect of cholecystokinin and of its antagonist, of atropine, and of food on the release of immunoglobulin A and immunoglobulin G specific antibodies in the rat intestine." *Gastroenterology* 93(6): 1242-6.

Fujishima, S., H. Takeda, et al. (2009). "The relationship between the expression of the glucocorticoid receptor in biopsied colonic mucosa and the glucocorticoid responsiveness of ulcerative colitis patients." *Clin Immunol* 133(2): 208-17.

Fukuzuka, K., C. K. Edwards, 3rd, et al. (2000). "Glucocorticoid and Fas ligand induced mucosal lymphocyte apoptosis after burn injury." *J Trauma* 49(4): 710-6.

Gallowitsch-Puerta, M. and V. A. Pavlov (2007). "Neuro-immune interactions via the cholinergic anti-inflammatory pathway." *Life Sci* 80(24-25): 2325-9.

Garvy, B. A. and P. J. Fraker (1991). "Suppression of the antigenic response of murine bone marrow B cells by physiological concentrations of glucocorticoids." *Immunology* 74(3): 519-23.

Gonzalez-Ariki, S. and A. J. Husband (1998). "The role of sympathetic innervation of the gut in regulating mucosal immune responses." *Brain Behav Immun* 12(1): 53-63.

Gonzalez-Ariki, S. and A. J. Husband (2000). "Ontogeny of IgA(+) cells in lamina propria: effects of sympathectomy." *Dev Comp Immunol* 24(1): 61-9.

Graybiel, A. M. (1995). "The basal ganglia." *Trends Neurosci* 18(2): 60-2.

Green, B. T. and D. R. Brown (2010). Interactions Between Bacteria and the Gut Mucosa: Do Enteric Neurotransmitters Acting on the Mucosal Epithelium Influence Intestinal Colonization or Infection? *Microbial Endocrinology*. M. Lyte and P. Freestone. Mew York, Springer: 89-109.

Green, B. T., M. Lyte, et al. (2003). "Neuromodulation of enteropathogen internalization in Peyer's patches from porcine jejunum." *J Neuroimmunol* 141(1-2): 74-82.

Gunnar, M. and K. Quevedo (2007). "The neurobiology of stress and development." *Annu Rev Psychol* 58: 145-73.

Halliday, W. J. and J. S. Garvey (1964). "Some Factors Affecting the Secondary Immune Response in Tissue Cultures Containing Hydrocortisone." *J Immunol* 93: 756-62.

Hasko, G. and C. Szabo (1998). "Regulation of cytokine and chemokine production by transmitters and co-transmitters of the autonomic nervous system." *Biochem Pharmacol* 56(9): 1079-87.

Heijnen, C. J., A. Kavelaars, et al. (1991). "Beta-endorphin: cytokine and neuropeptide." *Immunol Rev* 119: 11-63.

Jankovic, B. D. and K. Isakovic (1973). "Neuro-endocrine correlates of immune response. I. Effects of brain lesions on antibody production, Arthus reactivity and delayed hypersensitivity in the rat." *Int Arch Allergy Appl Immunol* 45(3): 360-72.

Jarillo-Luna, A., V. Rivera-Aguilar, et al. (2007). "Effect of repeated restraint stress on the levels of intestinal IgA in mice." *Psychoneuroendocrinology* 32(6): 681-92.

Jarillo-Luna, A., V. Rivera-Aguilar, et al. (2008). "Effect of restraint stress on the population of intestinal intraepithelial lymphocytes in mice." *Brain Behav Immun* 22(2): 265-75.

Jenkins, M. K., A. Khoruts, et al. (2001). "In vivo activation of antigen-specific CD4 T cells." *Annu Rev Immunol* 19: 23-45.

Jones, B. D. and S. Falkow (1996). "Salmonellosis: host immune responses and bacterial virulence determinants." *Annu Rev Immunol* 14: 533-61.

Jones, W. G., 2nd, J. P. Minei, et al. (1990). "Pathophysiologic glucocorticoid elevations promote bacterial translocation after thermal injury." *Infect Immun* 58(10): 3257-61.

Kaetzel, C. S. (2005). "The polymeric immunoglobulin receptor: bridging innate and adaptive immune responses at mucosal surfaces." *Immunol Rev* 206: 83-99.

Kincy-Cain, T. and K. L. Bost (1996). "Increased susceptibility of mice to Salmonella infection following in vivo treatment with the substance P antagonist, spantide II." *J Immunol* 157(1): 255-64.

Kincy-Cain, T. and K. L. Bost (1997). "Substance P-induced IL-12 production by murine macrophages." *J Immunol* 158(5): 2334-9.

Kincy-Cain, T., J. D. Clements, et al. (1996). "Endogenous and exogenous interleukin-12 augment the protective immune response in mice orally challenged with Salmonella dublin." *Infect Immun* 64(4): 1437-40.

Kirby, A. C., U. Yrlid, et al. (2002). "The innate immune response differs in primary and secondary Salmonella infection." *J Immunol* 169(8): 4450-9.

Klein, J. R. (2003). "Physiological relevance of thyroid stimulating hormone and thyroid stimulating hormone receptor in tissues other than the thyroid." *Autoimmunity* 36(6-7): 417-21.

Koebernick, H., L. Grode, et al. (2002). "Macrophage migration inhibitory factor (MIF) plays a pivotal role in immunity against Salmonella typhimurium." *Proc Natl Acad Sci U S A* 99(21): 13681-6.

Kohm, A. P. and V. M. Sanders (2001). "Norepinephrine and beta 2-adrenergic receptor stimulation regulate CD4+ T and B lymphocyte function in vitro and in vivo." *Pharmacol Rev* 53(4): 487-525.

Kulkarni-Narla, A., A. J. Beitz, et al. (1999). "Catecholaminergic, cholinergic and peptidergic innervation of gut-associated lymphoid tissue in porcine jejunum and ileum." *Cell Tissue Res*298(2): 275-86.

Kvetnansky, R., E. L. Sabban, et al. (2009). "Catecholaminergic systems in stress: structural and molecular genetic approaches." *Physiol Rev* 89(2): 535-606.

Li, M. O., Y. Y. Wan, et al. (2006). "Transforming growth factor-beta regulation of immune responses." *Annu Rev Immunol* 24: 99-146.

Li, T. W., J. Wang, et al. (1999). "Transcriptional control of the murine polymeric IgA receptor promoter by glucocorticoids." *Am J Physiol* 276(6 Pt 1): G1425-34.

Luger, T. A., T. E. Scholzen, et al. (2003). "New insights into the functions of alpha-MSH and related peptides in the immune system." *Ann N Y Acad Sci* 994: 133-40.

Lyte, M., L. Vulchanova, et al. (2011). "Stress at the intestinal surface: catecholamines and mucosa-bacteria interactions." *Cell Tissue Res* 343(1): 23-32.

Maaser, C., L. Eckmann, et al. (2002). "Ubiquitous production of macrophage migration inhibitory factor by human gastric and intestinal epithelium." *Gastroenterology* 122(3): 667-80.

Macpherson, A. J., K. D. McCoy, et al. (2008). "The immune geography of IgA induction and function." *Mucosal Immunol* 1(1): 11-22.

Madden, K. S. and D. L. Felten (1995). "Experimental basis for neural-immune interactions." *Physiol Rev* 75(1): 77-106.

Martinez-Carrillo, B. E., M. Godinez-Victoria, et al. (2011). "Repeated restraint stress reduces the number of IgA-producing cells in Peyer's patches." *Neuroimmunomodulation* 18(3): 131-41.

Matsui, H., M. Suzuki, et al. (2003). "Oral immunization with ATP-dependent protease-deficient mutants protects mice against subsequent oral challenge with virulent Salmonella enterica serovar typhimurium." *Infect Immun* 71(1): 30-9.

McGeer, E. G. and P. L. McGeer (1976). "Duplication of biochemical changes of Huntington's chorea by intrastriatal injections of glutamic and kainic acids." *Nature* 263(5577): 517-9.

Medina, F. A., T. M. Williams, et al. (2006). "A novel role for caveolin-1 in B lymphocyte function and the development of thymus-independent immune responses." *Cell Cycle* 5(16): 1865-71.

Meli, R., G. M. Raso, et al. (1996). "Recombinant human prolactin induces protection against Salmonella typhimurium infection in the mouse: role of nitric oxide." *Immunopharmacology* 34(1): 1-7.

Michetti, P., M. J. Mahan, et al. (1992). "Monoclonal secretory immunoglobulin A protects mice against oral challenge with the invasive pathogen Salmonella typhimurium." *Infect Immun* 60(5): 1786-92.

Michetti, P., N. Porta, et al. (1994). "Monoclonal immunoglobulin A prevents adherence and invasion of polarized epithelial cell monolayers by Salmonella typhimurium." *Gastroenterology* 107(4): 915-23.

Mittrucker, H. W. and S. H. Kaufmann (2000). "Immune response to infection with Salmonella typhimurium in mice." *J Leukoc Biol* 67(4): 457-63.

Moreira, C. G., D. Weinshenker, et al. (2010). "QseC mediates Salmonella enterica serovar typhimurium virulence in vitro and in vivo." *Infect Immun* 78(3): 914-26.

Mowat, A. M. (2003). "Anatomical basis of tolerance and immunity to intestinal antigens." *Nat Rev Immunol* 3(4): 331-41.

Munck, A. and A. Naray-Fejes-Toth (1992). "The ups and downs of glucocorticoid physiology. Permissive and suppressive effects revisited." *Mol Cell Endocrinol* 90(1): C1-4.

Nagy, E. and I. Bercz (1994). Prolactin as immunomodylatory hormone. *Advances in Psychoneuroimmunology, Hans Selye Symposiym on Neuroendocrinology and Stress.* i. Berczi and J. Szélenyi. New York, Plenum Press. 3: 110-123.

Nagy, E. and I. Berczi (1981). "Prolactin and contact sensitivity." *Allergy* 36(6): 429-31.

Nagy, E. and I. Berczi (1991). "Hypophysectomized rats depend on residual prolactin for survival." *Endocrinology* 128(6): 2776-84.

Nanda, N., G. K. Pal, et al. (2005). "Effect of dopamine injection into caudate nucleus on immune responsiveness in rats: a pilot study." *Immunol Lett* 96(1): 151-3.

Navolotskaya, E. V., N. V. Malkova, et al. (2002). "Effect of synthetic beta-endorphin-like peptide immunorphin on human T lymphocytes." *Biochemistry (Mosc)* 67(3): 357-63.

Nijhuis, L. E., B. J. Olivier, et al. "Neurogenic regulation of dendritic cells in the intestine." *Biochem Pharmacol* 80(12): 2002-8.

Nilssen, D. E., O. Oktedalen, et al. (1998). "Intestinal IgA- and IgM-producing cells are not decreased in marathon runners." *Int J Sports Med* 19(6): 425-31.

Norderhaug, I. N., F. E. Johansen, et al. (1999). "Regulation of the formation and external transport of secretory immunoglobulins." *Crit Rev Immunol* 19(5-6): 481-508.

Organista-Esparza, A., M. , M. Tinajero-Ruelas, et al. (2003). Effects of neurointermediate pituitary lobectomy and hypophysectomy on humoral immune response in the Wistar rat. *XLVI National Congress of Physiological Sciences.* Aguascalientes, Mexico, Mexican Society of Physiological Sciences.

Ottaway, C. A. (1991). "Neuroimmunomodulation in the intestinal mucosa." *Gastroenterol Clin North Am* 20(3): 511-29.

Pavlov, V. A. and K. J. Tracey (2005). "The cholinergic anti-inflammatory pathway." *Brain Behav Immun* 19(6): 493-9.

Pavlov, V. A. and K. J. Tracey (2006). "Controlling inflammation: the cholinergic anti-inflammatory pathway." *Biochem Soc Trans* 34(Pt 6): 1037-40.

Pavlov, V. A., H. Wang, et al. (2003). "The cholinergic anti-inflammatory pathway: a missing link in neuroimmunomodulation." *Mol Med* 9(5-8): 125-34.

Payan, D. G., J. P. McGillis, et al. (1986). "Neuroimmunology." *Adv Immunol* 39: 299-323.

Pestka, S., C. D. Krause, et al. (2004). "Interleukin-10 and related cytokines and receptors." *Annu Rev Immunol* 22: 929-79.

Pothoulakis, C. and I. Castagliuolo (2003). Infectious pathogen and the neuroenteric system. *Autonomic Neuroimmunology.* J. Bienenstock, E. Goetzl and M. Blennerhassett. London & New York, Taylor & Francis: 251-277.

Pycock, C. J. (1980). "Turning behaviour in animals." *Neuroscience* 5(3): 461-514.

Quintanar-Stephano, A., E. Abarca-Rojano, et al. (2010). "Hypophysectomy and neurointermediate pituitary lobectomy decrease humoral immune responses to T-independent and T-dependent antigens." *J Physiol Biochem* 66(1): 7-13.

Quintanar-Stephano, A., R. Chavira-Ramirez, et al. (2005). "Neurointermediate pituitary lobectomy decreases the incidence and severity of experimental autoimmune encephalomyelitis in Lewis rats." *J Endocrinol* 184(1): 51-8.

Quintanar-Stephano, A., K. Kovacs, et al. (2004). "Effects of neurointermediate pituitary lobectomy on humoral and cell-mediated immune responses in the rat." *Neuroimmunomodulation* 11(4): 233-40.

Quintanar-Stephano, A., A. Organista-Esparza, et al. (2005). Protection against adjuvant-induced arthritis (AIA) by the excision of the neuro-intermediate pituitary lobe (NIL) in Lewis rats. . *Experimental Biology meeting abstracts- FASEB Meeting.* San Diego, California. USA.

Quintanar-Stephano, A., A. Organista-Esparza, et al. (2004). Effects of Neurointermediate Pituitary Lobectomy and Desmopressin on Experimental Autoimmune Encephalomyelitis in Rats. *Experimental Biology. 2004. FASEB Meeting.* Washington. D.C. USA.

Rasko, D. A., C. G. Moreira, et al. (2008). "Targeting QseC signaling and virulence for antibiotic development." *Science* 321(5892): 1078-80.

Reichlin, S. (1993). "Neuroendocrine-immune interactions." *N Engl J Med* 329(17): 1246-53.

Renoux, G., K. Biziere, et al. (1989). "Sodium diethyldithiocarbamate protects against the MPTP-induced inhibition of immune responses in mice." *Life Sci* 44(12): 771-7.

Resendiz-Albor, A. A., H. Reina-Garfias, et al. (2010). "Regionalization of pIgR expression in the mucosa of mouse small intestine." *Immunol Lett* 128(1): 59-67.

Reyna-Garfias, H., A. Miliar, et al. (2010). "Repeated restraint stress increases IgA concentration in rat small intestine." *Brain Behav Immun* 24(1): 110-8.

Rivera-Aguilar, V., E. Querejeta, et al. (2008). "Role of the striatum in the humoral immune response to thymus-independent and thymus-dependent antigens in rats." *Immunol Lett* 120(1-2): 20-8.

Sanders, V. and J. A. Kavelaars (2007). Adrenergic regulation of immunity. *Psychoneuroimmunology.* R. Ader. Amsterdam, Elsevier. I: 63-83.

Sapolsky, R. M., L. M. Romero, et al. (2000). "How do glucocorticoids influence stress responses? Integrating permissive, suppressive, stimulatory, and preparative actions." *Endocr Rev* 21(1): 55-89.

Schmidt, L. D., Y. Xie, et al. (2007). "Autonomic neurotransmitters modulate immunoglobulin A secretion in porcine colonic mucosa." *J Neuroimmunol* 185(1-2): 20-8.

Schmidt, P. T., L. Eriksen, et al. (1999). "Fast acting nervous regulation of immunoglobulin A secretion from isolated perfused porcine ileum." *Gut* 45(5): 679-85.

Schreiber, K. L., L. D. Price, et al. (2007). "Evidence for neuromodulation of enteropathogen invasion in the intestinal mucosa." *J Neuroimmune Pharmacol* 2(4): 329-37.

Schwarcz, R., W. O. Whetsell, Jr., et al. (1983). "Quinolinic acid: an endogenous metabolite that produces axon-sparing lesions in rat brain." *Science* 219(4582): 316-8.

Shen, Y. Q., G. Hebert, et al. (2005). "In mice, production of plasma IL-1 and IL-6 in response to MPTP is related to behavioral lateralization." *Brain Res* 1045(1-2): 31-7.

Spiller, R. C. (2002). "Role of nerves in enteric infection." *Gut* 51(6): 759-62.

Spitz, J. C., S. Ghandi, et al. (1996). "Characteristics of the intestinal epithelial barrier during dietary manipulation and glucocorticoid stress." *Crit Care Med* 24(4): 635-41.

Sternberg, E. M. (2001). "Neuroendocrine regulation of autoimmune/inflammatory disease." *J Endocrinol* 169(3): 429-35.

Sternberg, E. M. (2006). "Neural regulation of innate immunity: a coordinated nonspecific host response to pathogens." *Nat Rev Immunol* 6(4): 318-28.

Stevens, A., D. W. Ray, et al. (2001). "Polymorphisms of the human prolactin gene--implications for production of lymphocyte prolactin and systemic lupus erythematosus." *Lupus* 10(10): 676-83.

Taylor, A. W. (2003). "Modulation of regulatory T cell immunity by the neuropeptide alpha-melanocyte stimulating hormone." *Cell Mol Biol (Noisy-le-grand)* 49(2): 143-9.

Thomas, M. D., B. Srivastava, et al. (2006). "Regulation of peripheral B cell maturation." *Cell Immunol* 239(2): 92-102.

Tracey, K. J. (2002). "The inflammatory reflex." *Nature* 420(6917): 853-9.

Tracey, K. J. (2007). "Physiology and immunology of the cholinergic antiinflammatory pathway." *J Clin Invest* 117(2): 289-96.

Tracey, K. J. (2010). "Understanding immunity requires more than immunology." *Nat Immunol* 11(7): 561-4.

Tracey, K. J., C. J. Czura, et al. (2001). "Mind over immunity." *FASEB J* 15(9): 1575-6.

Trinchieri, G. (1998). "Interleukin-12: a cytokine at the interface of inflammation and immunity." *Adv Immunol* 70: 83-243.

Van Der Zanden, E. P., G. E. Boeckxstaens, et al. (2009). "The vagus nerve as a modulator of intestinal inflammation." *Neurogastroenterol Motil* 21(1): 6-17.

Vargas, L., B. F. Nore, et al. (2002). "Functional interaction of caveolin-1 with Bruton's tyrosine kinase and Bmx." *J Biol Chem* 277(11): 9351-7.

Wang, X., B. R. Wang, et al. (2002). "Evidences for vagus nerve in maintenance of immune balance and transmission of immune information from gut to brain in STM-infected rats." *World J Gastroenterol* 8(3): 540-5.

Warwick-Davies, J., D. B. Lowrie, et al. (1995). "Growth hormone activation of human monocytes for superoxide production but not tumor necrosis factor production, cell adherence, or action against Mycobacterium tuberculosis." *Infect Immun* 63(11): 4312-6.

Warwick-Davies, J., D. B. Lowrie, et al. (1995). "Growth hormone is a human macrophage activating factor. Priming of human monocytes for enhanced release of H2O2." *J Immunol* 154(4): 1909-18.

Webster, J. I., L. Tonelli, et al. (2002). "Neuroendocrine regulation of immunity." *Annu Rev Immunol* 20: 125-63.

Weinstock, J. V. (2003). Substance P and the immune system. *Autonomic Neuroimmunology*. J. Bienenstock, D. T. Golenbock and M. Blennerhassett. London & New York, Taylor & Francis: 111-137.

Wick, M. J. (2004). "Living in the danger zone: innate immunity to Salmonella." *Curr Opin Microbiol* 7(1): 51-7.

Wiegers, G. J. and J. M. Reul (1998). "Induction of cytokine receptors by glucocorticoids: functional and pathological significance." *Trends Pharmacol Sci* 19(8): 317-21.

Wilder, R. L. (1995). "Neuroendocrine-immune system interactions and autoimmunity." *Annu Rev Immunol* 13: 307-38.

Wilson, I. D., R. D. Soltis, et al. (1982). "Cholinergic stimulation of immunoglobulin A secretion in rat intestine." *Gastroenterology* 83(4): 881-8.

6

Serology as an Epidemiological Tool for *Salmonella* Abortusovis Surveillance in the Wild-Domestic Ruminant Interface

Pablo Martín-Atance, Luis León and Mónica G. Candela
University of Murcia
Spain

1. Introduction

Salmonella sp, are opportunistic pathogens that can infect a wide range of hosts, including man (Murray, 1991). The increasing numbers of *Salmonella* infections reported in the last decades reveal an important health problem of considerable socio-economic impact (Kapperud et al., 1998). Salmonellosis has been reported in 85% of food-borne bacterial enteritis in humans from Spain (Pérez-Ciordia, et al; 2002), and *Salmonella* sp, are increasingly recorded in animals (Echeita et al.; 2005).

Unlike other *Salmonella* species, *Salmonella enterica* subspecies *enterica* serovar Abortusovis (S. Abortusovis) is adapted to sheep, and considered to be host specific (Jack, 1971). Discarded as a zoonotic pathogen, its importance lies in the economic losses that occur in ovine production systems in regions that depend on sheepherding (Pardon et al., 1988; Sojka, et al., 1983). It has been most frequently associated with ovine salmonellosis in ovine flocks from Europe and the Middle East, causing abortion outbreaks, stillbirths, and illness in lambs infected at birth (Jack, 1968; Pardon et al., 1988). These mainly result from the epidemic behavior of the disease, which is most recognized when the organism is newly introduced into a flock, because abortion storms reach high proportions. In endemic scenarios it also causes abortions in up to 50% of the ewes in a flock, usually during the first pregnancy, as in newly introduced (González 2000).

Available epidemiological data show a limited distribution for Abortusovis serovar. It is considered rare in most countries and regions of the world except in Europe, where it is particularly common, with reported cases in France, Spain, Germany, Cyprus, Italy, Switzerland, Russia, and Bulgaria, southwest England and Wales and also in Western Asia (Jack., 1968; Echeita et al., 2005; Valdezate et al., 2007). In northern Spain it has been considered to be among the major etiological agents of ovine abortion (González, 2000) , but it is also spread through 11 Spanish provinces, where 20 different clones have been identified in fifty-five field strains collected from epidemic abortions or neonatal mortality episodes affecting different ovine flocks during the period 1996–2001 (Valdezate et al., 2007).

The infection can appear in naive flocks by means of animal carriers such as new sheep replacements, contact with other animals in seasonal migration, wild and carrion birds, or rodents (Valdezate et al., 2007). Sensitive animals acquire the infection by ingestion of food and water contaminated by vaginal discharges, placenta, aborted foetus (liver and stomach contents), and infected newborn. Furthermore, in some conditions, faeces, milk and respiratory secretions can correspond to infectious material. Other routes of acquisition include respiratory and conjunctival routes (Jack., 1971).

From the third month of pregnancy, this pathogen induces abortion, in the absence of other clinical symptoms (Jack, 1971), but it is sometimes preceded by depression, uncertain walking, mucous vaginal discharge and diarrhoea. Following this, ewes seem to be healthy or show transient fever, but sometimes ewe mortality occurs from septicaemic complications like anorexia, acute metritis, enteritidis and peritonitis that result from placental retention (5-7% of cases) (Astorga et al., 2000), differing from infection causes by Dublin and Typhimurium serovars. In addition, neonatal mortality of lambs is frequent with living muttons at term which are non-viable and die within a few hours of birth from septicaemia. Occasionally, lambs appear to be healthy but die during the first month, showing signs of enteritis, pneumonia or polyarthritis. Conversely, the infection is asymptomatic in non-pregnant ewes and rams (Uzzau et al., 2001).

Within a flock, S. Abortusovis is maintained by effective transmission from infected to susceptible sheep through the oral, conjunctival, or respiratory routes, while venereal infections appear to be of minor importance (Uzzau et al., 2001). Spread to other susceptible populations is mostly the consequence of commercial translocations of asymptomatic carriers. The dissemination of S. Abortusovis by food, water, birds or other mammals, has been traditionally considered as negligible. But as well as the host-specificity to sheep of Salmonella Abortusovis, its adaptation to other mammals can be discussed: mice and rabbits can be experimentally infected, and it has occasionally been isolated from goats and rabbits.

Wild ruminants can act as asymptomatic carriers of pathogenic Salmonella serovars (Cubero et al., 2002; Renter et al., 2006), and some serovars can also cause clinical disease in deer species (Foreyt et al., 2001; McAllum et al., 1978). However, this bacteria has not been related to abortion in non domestic species.

Many European countries face difficulties in controlling S. Abortusovis disease because there are no ways to diagnose all infected animals (Lantier et al., 1983). In this sense, it is essential to consider sampling procedures (e.g. type of samples, sampling frequency) according to the objectives of the testing program, clinical findings, level of detection or precision of prevalence estimates required, cost and availability of sampling resources and laboratory facilities. In recent years a standard method for detecting Salmonella from primary animal production has been developed and evaluated, and an ISO-method (ISO 6579:2002 Microbiology of food and animal feeding stuffs - Horizontal method for the detection of Salmonella spp., Annex D) has now been adopted), and diagnosis procedures have been well defined (OIE, 2010).

The identification of the S. Abortusovis is based on the isolation of the organism; when infection of the reproductive system, abortion or conceptus occurs, it is necessary to culture fetal stomach contents, placenta and vaginal swabs. Nevertheless, few epidemiological surveys have been able to be carried out owing to the low number of available S. Abortusovis isolates (Valdezate et al., 2007).

In this sense, serology can improve the study of natural infections through identification of infected herds, rather than to confirm infected individual animals; although repeated herd tests can be used as an aid to detect chronic carriers. Antibodies to *S.* Abortusovis may become undetectable in some sheep 2–3 months after abortion. Flock diagnosis of *S.* Abortusovis in sheep can be performed by serological tests conducted on a statistically representative sample of the population, but results are not always indicative of active infection (OIE, 2010). These include serum agglutination test (SAT), hemagglutination inhibition, complement fixation, indirect immunofluorescence, gel immunodiffusion, and enzyme–linked immunosorbent assay (ELISA) (Davies, 2004).

In Spanish mountain rural areas extensive-grazing farming and game management practices favor the establishment of multispecies assemblies with a high diversity of herbivore species that share pastures and forests. There, sheep can be considered as primary hosts of *Salmonella* Abortusovis, but the potential dissemination of this agent to other possible sensible ruminant species has not been considered in epidemiological studies, even though specific antibodies have been found in Spanish ibex (*Capra pyrenaica*), fallow deer (*Dama dama*), European mouflon (*Ovis aries*) and red deer (*Cervus elaphus*) from Southern Spain (Pérez, 2007; León et al., 2002). In the Serranía de Cuenca domestic ruminant flocks find sympatric conditions for grazing with wild populations of red deer, fallow deer, roe deer (*Capreolus capreolus*), European mouflon and Spanish ibex, were introduced for big game hunting purposes from 1960 to 1979, and have successfully colonized the territory 30 years later.

2. Objectives

The aim of this study was to identify antibody responses against *Salmonella* Abortusovis in sheep flocks by means of the serum agglutination test (SAT), and evaluate the possibility of adapting it to official veterinary programs sampling efforts. A further aim was to evaluate the potential application of this test for screening *Salmonella* infections in cattle herds and wild ruminant populations. We therefore estimated the risk factors of sero-conversion against specific antigens of *Salmonella* Abortusovis in wild and domestic ruminants, and analyzed the statistical association between the sero-epidemiological indexes of each population and the related environmental conditions in the grazing pastures of the "Serranía de Cuenca Regional Game Reserve", where traditional shepherding constituted the core of the economic activity as in other many regions of Castile La-Mancha, Spain.

3. Material and methods

3.1 Study area

The Serranía de Cuenca Natural Park is one of the best conserved mountainous areas of Castile-La Mancha, Spain. The study area is situated in the centre-east of the Iberian Peninsula (40° 12´- 40° 28´ N / 1° 51´- 2° 03´ W), attached to the limits of the "Serranía de Cuenca Regional Game Reserve". It has a Mediterranean climate tempered by the altitude and barrier effect from the local orography to the wet winds from the west that confer a temporal semi-damp hydrologic pattern (annual rainfall varies between 600 and 1.000 mm, which is more intense from November to January), with long periods of drought during the hottest months (normally July and August). The climatic conditions that occur are typical for the Mediterranean Mountains, with temperate weather (Mean temperatures between 7'5 and

12'5°C), cold winters (Mean temperatures fall below the 0°C between 60 to 120 days) and dry, but not excessively hot, summers (law-ranking decree 99/2006). Vegetation is dominated by forests: Special importance is given to those constituted primarily by *Pinus nigra* Subsp. *Sazlmanii*, and the Oromediterranean forests of *Pynus sylvestris*. The latest, integrate the climax vegetal community at altitudes that limit the Mediterranean forests (*Quercus faginea* and *Q. ilex*), or on not suitable soils for the Sabin (*Juniperus sabina*), while Mediterranean scrubland (*Crataegus monogyna*, *Ligustrum vulgare*, *Viburnum lantana*, *Rhamnus saxatilis*) and pastures (*Poo-Festucetum istricis* association) present limited extensions (Peinado et al., 1985).

The "Serranía de Cuenca Regional Game Reserve"is an open area of 25.850'9 hectares covering eleven public mountains with free access to people, an open private property (El Maillo) and a fenced private property (Valsalobre), where human activities are restricted to traditional non-intensive use, such as eco-tourism, wood production, extensive farming, and hunting. In the core of this territory the human activities, including farming, are not allowed in a valley of 910 hectares, known as "El Hosquillo", which is partially fenced for its administration as Experimental Game Park. In contrast, game ungulates share the forest's pasturelands with cattle and small-ruminants herds in the surrounding protected areas.

3.2 Ruminant populations

The bio-climatic conditions in the region traditionally favor extensive ruminant farming system, specially focused on summer-grazing migratory merino-sheep flocks, for wool production (Cava, 1994). During recent decades this system has been modified by different social and economic influences that have led to a current orientation of the production to lamb meat, employing mixed raced animals for such purpose. On the other, goats have always played a self-sufficient role in the human communities of the Serranía de Cuenca and are usually maintained in the sheep flocks as guides or stepmothers, but are not of relevant economic importance. In contrast, beef productions are gaining importance, and the traditional local use of bovines as working elements is no longer practised. During the years 2.003-2.005 permissions for grazing in the "Serranía de Cuenca Regional Game Reserve" allowed the herding of 12.881 domestic ruminants in flocks classified by species composition in: 19 ovine, 14 ovine-caprine, and 6 bovine (Martín Atance, 2.009).

Big game hunting was historically practised in the area until the autochthonous wild ruminant populations became extinct, at the end of the 19th century. In order to encourage this activity, red and fallow deer were introduced in the "Serranía de Cuenca Regional Game Reserve" from 1960 to 1966, the European mouflon from 1974 to 1977, and Spanish ibex from 1976 to 1979 (Rojo Arribas, 2007). In 2001 the free ranging population sizes estimated in the Game Program of the "Serranía de Cuenca Regional Game Reserve" were: 850 red deer, 617 fallow deer, 300 mouflon and 230 Spanish ibex, in 25,724 hectares; and in semi-captivity: 209 red deer, 32 fallow deer, 91 mouflon, and more than 11 Spanish ibex at the fenced hectares in "El Hosquillo" Game Park (Martínez & Verona, 2002).

3.3 Collection and preparation of serum samples

The sampling of domestic ruminants herds that grazed at the study area were collected in two annual Livestock Sanitary Surveys: In 2,004, 241 blood samples were collected from cattle (5 herds), and 1,196 from sheep (13 herds), and in 2.005 samples were taken from 166 cattle (3 herds), and 2,543 from sheep (27 herds) in the course of Official Veterinary Programs. Sampling of wild ruminants was performed from 2,003 to 2,006 on 885 free ranging animals hunted in the "Serranía de Cuenca Regional Game Reserve", and from 225 live animals captured in the Game Park.

Blood samples were taken from the jugular veins in live animals, and from the heart in hunter-harvested wild ruminants. The samples were placed in test tubes and sent to the laboratory under refrigeration (4°C). They were then centrifuged and the serum was frozen at -80°C.

Animals were classified into four age groups: young (under 1 year), juveniles (1-2 years), adults (2-9 years), and old (over 9 years).

3.4 Serological procedures

We employed the Serum agglutination test (SAT) to evidence specific antibodies to *Salmonella* Abortusovis O antigen in 5,256 sera from 7 ruminant species: 3,739 from sheep, 556 from fallow deer, 407 from cattle, 314 from red deer, 211 from European mouflon, 21 from Spanish ibex and 8 from roe deer.

SAT were performed according to standard procedures (Pardon et al., 1983; Lindberg, & Le Minor, 1984; Sanchis et al., 1985; Sanchis et al., 1991; OIE, 2010). They were adapted to the microtitre format and used to determine somatic and flagellar titres of specific antibodies induced by *Salmonella* enteritidis subsp. enteritidis serovar Abortusovis natural infections. Antigens used were a serogroup B (O) antigen (Salmonella O Antigen -2840-56-3 Difco®) and a serogroup C (H) antigen (Salmonella Flagelar H Antigen -2846-56-7 Difco®). Negative (serum and normal saline) and positive (Standard sera Salmonella O Anti-sera Grupo B, Difco®, and Salmonella Flagellar Poli Anti-sera Difco®, respectively) controls were included in each test run as confirmatory method for quality control of SAT. Normal saline solution (0'85%) and bi-distilled water Mili-Q were also employed in test procedures.

Sera were screened at dilutions of 1:10 to 1:1.280; 50 µl ml of (O) antigen and added 100 µl of serum pre-diluted to 1:10. The plates were covered by a film and shacked automatically (40-50 rpm) for 5 minutes at 37° C. They were then incubated at the same temperature, without movement, for 18 hours. Sera that presented a positive reaction (from 1:20) were retested with (H) antigen.

To interpret the results of each test run the control wells were examined first, in order to confirm absence of agglutination in the negative controls, and agglutination in the positive controls. Agglutinations appeared as a "matt" or "carpet" at the bottom of the positive O antigen control wells and loose, woolly or cottony in the case of H antigens. In each sample, the highest dilution of serum that produced a positive agglutination was taken as titre.

Samples that held titres over 1:20 to both antigens were considered as sero-positive to *Salmonella* Abortusovis. The sera that reacted at 1:10 against any of the antigens were considered as doubtful, and as negative when agglutination was not observed.

3.5 Statistical tools and definition of epidemiological indexes

In order to describe the immune reactions observed, interpreted as previous exposure to *Salmonella* Abortusovis, and analyze the factors that could contribute to explain the variations observed between animals and populations, we performed statistical analysis (Caughley, 1977; Crawley, 1993; Petrie & Watson, 1999; Siegel, 1956; Daniel, 1993) suited to the methodological principles of Epidemiology (Thrusfield, 1990; Goldberg, 1994).

As a descriptive study, the main objective was to determine the patterns of sero-conversion presentations. The information obtained from every sample was registered in qualitative terms (presence or absence of sero-conversion), according to the evidence of previous exposure to *Salmonella* Abortusovis; and in quantitative terms related to the intensity of immune recognitions (titre values). The intensity of immune responses in animal groups was expressed by means of the Geometric Mean Titre (G.M.T.), calculated as described by Thrusfield (1995), and in order to measure the variability within groups we have used the standard deviations of the Inverse function of the titres.

The temporal relationships between host, agent, and environmental factors influence the risk of disease. They have been considered here in association to the Prevalence term, defined as the proportion of a population affected by a disease at a given point in time, and interpreted as the probability of a subject in a defined population being diseased at a particular annual sampling campaign. Prevalence is a function of both the incidence (frequency of new cases in a population) and duration (time to recovery of a disease). Also, the relationship between prevalence and incidence will be greatly influenced by the persistence of a detectable antibody following infection, where sero-positivity defines cases of disease.

As an analytical study, our objective was to understand the "natural history" of *Salmonella* Abortusovis. For this purpose we investigated the interrelatedness of the humoral response phenomenon within the biological system, selecting the segments of the chain of infection where the interactions performed betwen hosts and the environment could be measured.

We have used the risk to express the probability of sero-conversion occurrence following a particular exposure to S. Abortusovis. The differences in immune recognition between groups were respected to the frequency of the presence or absence of potential risk factors by means of a cohort study to evaluate the hosts' risk factors. Here, we considered those endogenous characteristics that could influence the immune response of individual ruminants: species and breed (genetic constitution); age and gender (Goldberg, 1994).

Our field study in th e"Serranía de Cuenca Natural Park" measured sero-positive and sero-negative animals considering their exposure to all the known and unknown environmental factors present in their natural environment. The quantification of sero-conversion cases occurrence was performed by counts of sero-positive individuals and expressed as a fraction of the number of animals that held a similar condition (= population at risk). From a mathematical perspective, frequencies were expressed through static measures as proportions and ratios (Thrusfield, 1990; Goldberg, 1994).

Finally,to obtain further data to understand the epidemiology of this infectious disease we evaluated the distribution of the agent in the ecosystem and the factors that could influence the modes of transmission between its compartments (Scott & Smith, 1994). As environmental risk factors we considered those not identifiable with the host or agent but related to the animals comprising individual populations, where they live under similar conditions of clime, management and nutrition. As empirical research, it involved the measurement of variables, estimation of population parameters and statistical testing of hypotheses by comparisons between groups.

In order to estimate the magnitude of an association between a putatively causal factor and sero-conversion and to assess if there was potential for a cause-effect relationship between a single or multiple risk factors and sero-conversion, we used Generalized linear models (GLM; McCullagh & Nelder 1989) to measure and represent the statistical interaction of response variables (dependents), such as abundance of sero-conversion (prevalences and frequencies),with environmental predictors (independent variables).

Statistical analysis were performed with Microsoft EXCEL 2000© (1985-1999, Microsoft Corporation, USA), Statistica 6.0® (1984-2001 Statsoft, EE.UU.), and Epilnfo 3.3.2 (Center for Disease Control, USA, 2005) integrated epidemiological statistics package. The analysis of the risk factors for the infection was calculated by the Pearson Chi-square test without correction and the Fisher exact test. We considered the value of two tailed P in all analyses. The level of significance was set at $P \leq 0.05$. The association of risk factors and infection were quantified by the analysis of the odds ratio (OR) using Cornfield 95 % confidence limits. Finally, we included the statistically significant factors in General Linear Models to establish which variables act as predisposing factors.

4. Results

Antibodies to *Salmonella enterica* Serogorup B Somatic (O) were not found in 4,318 (82.1 %) samples and were classified as sero-negative SAT reactions. We observed coloured films on the surfaces of the well at 1/10 dilutions in 716 (13.6 %) sera, that were considered as doubtful and, at higher titres 222 (4.2 %) were identified as sero-positives.

Results showed that frequency of immune reactions (titres $\geq$ 1:10) against somatic antigen of *Salmonella* Abortusovis (applicable to others included in B serogroup) were higher bovines (RF = 36.8%) than in sheep (RF = 19.73%), fallow deer (RF = 7.55%), mouflon (RF = 9.47%), or red deer (RF = 1.59%) (Table 1; Figure 1).

Moreover, specific antibodies to *Salmonella* Abortusovis H antigen were revealed by SAT among the 939 sera that showed titres of 1:10 or higher to the O antigen (Cattle = 150; sheep = 738; fallow deer = 42; red deer = 5; mouflon = 4). Results showed that frequency of immune reactions (titres $\geq$ 1:10) against flagellar antigen of *Salmonella* Abortusovis were higher in mouflon (RF = 75%) and domestic ovine (RF = 74.39%), than in fallow deer (RF = 54.76%), cattle (RF = 30-66%), or red deer (RF = 0%) (Table 1; Figure 2).

The relative intensity of reactions in this species against each antigen was estimated by the Mean Geometric Titer, and only considering sero-positives (titers $\geq$ 1:20). Immune responses to O antigen were higher in cattle (MGT = 1:39.2), than in fallow deer (MGT = 1:23.7), sheep (MGT =1:22), or red deer and European mouflon (MGT =1:20). Other ways, responses to H antigen were higher in ovines (MGT = 1:30.6) and mouflon (MGT= 1:25.2), than cattle or fallow deer (MGT = 1:20), and were not detected in red deer (Table 1).

The interpretation of serological results obtained in both techniques for *Salmonella* Abortusovis serological diagnosis was performed with attention to positive reactions (titres

≥ 1:20) to both antigens O and H. This condition was only found in 150 (2.8%) serum samples: 145 (3.85%) ovine sera, 3 mouflon sera (0.95%), and 2 fallow deer sera (0.35%) (Table 1; Figure 3).

Species		Somatic antigen "O"(group B)					Flagellar antigen H (group C)			
		Titre						Titre		
	n	1:10	> 1:10			n	1:10	> 1:10		
		AF	AF	%	MGT		AF	AF	%	MGT
C	107	111	39	9,58	39,2	150	0	46	30,6	20
S	3739	567	171	4,57	22,0	738	128	421	74,3	30,68
RD	314	4	1	0,32	20	5	0	0	0	-
FD	556	34	8	1,44	23,7	42	21	2	54,7	20
RoD	8	0	0	0,00	-	0	0	0	-	-
SI	21	0	0	0,00	-	0	0	0	-	-
M	211	1	3	1,42	20	4	0	3	75,0	25,19
T	5256	716	222	4,22	-	939	149	490	51'7	-

AF: absolute frequency; GMT: mean geometric titre; C: cattle; S: sheep; RD: red deer; FD: fallow deer; RoD: roe deer; SI: Spanish ibex; M: mouflon; T: Total.

Table 1. Frequencies of sero-positives and Mean Geometric Titres aganist S Abortusovis Somatic and flagellar antigens obtained by Serum Agglutination Test

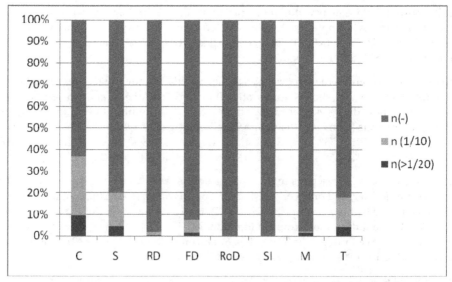

Fig. 1. Percentage of sero-positive, doubtful and sero-negative reactions against O Antigen

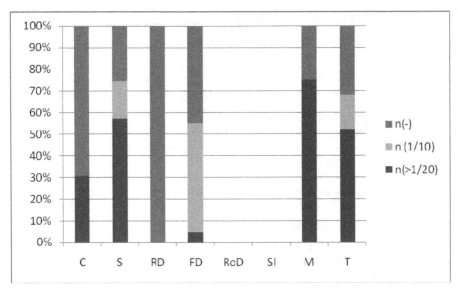

Fig. 2. Percentage of sero-positive, doubtful and sero-negative reactions against F Antigen

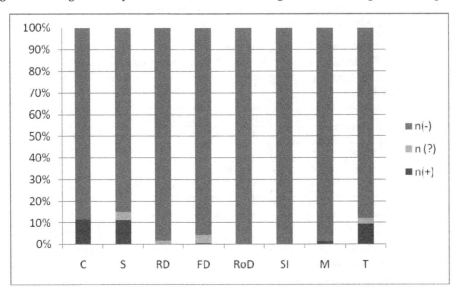

Fig. 3. Percentage of sera catalogued as sero-positive, doubtful and sero-negative against
Salmonella Abortusovis

Statistical relationship between the sero-conversion cases and individual factors of the animals was explored by means of risk factors analysis. Statistical risk to sero-conversion was observed between sexes in domestic sheep (X^2 = 42.37; p < 0.001), that was higher for males compared to females (Relative risk = 3.85; IC95%: 2.5-5.8 vs. RR = 0,26; IC95%: 0.1-0.3) (Table 2).

Species	TOTAL			♀		♂		φ (IC 95%) ♀		X²	p
	N°	(+)	(-)	(+)	(-)	(+)	(-)				
C	361	0	361	0	290	0	71	-	-	-	-
S	3249	139	3110	116	2974	23	136	0,26 (0,1-0,3)	3,85 (2,5-5,8)	42,37	0,00
RD	314	0	314	0	184	0	130	-	-	-	-
FD	533	2	531	2	307	0	224	-	-	-	-
RoD	8	0	8	0	5	0	3	-	-	-	-
SI	21	0	21	0	9	0	12	-	-	-	-
M	211	3	208	1	141	2	67	0,24 (0,02-2,6)	2,12 0,3-44,6)	1,60	0,20

(+) Positive. (-) Negative. Risk *φ.- Odds ratio.* AF: absolute frequency; GMT: mean geometric titre; C: cattle; S: sheep; RD: red deer; FD: fallow deer; RoD: roe deer; SI: Spanish ibex; M: mouflon; T: Total.

Table 2. Risk associated to sex in the studied species.

Among domestic ruminants, yearling sheep showed a high risk of sero-conversion to *S.* Abortusovis (p=0'01, φ=2'27) respect to older animals (p=0'14, φ=0'6). In the same way. the results for mouflon sheep, revealed a significantly high *Odds ratio* in yearling animals (φ = 7'58), that was followed by a decreasing value (φ = 3'12 in the next age group (2 to 4 years); estimates in the rest of the cohorts considered were not able to be performed due to the scarce absolute number of sero-positive animals. (Table 3).

By the same way, the juvenile age group seems to be an important risk factor for sero-conversion in domestic sheep (OR = 1'39; $IC_{95\%}$:1'13-2'71) and mouflon (OR = 7,6; $IC_{95\%}$:0,73-77,18), compared to other age categories (X^2 = 5'6; p = 0'01 and X^2 = 3,8; p = 0,049, respectively) (Table 3).

Group age	Specie	Cases			Probability		Risk	
		+	-	RF (%)	X²	p	φ	IC 95%
Young	Sheep	8	77	9'4	5'6	0'01	2'27	1'15-4'49
	Cattle	0	66	0'0	-	-	-	-
Adult	Sheep	131	3004	4'3	2'16	0'14	0'6	0'3-1'19
	Cattle	0	177	0	-	-	-	-
Old	Sheep	0	29	0	-	-	-	-
	Cattle	0	118	0	-	-	-	-

RF: Relative frquency. *Odds ratio*

Table 3. *Odds ratio* (φ) for sero-conversion to *S.* Abortusovis in the different age groups.

Sero-prevalence values in sheep flocks were estimated in 2004 and 2005. The frequency of affected flocks in the first year reached 100% (13/13), with sero-prevalences ranging from 4.3% to 22.0% (12-33± 5.9%) and also low MGT (1:23.25 ± 1:5.57). The following year sero-positive sheep were found only in the 33.33% of the sampled flocks (9/27), and sero-prevalence values in the same flocks ranged from 1.06% to 7.9% (3.0±2.14%) with very low MGT (1:0.99±7.91). The results obtained in the Serranía de Cuenca show a decline phase of an epidemic outbreak.

Mean sero-prevalence values in each flock and grazing area were estimated and used as indexes to evaluate association of sero-conversion with management factors, and spatial relationships with environmental conditions, respectively. In the correlation matrix analysis built to evaluate management factors in sheep flocks a significant trend to seroconversión was observed in association with lambing intensities of 3 parturitions each 2 years (r = 0.4013; p = 0-028), but this could not be confirmed by ANOVA ($F_{(7,23)}$= 1.19; p = 0.34).

The spatial distribution of infected flocks in the "Serranía de Cuenca Regional Game Reserve" Pasture Areas conditioned a wide distribution of sero-positive sheep, restricted only by their own flock permissions, with mean sero-prevalences of 4.2 ± 3.04% (Range: 0 to 9.57%). These sero-prevalence values were associated with the relative extension of *Pinus nigra* within each area, by means of Lineal Regression Models: "PRV OA Salm" = 1.852 + 0.07 * "Pn" (R^2 = 0.33; p = 0.039).

The scarce evidence of specific antibodies against *Salmonella* Abortusovis in other ruminant populations prevents evaluating the potential host-pathogen interactions in the lesser sampled species (roe deer and Spanish ibex). Furthermore, frequencies of sero-positives in mouflon (3/211, 1.4 %) and fallow deer (2/555. 0.3%) populations were low and not intense (Range of titres: 1:20-1:40), and immune responses were absent in red deer, as in cattle (Table 4).

RD (n = 314)	FD (n = 555)	RoD (n= 8)	SI (n = 21)	M (n = 211)
0	2 (0.36)	0	0	3 (1.4)

RD: red deer; FD: fallow deer; RoD: roe deer; SI: Spanish ibex; M: mouflon.

Table 4. Frequency of sero-positives in wild ruminant species.

5. Discussion

Different *Salmonella* serovars have been isolated from a wide variety of vertebrates, including European wild ungulates, but bacterial isolation is not sufficient, in itself, for a diagnosis of salmonellosis (Mörner, 2001; Nielsen et al, 1981).

Diagnosis of salmonellosis should be based on culture and identification of the bacteria, together with clinical and pathological evidence (Olsen et al, 2003; Threlfall & Frost, 1990; Valdezate et al., 2007; Van der Zee & Huis In't Veld, 2000). In the case of paratyphoid abortion diagnosis, this can be compromised, because the use of common bacteriological procedures is not an advisable option, and clinical signs or lesions are not specific enough (González, 2000; Beuzón et al., 1997; Linklater, 1983). In such situations, serological methods, such as SAT on microplates, may also be used to identify infected flocks or herds, rather than to identify individual infected animals (Pardon et al., 1988; González, 2000; OIE, 2010).

For detailed epidemiological investigations strain identifications are necessary (OIE, 2010). Traditionally, the identification of S. Abortusovis has relied on the use of antisera against O and H antigens (Brenner, 1984; Vodas and Martinov, 1986), and currently molecular characterization techniques like ribotyping, plasmid profiling, and IS200 fingerprinting can be performed successfully to identify different and predominant Abortusovis genotypes (Nikbakht, et al.2002; Schiaffino et al. 1996; Nastasi, et al. 1992). The use of this method led to the identification of two different strains in sheep flocks from Cuenca province between

1996 and 2001, but this epidemiological surveillance system is limited by the low number of Abortusovis isolates available (Valdezate et al., 2007).

On the other hand, detection of specific immune responses to S. Abortusovis can provide further evidence of infection, but little is known about the duration of effective immunity following *Salmonella* infections, and positive results cannot always be interpreted as indicative of active infections (OIE, 2010; Brennan et al., 1994; 1995). Animals that have been infected recently would, in all probability, eventually be detected serologically by an appropriate monitoring programme throughout the life of the flock/herd, but there are often cost limitations to the application of effective monitoring programmes (OIE, 2010).

The agglutination test is considered the preferred method in export and diagnostic purposes for samples from all species of farm animals (Davies, 2004; OIE, 2010). Our research demonstrated the possibility of performing it alongside the analysis scheme established in the brucellosis official eradication program. SAT proved to be easily adaptable to the routine diagnostic procedures in "Albaladejito" Veterinary Laboratory as well as being an economical method of performing simultaneous analyses of large numbers of samples. In this sense, we must argue for the potential adoption of this method by official veterinary programs if further studies are to be carried on.

In order to define the level of detection or accuracy of prevalence in the testing program we must consider the lower sensitivity of SAT in comparison with ELISA tests (Berthon, et al., 1994, Sting et al., 1997; Veling et al., 2000), and the conditions relating to sampling procedures regarding the dynamics of immune responses aganist *Salmonella* Abortusovis (OIE, 2010). For most animals, a significant increase in agglutination titres could be observed from day 5 after inoculation (Lantier, 1987), but care in interpreting the serological results has to be taken if this test is performed after abortion, as antibody levels fall and may become undetectable 2-3 months later (Davies, 2004; González, 2000). Thus, our serological results should be interpreted as punctual sero-frequencies or sero-prevalences.

The wide frequency of O-agglutinating sera found among the domestic (bovines RF = 39.2%; ovines RF=22 %) and wild ruminants (fallow deer RF =23.7%; mouflon RF = 20%: red deer RF = 20%), sampled in the "Serranía de Cuenca Regional Game Reserve" indicated wide immune recognition of *Salmonella* between 2003 and 2006, but these immune responses may respond to other group B *Salmonella* infections, whatever the O antigen (Bernard et al., 2002). In fact, results obtained in cattle and red deer against group C H antigen were not correlated, in any case, to the respecting anti-O titres. In this sense, although serological responses can be demonstrated against both flagellar and somatic antigens, it has been stated that it is advisable to restrict the search to anti-H agglutinins, as these are more specific, reach higher titres and have a precocity and persistence similar to those against O antigen (Pardon et al., 1988; González, 2000).

The frequencies obtained in the group C H-antigen SAT showed very high values among the ovines (mouflon RF = 75%; sheep RF = 74.39%), but were also found in a high percentage of fallow deer (RF = 54.76 %), and cattle (RF= 30.66%). However, many samples presented a low titre to group B O antigen (1:10), but reached high agglutination titres to H antigen, and were not considered as specific responses against S. Abortusovis. These were recorded in a limited percentage of the sheep analysed (3.85%), mouflon (0.95%), and fallow deer (0.35%), thus restricting the potential host range. In order to improve the knowledge of *Salmonella* infections in each species it would be advisable to include other antigens in the serological screening.

The presence of humoral immune specific recognition of *S.* Abortusovis in sheep from the "Serranía de Cuenca Regional Game Reserve" is further evidence of paratyphoid abortion in the province of Cuenca (Valdezate et al., 2007). Besides possible controversial discussions about prevalence values estimated under different methods (Jack., 1968; Uzzau et al., 2000; Valdezate et al., 2007), the characteristic epidemic behaviour of paratyphoid abortions leads to differences of incidence and importance of S. Abortusovis infection between sheep flocks, countries and periods of time (González, 2000; Giannati-Stefanou et al., 1997). In this sense, the use of SAT can improve sheep abortion surveillance by identifying infected flocks (OIE, 2010), but also by indicating epidemiological trends, as may be suggested by the significant differences found among sero-epidemiological indexes (frequency of affected flocks, MGT, and sheep sero-prevalence) estimated in consecutive sampling campaigns, which clearly indicate a decreased epidemic phase of infection.

In addition, the main affection of yearlings and recently purchased sheep is usual after paratyphoid abortion storms (González, 2000). This situation is also suggested by the results obtained in the risk factor analysis. The odds ratio for sero-conversion in sheep showed a statistical association to host factors such as sex or age, with higher risk in males and yearlings. These findings could respond to different predisposing physiological conditions of lambs, ewes, and rams, but also to environmental factors related to management practices. At this point, it is necessary to consider that in the "Serranía de Cuenca Regional Game Reserve" extensive grazing system these sheep groups maintain different farming conditions, such as transport, herding, housing or nutrition, that respond to the conditions of the reproduction scheme in each flock (Martín Atance, 2009).

The communal grazing practices in the "Serranía de Cuenca Regional Game Reserve" may allow a wide dissemination of infection between flocks, but differences in the extension of Pinus nigra forests within each grazing area can modulate the variance of the mean herd-sero-prevalence values estimated ($R^2 = 0.33$; $p = 0.039$). In our opinion, this finding could indicate the effect of the environment on the host conditions, such as nutritional stress in carrier hosts, but this hypothesis needs to be confirmed by specific research (González, 2000; Linklater et al., 1991).

The detection of SAT specific responses in mouflon and fallow deer allows us to hypothesize about the possible sensitivity to infection of both species, or even a potential adaptation of similar strains to this hosts. In this sense, it would be recommendable to favour research into the genetic and physiological differences of *Salmonella* isolates from sheep and compare them to others of wildlife origin in order to test the basis for *Salmonella enterica* evolution in relation to hosts (Bäumler et al., 1997; Hensel, 2004; Popoff & Le Minor, 2005). Nevertheless, the scarce amount of evidence suggests that wild ruminants play a minor role in the epidemiological cycle of paratyphoid abortion in the "Serranía de Cuenca Regional Game Reserve", as has been suggested in Andalusia, where sero-positive Spanish ibex, mouflon, fallow deer and red deer have been detected at higher rates (Arenas et al., 1993; Cubero et al., 2002; Pérez, 2007; León Vizcaíno et al., 1980; León Vizcaíno et al., 1992; León Vizcaíno et al., 1994; León Vizcaíno et al., 2002).

These results indicate the epidemiological role of sheep as primary hosts of S. Abortusovis, and the absence of natural niches of infection among other ruminant populations. However, the presence of specific antibodies in European mouflon and fallow deer could be indicative of the infection in this populations and the presence of chronic carriers that might help to disperse this agent.

6. References

Arenas, A. & Perea, A. (1993). *El Ciervo en Sierra Morena*. Servicio de Publicaciones de la Facultad de Veterinaria. Universidad de Córdoba. Córdoba. España. ISBN 84-600-8657 7.

Astorga R, Gomez JC, Arenas A, Perea A. (2000). Patologia de los pequeños rumiantes en imágenes. Sindromes de mortalidad perinatal y mamitis-agalaxia. Revista del Consejo General de Colegios Veterinarios de España, http://www.colvet.es/infovet.

Bäumler A, Gilde A, Tsolis R, Velden A, Ahmer B, Heffron F. (1997). Contribution of horizontal genes transfer and deletion events to development of distinctive patterns of fimbrial operons during evolution of *Salmonella* serovars. *The Journal of. Bacteriology*. Vol.179, No.2 (January 1997), pp.:317-322

Berthon, P. Gohin, I. Lantier, I. & Olivier, M. (1994). Humoral immune response to *Salmonella* Abortusovis in sheep: in vitro induction of an antibody synthesis from either sensitized or unprimed lymph node cells. Veterinary Immunology and Immunopathology. Vol.41, No.3-4, (June 1994), pp. 275-294

Beuzón CR, Schiaffino A, Leori G, Cappuccinelli P, Rubino S, Casadesús J. (1997). Identification of *Salmonella* Abortusovis by PCR amplification of a serovar-specific IS200 element. *Applied* Environmental Microbiology. Vol. 63, No.5, (May; 1997).pp.2082-5.

Brennan, F. R., J. J. Oliver, & G. D. Baird. (1994). Differences in the immune responses of mice and sheep to an aromatic-dependent mutant of *Salmonella typhimurium*. *Journal of Medical Microbiology*. Vol.41 (July 1994), pp. 20–28.

Brennan, F. R., J. J. Oliver, & G. D. Baird. (1995). In vitro studies with lymphocytes from sheep orally inoculated with an aromatic-dependent mutant of *Salmonella typhimurium*. *Research in Veterinary Science*. Vol.58, No.2 (March 1995), pp.152–157.

Brenner, D. J. (1984). Family I. *Enterobacteriaceae,*. In: *Bergey's manual of systematic bacteriology*, N. R. Krieg and J. G. Holt (Ed.), vol. 1. pp. 408–516- Baltimore, Maryland. United States of America.

Caughley G. (1977). *Analysis of vertebrate populations*. The Blackburn Press. John Wiley and Sons.Ltd (eds). Caldwell, New Jersey United States of America pp.234

Cava LE. 1994. *La Serranía Alta de Cuenca: Evolución de los usos del suelo y problemática socioterritorial*. Facultad de Geografía e Historia. Universidad Internacional Menéndez Pelayo. Programa LEADER ".Serranía de Cuenca". Tesis Doctoral. Universidad Nacional de Educación a Distancia (UNED), Madrid, Spain. pp 588. ISBN: 84-605-1514-1.

Crawley, M.J. (1993) *GLIM for Ecologists*. Blackwell Science, Oxford, United Kingdom.

Cubero Mª-J. González, M., & León, L. (2002). Enfermedades infecciosas en las poblaciones de cabra montés. In: *Control de la sarna sarcóptica de la cabra montés (Capra pyrenaica hispanica) en Andalucía.* (J. Pérez, Ed.). Consejería de Medio Ambiente de la Junta de Andalucía, Sevilla. pp 201-256. ISBN: 84-8439-098-5

Daniel, W. W. 1993. *Bioestadística: base para el análisis de las ciencias de la salud.* Limusa, México, 667pp

Davies, R.H., Dalziel, R., Gibbens, J.C., Wilesmith, J.W., Ryan, J.M.B., Evans, S.J., Byrne, C., Paiba, G.A., Pascoe1, S.J.S. & Teale, C.J. (2004). National survey for *Salmonella* in pigs, cattle and sheep at slaughter in Great Britain (1999–2000). *Journal of Applied Microbiology.* Vol.96, No 4, (April 2004) pp750–760

Decreto 99/2006, de 1 de agosto, por el que se aprueba el Plan de Ordenación de los Recursos Naturales de la Serranía de Cuenca. Boletín Oficial del Estado, 28 de mayo de 2007. No. 119. 10024.

Echeita, M.A., Aladueña, A. M., Díez, R., Arroyo, M., Cerdán, F., Gutiérrez, R., de la Fuente, M., González-Sanz, R., Herrera-León, S. & Usera, M. A. (2005). Distribución de los serotipos y fagotipos de *Salmonella* de origen humano aislados en España en 1997-2001 *Enfermedades Infecciosas y Microbiología Clínica* Vol. 23, No. 3 (Diciembre 2005), pp. 127-34.

Foreyt, W.J., Besser, T. E., & Lonning, S. M. (2001). Mortality in captive elk from salmonellosis. *Journal of Wildlife Diseases* Vol.37, No. 2 (2001), pp. 399–402.

Giannati-Stefanou, A.;Bourtzi-Hatzopoulou, E.;Sarris, K.;G, Xenos. (1997). Epizootiological. Study of sheep and goat abortion by *Salmonella* abortus ovis in Greece. Journal of the Hellenic Veterinary Medical Society, Vol.48, No. 2 (April-June 1997) pp. 93-98

Goldberg M. 1994. *La epidemiología sin esfuerzos.* Ed. Díaz de Santos, Madrid, pp 195.

González L (2000). *Salmonella abortus ovis* infection, In: *Diseases of sheep.* Martin & Aitken, (Eds.), 102-107, Blackwell Science, 3rd ed. ISBN 0-632-05139-6, Oxford, United Kingdom

Hensel, M. (2004) Evolution of pathogenicity islands of *Salmonella enteric.* International Journal of Medical Microbiology. Vol. 294, No. 2-3, (September 2004), pp.95-102

Jack, E. J. 1971. *Salmonella* abortion in sheep. *Veterinary annual.* Vol.12, pp, 57-63.

Jack, K. J. 1968. *Salmonella Abortusovis*: an atypical *Salmonella. Veteriary Research.* Vol.82. pp. 1168–1174.

Kapperud, G., H. Stenwing, Lassen, G. (1998). Epidemiology of *Salmonella typhimurium* O:4–12 infection in Norway: Evidence of transmission from an avian wildlife reservoir. *American Journal of Epidemiology.* Vol. 147, No.8 (April 1998), pp.774–782.

Lantier, F. (1987). Kinetics of experimental *Salmonella* abortus ovis infection in ewes. *Annales de Recherches Vétérinaires* Vol.18 No.4 (October 1987) pp.393-396.

Lantier F, Pardon P, Marly J. (1983). Immunogenicity of a low-virulence vaccinal strain against *Salmonella* abortus ovis infection in mice. *Infection and Immunity.* Vol. 40 (May 1983) pp.601–607.

León Vizcaíno, L.; Alonso De Vega, F.; Garrido Abellán, F.; González Candela, M.; Martínez Carrasco-Pleite, C.; Pérez Béjar, L.; Cubero Pablo, M.J.; Ruiz De Ybáñez Carnero, R. & Arenas Casas, A. (2002). Estudio en masa sobre infecciones que causan mortalidad perinatal congénita entre rumiantes domésticos y silvestres en las sierras béticas.In: XXXIII Jornadas Científicas y XII Internacionales de la Sociedad Española de Ovinotecnia y Caprinotecnia. Consejería de Agricultura y Pesca. Junta de Andalucía. pp. 325-330.

León, L., Astorga, R. & Cubero Mª. J. (1994). Las enfermedades del ciervo: estudio serológico. In: *El ciervo en Andalucía*. (R. Soriguer, P. Fandos, E. Bernaldez y J. Delibes, eds.). Consejería de Agricultura y Pesca de la Junta de Andalucía, Sevilla. Pp. 195-203.

León, L., De Menegui, D., Meneguz, P.G., Rosati, S. & Rossi, L. (1992). Encuesta seroepidemiológica de infecciones en la población de cabra montés del Parque Natural de la Sierra de las Nieves (Ronda). In. Proceedings of *V Internatinal Congress Génus Capra*. (20-22 Octubre 1992), Ronda. Málaga.

León, L., Miranda A., Perea, A., Carranza J. & Hermoso, M. (1980). Investigación inmunológica de diversos agentes infecciosos en ciervos y jabalíes de Sierra Morena. 490 501. In *Proceedings of II Reunión Iberoamericana. Conservación de Zoología de Vertebrados*. Cáceres, Spain.

Lindberg, A.A. & Le Minor L. (1984). Serology of *Salmonella*. In: Methods in Microbiology, Vol. 15, Bergman T.E., ed. Academic Press London, United Kingdom, 1–14.

Linklater, K. A. (1983): Abortion in sheep associated with *Salmonella montevideo* infection. *Veterinary Record*. Vol. 112, No. 16 (April 1983), pp. 372-374.

Linklater, K. A. (1991) Salmonellosis and *Salmonella* Abortion. In: *Diseases of Sheep*. 2nd edition. Ed.W. B. Martin & I. D. Aitken. pp. 65-70. Blackwell Scientific Publications, Oxford. United Kingdom.

Martín Atance. P. 2009. Seroepidemiología de infecciones asociadas al síndrome de mortalidad perinatal congénita e interacciones entre rumiantes silvestres y domésticos en la serranía alta de Cuenca. Tesis Doctoral. Universidad de Murcia.

Martínez A, Verona MA. 2002. Reserva de Caza "Serranía de Cuenca". Plan Téncico de caza para las temporadas 2003/2006 al 2007/2008. Consejería de Agricultura y Medio Ambiente. Delegación Provincial de Cuenca. Servicio Medio Ambiente Natural., Cuenca, pp. 77

Mcallum, H. J., A. S. Familton, R. A. Brown,& P. Hemmingsen. (1978). Salmonellosis in red deer calves (Cervus elaphus). *New Zealand Veterinary Journal*. Vol.26, pp 130–131.

McCullagh, P. & Nelder. J. A. (1989). *Generalized Linear Models*. Second edition. Chapman and Hall. London.

Mörner, T. (2001). Salmonellosis. Pp 505-507. In ES Williams and I.K. Barker (eds). Infectious diseases of wild mammals. 3rd Ed. Iowa State University Press, Ames, IA.

Murray, C. J. 1991. *Salmonellae* in the environment. *Scientific and Technical Review*. Vol. 10, No.3, (September 1991), pp. 765-85.

Nastasi A. Mammina C, Villafrate MR, Caracappa S, Di Noto AM, & Balbo R. (1992) Epidemiological evaluation by rRNA-DNA hybridation of strains of *Salmonella enterica* subsp. enterica serovar Abortusovis isolated in southern Italy in the years 1981–1989. *Bollettino dell Istituto Sieroterapico Milanese* (Milano) 1991; Vol. 70, No. 1-2, (1991-1992), pp. 475–481.

Nielsen, B. B., B. Clausen & Elvestad, K.. (1981). The incidence of *Salmonella* bacteria in wild-living animals from Denmark and in imported animals. *Nordisk veterinærmedicin*. Vol. 33, No.9-11, (September-November 1981), pp. 427-33.

Nikbakht GH, Raffatellu M, Uzzau S, Tadjbakhsh H, & Rubino S.(2002). Fingerprinting of *Salmonella enterica* subsp. enterica serovar Abortusovis in Irán. *Epidemiology and Infection*. 2002; Vol. 128, No.2, (April 2002), pp 333–336.

OIE. 2010. Terrestrial Manual. World Organization for Animal Health.

http://www.oie.int/fileadmin/Home/eng/Health_standards/tahm/2.09.09_SAL
MONELLOSIS.pdf

Olsen E.V., Pathirana S.T., Samoylov A.M., Barbaree J.M., Chin B.A., Neely W.C. &
Vodyanoy V. (2003). Specific and selective biosensor for *Salmonella* and its detection
in the environment. *Journal of Microbiological Methods*, Vol. 53, No.2 (May 2003), pp.
273-285.

Pardon P, Marly J, Sanchis R, Fensterbank R, 1983. Influence des voies et doses d'inoculation
avec *Salmonella* Abortusovis sur l'effet abortif et la réponse sérologique des brebis.
Annales de Recherches Vétérinaires Vol.14, pp. 129-139

Pardon, P., Sanchis, R., Marly, J., Lantier, F., Pepin ,M., & Popoff, M. (1988). Salmonellose
ovine due à *Salmonella abortus ovis*. *Annales de Recherches Vétérinaires*, Vol. 19, No.3,
(December1988), pp. 221-235.

Peinado M Martínez J-Mª. (1985). El paisaje vegetal de Castilla-La Mancha. Servicio de
Publicaciones. Junta deComunidades de Castilla-La Mancha. Toledo, pp 230.

Pérez Béjar Linarejos, R. (2007). *Aspectos epidemiológicos de las enfermedades contagiosas de la
reproducción en las poblaciones de rumiantes silvestres y domésticos del Parque Natural de
las Sierras de Cazorla, Segura y Las Villas (Jaén)*. Tesis Doctoral. Universidad de
Murcia. Spain.

Pérez-Ciordia I., Ferrero M., Sánchez E., Abadías M., Martínez-Navarro F. & Herrera D.
(2002). Enteritis por *Salmonella* en Huesca. 1996-1999. *Enfermedades Infecciosas y
Microbiología Clínica*. Vol. 20, No.1, (August-September 1991), pp. 765-85.

Petrie, A., and P. Watson. (1999). Statistics for Veterinary and Animal Science. Blackwell
Science, Malden, Massachusetts, United States of America.

Popoff, M.Y. & Le Minor L.E. (2005). Genus XXXIII. *Salmonella*. *Bergey's manual of systematic
bacteriology*. D. J. Brenner, N. R. Krieg & J. T. Stanley (Eds). New York, Springer
Science. Vol 2, Part B. *The Gammaproteobacteria*. pp 764-799.

Renter, D. G., D. P. Gnad, J. M. Sargeant, & Hygnstrom. S. E. (2006). Prevalence and serovars
of *Salmonella* in the feces of free-ranging whitetailed deer (*Odocoileus virginianus*) in
Nebraska. *Journal of Wildlife Diseases* Vol. 42. No. 3 (July 2006) Pp. 699–703.

Rojo Arribas, F.J. (2007). Memorandum de la caza del trofeo en la Reserva de Caza de la
Serranía de Cuenca.(May, 2007). Not indexed.

Sanchis R, Pardon P, Abadie G (1991) Abortion and serological reaction of ewes after
conjuctival instillation of *Salmonella* Abortusovis. *Annales de Recherches Vétérinaires*
Vol. 22, pp59-64

Sanchis R, Polveroni G, Pardon P (1985) Serodiagnostic de la Salmonellose a *Salmonella*
Abortusovis. Microtechnique de seroagglutination. Bull LabeVt 145,-9/1520

Schiaffino A, Beuzón CR, Uzzau S, Leori G, Cappuccinelli P, Casadesús J, Rubino S. (1996).
Strain typing with IS*200* fingerprints in *Salmonella Abortusovis*. *Applied and
Environmental Microbiology*. Vol. 62No. 7. (July 1996) pp. 2375–2380.

Scott, M. E. & Smith, G. (1994). *Parasitic and infectious diseases: epidemiology and ecology*.
Academic Press. San Diego.

Siegel, S. (1956). Non Parametric Statistics for the Behavioral Sciences. McGraw-Hill. New
York. United States of America.

Sojka, W.J., Wray, C., Shreeve, J.E., and Bell, J.C. (1983). The incidence of *Salmonella* infection
in sheep in England and Wales. 1975-81. *British Veterinary Journal* Vol.139:(June
1983), pp. 386-392.

Sting R, Nagel C, Steng G. (1997). Detection methods for *Salmonella* abortus ovis and examinations in sheep flocks in northern Baden-Württemberg. Zentralblatt fur Veterinarmedizin B. Vol.44 No.2. (April 1997) pp. 87-98.

Threlfall, E. J. & Frost, J. A. (1990). The identification, typing and fingerprinting of *Salmonella*: laboratory aspects and epidemiological applications. Journal of Applied Bacteriology. Vol.68 No.1 (January 1990)pp. 5-16.

Thrusfield M. (1990). *Veterinary Epidemiology.* Acribia Ed. Zaragoza, Spain, pp 483. ISBN 84-200-0674-2

Uzzau, S., Leori G.S.,[2] Petruzzi, V., Watson, P.R., Schianchi,G. Bacciu, D., Mazzarello, V., Wallis, T. S., & Rubino, S.(2001) *Salmonella enterica* Serovar-Host Specificity Does Not Correlate with the Magnitude of Intestinal Invasion in Sheep. *Infection and Immunity*, Vol. 69, No. 5 (May 200)1, pp. 3092-3099

Valdezate S, Astorga R, Herrera-León S, Perea A, Usera MA, Huerta B, Echeita A. (2007) Epidemiological tracing of *Salmonella enterica* serovar Abortusovis from Spanish ovine flocks by PFGE fingerprinting." *Epidemiology and Infection.* 2007. Vol. 135:(Online publication August 2006), pp. 695-702

Van der Zee, H. &Huis In't Veld,. J.H.J. (2000). Methods for the rapid detection of *Salmonella*. In:*Salmonella in Domestic Animals*. C. Wray and A. Wray. Wallingford (eds), Oxon, UK, CAB International. pp. 373-391.

Veling, J. Van Zijderveld, F. G., Van Zijderveld -Van Bemmel, A. M. Barkema, H. W. &. Schukken, Y. H. (2000). Evaluation of Three Newly Developed Enzyme-Linked Immunosorbent Assays and Two Agglutination Tests for Detecting *Salmonella enterica* subsp. *enterica* Serovar Dublin Infections in Dairy Cattle. *Journal of Clinical Microbiology*, Vol. 38. No.12 (December 2000), pp. 4402–4407

Vodas, K., & Martinov, S. (1986). Diagnostic value of serological and bacteriological procedures in sheep infected with *Salmonella Abortusovis* and *Chlamydia psittaci* var. *ovis. Veterinarno-meditsinski nauki*. Vol. 23, (1986), pp. 14–22

7

Porins in the Inflammatory and Immunological Response Following *Salmonella* Infections

Emilia Galdiero[2], Aikaterini Kampanaraki[1],
Eleonora Mignogna[1] and Marilena Galdiero[1]
[1]*Seconda Università degli Studi di Napoli, Facoltà di Medicina e Chirurgia*
Dipartimento di Medicina Sperimentale
Sezione di Microbiologia e Microbiologia Clinica, Napoli
[2]*Università Federico II di Napoli. Facoltà di Scienze Matematica e Fisica*
Dipartimento di Biologia Strutturale e Funzionale, Napoli
Italia

1. Introduction

Today numerous information are available on the molecular mechanisms activated by *Salmonella* and its components during the interaction with host cells and in determining the disease state. The molecular mechanisms of how *Salmonella* enter host cells and function as an intracellular pathogen are under intense investigation. Much progress has been made in identifying the bacterial factors that mediate invasion. Once *Salmonella* enters a cell, it remains within a membrane-bound vacuole and does not appear to fuse with lysosomes, the outcome of infection is determined both by bacterial and host factors, including the virulence of *Salmonella* strain, and the ability of the host to respond with an inflammatory and immunological reaction. The host response involves multiple cells that are resident at the site of infection or infiltrate from the circulation. Induction of an array of cytokines occurs in response to infection of macrophages with live *Salmonella* and after exposure to various *Salmonella* components including lipopolysaccharide (LPS), and porins. Of the different biologically active components present in Salmonella, LPS and porins are the most potent inducer of host response. LPS and porins present an intrinsic biological activity on cell involved in the inflammatory response and also on other types of cells; moreover they are immunogenic molecules against which the organism raises the humoral and cellular response. The LPS molecule is the most studied. Techniques previously used in the extraction of LPS from the endotoxin had greatly favoured the study of this portion of the macro-complex ignoring the protein fraction, allowing the identification of most of the effects of endotoxin with those of LPS. Subsequent extraction techniques for membrane proteins then allowed the study of the protein fraction, which was extracted globally in the endotoxin. Later experiments suggested that a chemical subunits of LPS, lipid A, was the actually toxic moiety and that the O-specific chain found on LPS was not involved in the toxic effect. The endotoxin-associated protein (EP) consists of a complex of 4-5 major proteins that range in size from 10 to 35 KDa. Originally considered to be a superfluous carrier of LPS, EP is now recognized to have potent biological activities, some of which are unique (Mangan et al., 1992). For example, EP is a powerful mitogen for C3H/HeJ mouse and human lymphocytes which are hyporesponsive to LPS. Among the proteins associated with LPS, the techniques actually used for the extraction of native proteins from the cellular membranes allowed the isolation of outer-membrane proteins (OMPs). LPS and OMPs are released by different bacteria during both in vitro and in vivo growth and this release is significantly enhanced when the bacteria are lysed following exposure to antibiotics or human serum. Molecular complexes of LPS and OMPs, together with the other molecules which constitute the external surface of Gram-negative bacteria, are released as outer membrane vesicles (OMVs). OMVs are formed by blebbing and pinching of segments of the bacterial outer membrane.

2. Outer membrane and porins

The cell envelope of Gram-negative bacteria (Figure 1) is composed of two distinct membranes, the inner plasma membrane and the outer membrane. The peptidoglican layer is located between the two membranes and this area between the plasma membrane and the outer membranes is referred to as the periplasmic region. The outer membrane is composed of phospholipid and protein as is the cytoplasmatic membrane. The outer leaflet is occupied by about 45% lipopolysaccharide (LPS). The phospholipid is localized almost exclusively in the inner layer of the outer membrane bilayer. In *Salmonella enteric serovar Typhimurium*, the phospholipid composition of the outer membrane consists predominantly of phosphatidylethanolamine, phosphatidylglycerol and cardiolipin. LPS consists of a hydrophobic membrane anchor, lipid A, a short core oligosaccharide, and an O antigen that may be a long polysaccharide. The lipid A is rather well conserved among Gram-negative bacteria. The core oligosaccharide and O-antigen, if present, is the most variable part of LPS and shows even a high degree of variability between different strains of the same species. In bacteria the number of porin copies was determined to be up to 100.000 for cell. Porins form β-barrels with 14, 16 or 18 strands, all of which are connected by extraplasmic loops and periplasmic turns with the particularly long loop L3 folded inside the barrel. Porins of 16 strands are called general or non-specific porins and form pores allowing the diffusion of hydrophilic molecules, showing no particular substrate specificity; while 18-strands porins are substrate specific porins. The best-studied examples is the sucrose-specific porin ScrY from *S. typhimurium*. The three-dimensional structures of this specific porin has been elucidated, (Forst et al., 1998). ScrY forms homotrimers whose monomers consist of 18-stranded antiparallel β-barrels. As was found with the general porins, the third loop, L3, folds back inside the β-barrel. A peculiar feature of ScrY is the presence of a 70-residue-long N-terminal extension, which hangs out into the periplasm in the form of a parallel triple-stranded coiled-coil.

An homology based 3D structural model for the porin OmpC from *S. typhimurium* was built to understand the possible unique conformational features of its antigenic loops with respect to other immunologically cross reacting porins. The homology model was built based on the known crystal structures of the *E. coli* porins OmpF and PhoE. The resulting model was compared with other porin structures, having β-barrel fold with 16 transmembrane β-strands, and found that the variable regions are unique in terms of sequence and structure (Arochiasamy et al., 2000).

Recently, a structural model for a 50kDa antigen protein of *S. enterica serovar Typhimurium* was also built by Siew-Choong et al. (Yee et al., 2011). The characteristic of the built model also resembles the structure of known transmembrane proteins in other Gram-negative bacteria (S.

Galdiero et al., 2003). It shows a similar structure as the TolC transmembrane channel protein with the combination of β-barrel domain projecting from the membrane, across the periplasmic space with α-helical domain and the equatorial domain (mixed of β-sheets/α-helices). The upper part of the structure is open and could provide solvent access, while the lower part is narrowed. The structure shows that it may be an ion channel whose conductance depends on the open or close conformation at the lower end, which is similar with the characteristic of TolC and its analogues. The surface exposed loops might act as a "lid" to access into the top end of the β-sheets domain. The β-barrel domain consists of 16 strands, which is within the number of strands that has been characterized for other bacterial outer membrane proteins (S. Galdiero et al., 2007). The 40 Å long axis of the β-barrel domain fits into a lipid bilayer membrane which is typically 30 Å. As for other porins the base of the β-barrel is mainly composed of aromatic residues, specifically phenylalanine and tyrosine. These residues are usually found in a typical β-barrel membrane protein to define the inner edge of a lipid bilayer of membrane. The lower part of the built structure is the left twisted antiparallel-helices barrel, which is involved in the control of the opening or closing of the access.

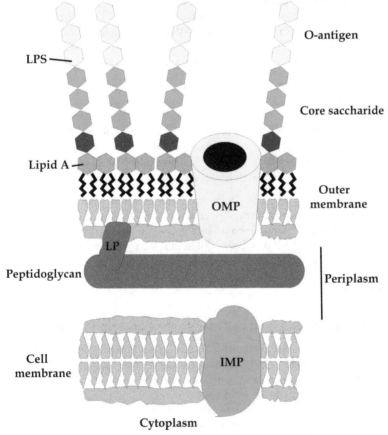

Fig. 1. Depiction of Gram-negative cell envelope. IMP, integral membrane protein; LP, lipoprotein; LPS, lipopolysaccharide; OMP, outer membrane protein.

3. Biological activity of porins

The release of porins at the infection site, whether secreted during growth or derived from the lysis of the bacterial cells, involve the host defence and influence the course of enterobacterial diseases.

Porins of *S. enterica serovar Typhimurium* added to macrophage cultures in vitro are able to modify several macrophage functions. The superficial hydrophobicity of macrophages adhering to the slide, as measured by the average contact angle monolayers, shows an increase of about 15-20%. The phagocitic index of macrophage treated with porins was found to be significantly lower. The same reduction was observed in the intracellular killing of macrophages treated with porins (Tufano et al., 1984). Porins inhibit phagocytosis by activating the adenylate cyclase system (Di Donato et al., 1986). The effect of *Salmonella typhimurium* porins was also studied on human polymorphonuclear leukocytes (PMNs). Labelled porins were shown to bind to the PMNs, and could be completely displaced by unlabeled porins. The binding caused modifications of membrane integrity and of the physico-chemical characteristics of the PMN surface, e.g. decreased oxidative burst, decreased hydrophobicity and altered cell morphology. The porins acted as both chemotaxins and chemotaxinogens. When PMNs were preincubated with porins their migration in the presence of commonly used chemoattractions (serum activated by zymosan or N-formyl-L-methionyl-L-leucyl-L-phenylalanine) was inhibited (Tufano et al., 1989).

S. enterica serovar Typhimurium porins injected into the paws of rats, induced a dose dependent edema which was maximal at 2 to 3h and still present at 5h. Edema was unaffected in animals which had their complement levels depleted, demonstrating that inflammation was not associated with complement activation; however it could be decreased by indomethacin and by dexamethasone. Rat peritoneal cell incubated with porins released histamine but little prostacyclin, suggesting that porins have little ability to induce the prostenoid producing enzyme cicloxygenase II (F. Galdiero et al., 1990).

Porins were also shown to kill D-glucosamine-sensitized LPS-responsive and LPS-unresponsive mice. A 100 µg amount of porins was sufficient to kill 80-90% of animals. But lethal effect of the porin preparation could be completely blocked by pre-administration of a neutralizing antiserum to TNF-α but was not abolished by polimixin B indicating that LPS did not contribute to the biological responses. Porins were also pyrogenic in rabbits and elicited a localized Swartzman reaction when used as the sensitizing and eliciting agent (F. Galdiero et al., 1994).

4. Porins interaction with host cell membranes and signal transduction

The cell ability to sense external stimuli and to react by initiating a program of expression often involves propagation of a cell surface-initiated signal along a specific pathway of protein kinases whose ultimate targets are nuclear transcription factors (Figure 2). These pathways consist of a cascade of biochemical events that include phosphorylation of a variety of kinases, that in turn modulate other factors that control gene expression. Porins purified from *S. enterica serovar Typhimurium* induce tyrosine-phosphorylation in THP-1 cells and in C3H/HeJ macrophages. After porin stimulation a pattern of tyrosine-phosphorilated proteins appeared in the soluble cytoplasmic fraction, the membrane fraction and in the insoluble protein fraction. The events of tyrosine protein phosphorylation were present in macrophage from LPS-hyporesponsive C3H/HeJ mice stimulated with porins, while they were markedly reduced where the macrophage where stimulated with LPS-(R). Among the most prominent tyrosine phosphorylated bands in porin-stimulated cells there is a number of proteins with a molecular mass that is similar to that of the family of tyrosine/serine/threonine protein kinases. Mitogen-activated protein kinase (MAPK) cascade are among the best known signal transduction systems and play a key role in the regulation of gene expression as well as cytoplasmic activities.

MAPKs have been shown to be involved in the regulation of cytokine responses. MAPKs are activated upon phosphorylation of both tyrosine and threonine residues by MAPK kinase (MEK). These enzymes participate in cell signalling pathways leading to AP-1 and NF-κb activation following porin stimulation of cells. Raf-1 was also phosphorylated in response to the treatment of U-937 cells with porins; the porin-mediated increase in Raf-1 phosphorylation is accompanied by the phosphorilation of MAPK kinase 1/2 (MEK 1/2), p38, ERK 1/2 and C-Jun N-terminal kinase. P38 signalling pathway mainly regulates AP-1 and NF-κb activation in cells treated with S. enterica serovar Typhimurium porins. The transcriptional factor AP-1 is composed of multiple protein complexes formed between the protein products of proto-oncogenes C-fos and C-jun and their related gene family members. The fos family consists of C-fos, the gene for Fos-related antigen-1 (Fra-1), Fra-2 and Fas-B and its naturally truncated form Fas B2; the Jun family consists of C- jun, Jun B and Jun-D. The AP-1 family of transcription factors consists of homodimers and heterodimers of these subunities; the different complexes regulate their abilities to transactivate or repress transcription. AP-1 composition may change in the cell as a function of time and stimules; the binding affinity for a given target DNA sequence is determined by the different AP-1 dimer combinations and the context of the surrounding sequences. In U937 cells treated with porins from Salmonella, different complexes including C-Jun and Fra-2 subunits appeared. While in cells treated with LPS, the stimulus leads to AP-1 complexes containing Jun D, C-Fos and C-Jun, stimulation by porins induces AP-1 complexes containing Fra-2 in addiction to the other subunits (M. Galdiero et al., 2002). The formation of different complex represent a further difference between stimulation with LPS and stimulation with porins to be added to other observations: that cytokine release after stimulation with porins begins after 120 min and continues for 5 to 6 hours, whereas cytokine release following LPS stimulation begins after 30 min. and decrease at 120 min (M. Galdiero et al., 1995).

Porins trigger multiple synergistic signal transduction pathways, including protein kinase A (PKA), proteine kinase C (PKC), NT proteine tyrosine kinase (NT-PTKs). The role of PKC in signal transduction in mouse macrophages stimulated by S. enterica serovar Typhimurium porins is reported by Gupta S et al., (Gupta et al., 1999). Their experiments showed that porin activation of macrophages results in the increased inositol-triphosphate and intracellular Ca2+ mobilization: there is a translocation of PKC to the membrane which is accompanied by nitric oxide release.

Several polypeptide ligands use the JAK-STAT molecules in signal transduction (Darnell et al., 1994). The family of transcription factors called STAT (signal transducers and activators of transcription) have been found to be activated by the Janus Kinases (JAKs) that are associated with the cytokine receptor components (Kisseleva et al., 2002) . In resting cells the

STATS, when phoshorylated by the JAKs, dimerize via their SH2 domains and traslocate in the nucleus, where they interact with specific DNA sequences and transactivate the associated genes. The Jak/Stat signalling pathway plays a fundamental role in response to infection and in sepsis (Scott et al., 2002) In vitro experiments on U937 cells demonstrate a complex indirect mechanism of STAT-1 and STAT-3 activation after stimulation of cells with porins. The treatment with porins did not results in increase of JAK phosphorylation although STAT-1 and STAT-3 activation was observed. The activation of STAT-1/STAT-3 by porins can occur through the activation of MAPK and possibly other PTKs but not through JAK activation (M. Galdiero et al., 2006).

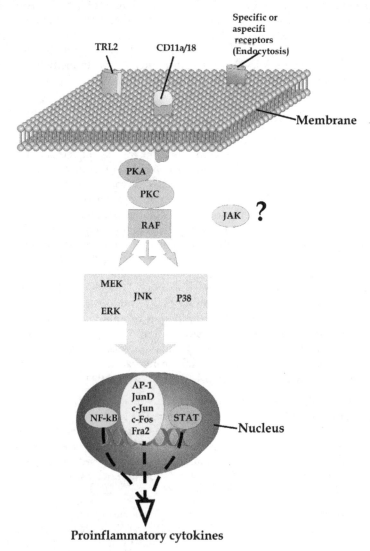

Fig. 2. Porin signal transduction pathways

5. Cytokines release by porins

Many of the pathophysiologic mechanisms of Gram-negative bacterial infections are due to bacterial surface components acting on cells directly or via mediators such as cytokines. Cytokines are polypeptides that exert a wide spectrum of biological effects, including haematopoietic, metabolic, inflammatory and immunologic homeostasis. Among the large family of cytokines, porins by S. enterica serovar Typhimurium induce the release of TNF-α, IL-1, IL-6, and TGF by macrophages and IL-4 and IFN-γ by lymphocytes. The role played by porins in the production of cytokines derives from the comparison with LPS-S e LPS-R extracted by the same strain of S. typhimurium. Porins at 1 μg/ml induce the greatest release of TNF-α, IL-1α and IL-6 by monocytes and IL-4 by lymphocytes, while porins at 5μg/ml induce the greatest release of IFN-γ by lymphocytes. The R-form of LPS (LPS-R) induces the greatest release of TNF-α and IL-1α by monocytes when used at 1 μg/ml concentration. At concentration of 5 and 10 μg/ml, respectively, LPS-R induce the maximal release of IL-6 from monocytes and the maximal release of IL-4 from lymphocytes. The S-form of LPS (LPS-S) induces the greatest release of TNF-α, IL-1α and IL-6 by monocytes and that of IL-4 by lymphocytes when used at a concentration of 1 μg/ml. Porins (5 μg/ml) induce the release of IL-8 by THP-1 cells after 24h of stimulation (Vitiello et al., 2004). The level of IL-8 in THP-1 cells stimulated with porins was comparable to that induced in response to 1 μg/ml of LPS-R.

While CD-14, CD-11/18 and Toll receptors 2 and 4 appears to be very important LPS signal transducer, porin-specific receptors are still unknown. Therefore, it is possible that porin stimulation is not due to binding to specific receptors, but the consequence of the perturbation of the cell membrane lipoproteic phase, induced during adsorption or porin penetration. CD-14 is a glycosyl-phosphatidyl inositol linked 55 kDa protein present on the surface of monocytes and polymorphonuclear leucocytes, and it function as the cell surface receptor for LPS and several surface components of Gram-positive bacteria. CD14 is also found as a soluble protein (sCD14) in human serum. CD14 lacks transmembrane and cytokine-binding domains and is not believed to have intrinsic signalling capabilities. Toll-like receptors 4 (TLR4) appears to be very important LPS signal transducer. It's thought that, also, Toll-like receptor 2 (TLR2) functions as a signal transducer upon LPS binding by CD14. TLRs make up a family of evolutionary conserved pattern recognition molecules that are important signal transducers for the induction of mammalian innate immunity responses, including cytokine responses. The best characterized TLRs to date are TLR2 and TLR4. TLR2 is involved in the recognition of a wide assay of bacterial products, including peptidoglycan, lipopeptides, zymosan and bacterial lipoproteins, whereas TLR4 is activated by LPS. CD 14 acts as a abroad specificity coreceptor that can enhance cell activation induced by TLR4 or TLR2 agonists. Data from Haemophilus influenzae (Hib) porin and from neisserial porin P or B indicate that porins from different bacteria may be recognized by TLR-2 (M. Galdiero et al., 2004). The Hib porin-induced TNF-α and IL-6 production was eliminated in macrophages from TLR2 or MyD88 deficient mice. In contrast, macrophages from LPS hyporesponsive C3H/HeJ mice which are defective in TLR4 function, responded normally to Hib porin. Neisserial porin adjuvant activity was mediated by surface expression of B7-2 and class 2 major histocompatibility complex on B cells by TLR-2-dependent mechanisms; the presence of the adaptor molecule MyD88 was also required.

CD11/18 (M. Galdiero et al., 2004) integrin may also participate in LPS signalling. This family of receptors are heterodimeric cell surface glycoproteins composed of a CD11 and a

CD18 subunit. The release of TNF-α, IL-6 and IL-8 by THP-1 cells stimulated by porins is independent of CD14, but is partially dependent on CD11/18 integrins. *S. enterica serovar Typhimurium* porins enhance the synthesis and release of IL-6 in U937 cells regulating the transcriptional activity of IL-6 gene by nuclear transduction of NF-κB. The characterization of the human IL-6 promoter revealed a highly conserved control region of 300bp upstream of the transcriptional initiation site that contains the elements necessary for its induction by a variety of stimuli commonly associated with acute inflammatory or proliferative states. In particular, electrophoresis mobility shift assay, as well as promoter deletion and point mutation analysis, revealed the presence of an NF-κB binding element. In U937 cells stimulated by *Salmonella* porins, NF-κB is able to enhance IL-6 gene promoter activity. Activation of this nuclear factor may be responsible for porin induced expression and release of IL-6 (Finamore et al., 2009).

These observations allow to outline a specific ability of porins that is expressed by stimulation of the cell surface, signal transmission, activation of nuclear factors, activation of gene promoters and finally release of cytokines.

6. Transmigration of leukocytes following porin activation

It has long been recognized that *Salmonella* provokes un intense intestinal inflammatory response, consisting largely of neutrophil migration across the epithelial lining of the intestine (M. Galdiero et al., 1999); this inflammatory event manifests as an epithelial dysfunction, namely diarrhea. In an in vitro model, Mc Cornick et al (McCormick et al., 1995) showed that *S. typhimurium* tran-epithelial signaling to polymorphonuclear neutrophils (PNM) plays a direct and substantial role in stimulating enteritis in humans. Leukocyte-endothelial cell interaction both in vivo and in vitro are active multistep processes. The initial adhesion of circulating leukocytes to vascular endothelium is induced by interaction of constitutively functional leukocyte homing receptors with regulated endothelial cell ligands. During inflammation a dramatic increase of endothelial cell surface molecule expression occurs that support the adhesion of circulating leucocytes. Bacteria or bacterial products may constitute important inducers of surface molecule expression on endothelial cells. LPS-S induce adhesion of leucocytes to endothelial cells as potent as that induced by IL-1β (Takeuchi et al., 1967) *S. typhimurium* porins and LPS-R are able to induce the release of s-E-selectin and sICAM-1 from human umbilical vein endothelial cells (HUVEC) and also were to up-regulate the surface expression of E-selectin and ICAM-1 on endothelial cells (Donnarumma et al., 1996).

Treatment of the HUVEC with either porins or LPS in the form S or R increased the transmigration of different leukocyte populations, in particular that of neutrophils; transmigration increased remarkably during the simultaneous stimulation of endothelial cells by IL-1β together with either porins or LPS (M. Galdiero et al., 1999). Porin treatment caused transmigration that lasted several hours longer than that caused by LPS and further increase the activation of those cells already activated by IL-1β. Consequently, the in vivo activity of the two molecules shows an effect prolonged in time. In vitro, the simultaneous stimulation of endothelial cells with IL-1β and either porins or LPS causes overlapping effect leading to a very high migration index. Neutrophil transmigration was partially inhibited by monoclonal antibodies (MoAb) binding to E-selectine; the transmigration of lymphocites and monocytes was partially inhibited by MoAbs anti-VCAM-1; the transmigration of neutrophils, lymphocytes and monocytes was partially inhibited by MoAb anti-ICAM1. Monocyte and granulocyte transmigration was, also, inhibited by MoAbs binding to CD11a/CD18 and CD11b/CD18. Lymphocyte transmigration was inhibited by MoAbs CD11a/CD18 and not by CD11b/CD18. Therefore, porins may constitute important inducers of surface molecule expression on endothelia cells. This ability makes these molecules particularly important in the inflammatory process during *Salmonella* infections.

7. Porin immunogenicity

Porins demonstrate immunogenic and adjuvant properties. Heat denaturable surface components play a role in inducing protection to *S.enterica serovar Typhimurium* infection in mice (Isibasi et al., 1994). The protective role of the outer membrane proteins or of porins from Neisseria (Melancon et al., 1983), *Salmonella* (Muthukkmar et al., 1993), Haemophilus and Vibrio genus has been shown. Antiporin antibodies have been demonstrated to be bactericidal and opsonic (Isibasi et al., 1994); patients with pelvic inflammatory disease (PID), who recover spontaneously have high levels of antiporins antibodies showing that the presence of serum antibodies to neisserial porins may correlate with protection against PID. Several studies have been recently performed with porins extracted from *N. meningitidis, N. gonorrhoeae* and *H. influenza* (Massari et al., 2003; Song et al., 1998; Wetzler et al., 1996). Porin vaccines have also been developed to protect against *S. typhimurium* infection. Porins are excellent antigens, efficiently stimulating humoral and cell-mediated immune response of the host immune systems which could play a role in the protection against the disease. *S. enterica serovar Typhimurium* porins are also able to induce expression of CD86 on antigen–presenting cells (Massari et al., 2002). Macrophages from mice immunized with porins and infected later with Salmonella, express 53% more B7 versus control macrophages and can therefore support a host–protective immune response.

Complement is an important arm of innate immune defenses against invading pathogens. Complement activation leads to the deposition of C3 fragments, which can enhance opsonophagocytosis of microbes. Porin contribute to complement activation mainly through the classical pathways in an antibody-indipendent manner. All of the porins tested to date have been shown to bind fragment C1q, the first component of the classical pathway of the complement system. The effect of porins purified from *S. enteric serovar Tiphimurium* on the complement system was investigated both in vitro and in vivo. Incubation of porins with either human or guinea pig serum resulted in the consumption of the total complement activity when an amount of porins ranging from 8 to 10 µg per 100 ml of serum was used. The activation of complement system was temperature dependent, suggesting an active process rather than passive adsorption of the complement components by porins. In addition, the activation had a fast kinetic and proceeded mainly through the classical pathway. This conclusion is supported by the consumption of C1s and C4 in normal human serum treated with porins and also by the depletion of C3 activity in the C1s-deficient serum which was marked only when purified C1s was added to the serum before incubation with porins. Injection of 100 µg of porins into guinea pigs induced profound complement consumption at 6 h post-injection that persisted up to 12 h (F. Galdiero et al., 1984). Also porin Omp K 36 from *Klebsiella pneumonia* interact with C1q.

The cloning of the genes encoding the outer membrane proteins has facilitated the production of pure porins free from other bacterial antigens for investigation as potential products for protective responses. The protection against bacterial infection is correlated to the presence of the antibodies with the ability to activate complement mediated killing of bacterial cells. The amminoacid sequences of the superficial loops of porin are highly variable in different strains and are the regions actually responsible for stimulation the production of bactericidal antibodies (Snapper et al., 1997). The role of surface loops of porin in immunological responses has been also studied in Haemophilus. Protein P2 of H. influenzae is a homotrimeric porin, which constitutes approximately one-half of the total outer membrane protein and it is an important target of the immune response to Haemophilus. P2 contains 16 transmembrane regions with β-sheet conformation and 8 suface-exposed loops. Analysis of sequences of P2 genes indicates that the transmembrane regions are relatively conserved among strains while considerable heterogenicity exists in surface-exposed loops (Qi et al., 1994). Challenging with whole bacterial cells resulted in a prominent antibody response directed at the P2 molecule. Analysis of the antibodies to whole organisms, and peptides corresponding to each of the eight loops of P2 by immunoassay revealed that bacterial antibodies were prevalently specific for loop 5, a highly variable region, and for loop 6, a conserved surface exposed loop.

Infection of mice with S. typhimurium is widely accepted as a valuable experimental model for human typhoid fever. Oral infection with S. enteric serovar Typhimurium induces a strong T helper 1 (Th1) response that is responsible for the CD4+ T-cell-mediated protection. Cytokines such as IFN-γ and TNF-α released by Th1 cells activate bactericidal pathway in macrophages. Moreover, CD 4+ T lymphocytes help B lymphocytes to produce antibodies and salmonella-specific CD8+T lymphocytes. Various studies have demonstrated that porins induce a Th1 response (Gupta et al., 1999, M. Galdiero et al., 1998) reported that porins of S. enteric serovar Typhimurium elicit Th1 response in the host. In fact, porins appear to stimulate T cell proliferation in the presence of macrophages incubated with dead bacteria, live Salmonella infected macrophages stimulated a minor proliferation compared to dead Salmonella incubated macrophages. Furthermore infection with live Salmonella induced the loss of accessory molecules, such as B7 and ICAM-1 on macrophages. OMPA of S. enteric serovar Typhimurium activate dendritic cells and enhances Th1 polarization (Lee et al., 2010). Others studies demonstrate that purified porins are able to induce a different response to that induced by the porins present on the S. typhimurium cell surface. Porin treated or orally infected mice show anti-porin antibodies with bactericidal activity. The complete adaptive transfer of resistance to S. typhimurium infection is achieved only using splenic T cells from survivor mice after experimental infection. After stimulation with specific antigen in vitro CD4+ cells from porin-immunized mice released large amounts of IL-4, while CD4+ cells from S. typhimurium infected mice predominantly secreted IFN-γ. Limiting dilution analysis showed that infection resulted in a higher precursor frequency of IFN-γ producing CD4+ T cells and a lower precursor frequency of IL-4 producing CD4+ T cells, while immunization with porins resulted in a higher precursor frequency of IL-4 producing cells and a low frequency of IFN-γ producing cells.

Analysis of polymerase chain reaction-amplified cDNA from the spleens of infected mice demonstrated that IFN-γ, IL-2 and IL-12 mRNA were found 5 days after in vitro challenge and increased after 15 days; IL-10 expression was rarely present after both 5 and 15 days, while IL-4 mRNA expression was not detected. In porin-immunized mice, the IL-4 mRNA expression increased after 15 days, IFN-γ mRNA expression decreased after 15 days, while IL-2, IL-10 and IL-12 mRNA remained relatively unchanged (M. Galdiero et al., 1998) Other works demonstrated that subletal doses of live *S. typhimurium* give rise to an IFN-γ dominant Th1–like immune response whereas heat-killed bacteria generate an IL-4 dominant Th2-like response (Thatte et al., 1993). Therefore, during experimental oral infection with *S. typhimurium* in mice a T-lynphocyte differentiation occurred, leading to prevalent TH1 response, while the immunization with isolated porins did not induce in vivo a similar pattern of differentiation. During the initial phase of infection with virulent strains which express large amounts of porins on their surface, *Salmonella* are present in the bloodstream and are resistant to complement-mediated lysis (Munn et al., 1982); bacterial cell is able to survive and entry into macrophages and therefore is resistant to bactericidal anti-porin antibodies. In this phase of infection, cell-mediated immunity is important for protection against typhoid fever. Transfer of immunity experiments have demonstrated that CD4+ cells, CD8+ cells and serum are all required to protect mice from challenge with virulent *S. typhimurium* (Mastroeni et al., 1993).

8. Conclusion

Porins from several Gram-negative bacteria, including Salmonella, play a fundamental role in the host-pathogen interaction, eliciting diverse biological proinflammatory activities and immune responses.

Porins present an intrinsic biological activity when interacting with eukaryotic cells, but also behave as antigens stimulating specific immune responses. Porins of *Salmonella* have endotoxin-like effects such as lethal action, the ability to elicit a local Shwartzman reaction, to activate the complement system and pyrogenicity. Furthermore, the porins stimulate pro-and anti-inflammatory cytokine synthesis and release. It has been established that protein tyrosine phosphorylation plays a central role in porin mediated transduction processes. Signal transduction pathways and transcriptional activation known to occur during immune cell activation have been widely investigated in cells stimulated by porins. Bacterial porins also may constitute important inducers of surface molecule expression on endothelial cells and contribute to endothelial transmigration of leucocytes. The protective role of porins from *Salmonella* and other bacteria has been demonstrated. Anti-porin antibodies have been shown to be bactericidal and opsonic; patients with pelvic inflammatory disease who self cure present high levels of antiporins antibodies. Porins also play an important role in the development of cellular immunity. During experimental oral infection with *S. typhimurium* in survivor mice a T-lymphocyte differentiation occurred leading to a prevalence of the Th-1 response, while the treatment with purified porins did not induce in vivo a similar pattern of differentiation. Transfer of immunity experiments have demonstrated that CD4+, CD8+ cells and serum are all required to protect naive mice from challenge with virulent *S. typhimurium*. These studies have led to the establishment of a multiplicity of targets for novel therapies.

9. References

Arockiasamy, A., & Krishnaswamy, S. (2000). Homology model of surface antigen OmpC from *Salmonella* typhi and its functional implications. *Journal of Biomolecular Structure & Dynamics.* Vol. 18, No. 2, pp. 261-271.

Darnell, J.E.Jr., Kerr, I.M., & Stark, G.R. (1994). Jak-STAT pathways and transcriptional activation in response to IFNs and other extracellular signaling proteins. *Science,* Vol. 264, No. 5164, pp. 1415-1421.

Di Donato, A., Draetta, G.F., Illiano, G., Tufano, M.A., Sommese, L., Galdiero, F. (1986). Do porins inhibit the macrophage phagocyting activity by stimulating the adenylate cyclase? *J Cyclic Nucleotide Protein Phosphor Res.*, Vol. 11, No. 2, pp. 87-97.

Donnarumma, G., Brancaccio, F., Cipollato de l'Ero, C., Folgore, A., A. Marcatili, A., & Galdiero, M. (1996). Release of GM-CSF, sEselectin and sICAM-1 by human vascular endothelium stimulated with Gram-negative and Gram-positive bacterial components. *Endothelium*, Vol. 4, pp. 11-12.

Finamore, E., Vitiello, M., D'Isanto, M., Galdiero, E., Falanga, A., Kampanaraki, A., Raieta, K., & Galdiero, M. (2009). Evidence for IL-6 promoter nuclear activation in U937 cells stimulated with *Salmonella* enterica serovar Typhimurium porins. *European Cytokine Network,*Vol. 20, No. 3, pp. 140-147.

Forst, D., Welte, W., Wacker, T., & Diederichs, K. (1998). Structure of the sucrose-specific porin ScrY from *Salmonella* typhimurium and its complex with sucrose. *Nature Structural Biology*, Vol. 5, No. 1, pp. 37-46.

Galdiero, F., Tufano, M. A., Sommese, L., Folgore, A., & Tedesco, F. (1984). Activation of complement system by porins extracted from *Salmonella* typhimurium. *Infectection and Immunity*, Vol.46, No. 2, pp. 559- 563.

Galdiero, F., Tufano, M.A., Galdiero, M., Masiello, S., & Di Rosa, M. (1990). Inflammatory effects of *Salmonella* typhimuriu porins. *Infection and Immunity*, Vol. 58, No. 10, pp. 3183-3186.

Galdiero, F., Sommese, L., Scarfogliero, P., & Galdiero, M. (1994). Biological activities - lethality, Shwartzman reaction and pyrogenicity- of *Salmonella* typhimurium porins. *Microbial Pathogenesis*, Vol. 16, No. 2, pp. 111-119.

Galdiero, M., Cipollaro de L'ero, G., Donnarumma, G., Marcatili, A., & Galdiero, F. (1995). Interleukin-1 and interleukin-6 gene expression in human monocytes stimulated with *Salmonella* typhimurium porins. *Immunology,*Vol. 86, No. 4, pp. 612-619.

Galdiero, M., De Martino, L., Marcatili, A., Nuzzo, I., Vitiello, M., & Cipollaro de l'Ero, G. (1998). Th1 and Th2 cell involvement in immune response to *Salmonella* typhimurium porins. *Immunology*, Vol. 94, No. 1, pp. 5-13.

Galdiero, M., Folgore, A., Molitierno, M., & Greco, R. (1999). Porins and lipopolysaccharide (LPS) from *Salmonella* typhimurium induce leucocyte transmigration through human endothelial cells in vitro. *Clinical &Experimental Immunoloy*, Vol. 116, No. 3, pp. 453-461.

Galdiero, M., Vitiello, M., Sanzari, E., D'Isanto, M., Tortora, A., Longanella, A., & Galdiero S. (2002). Porins from *Salmonella* enterica serovar Typhimurium activate the transcription factors activating protein 1 and NF-kappaB through the Raf-1-mitogen-activated protein kinase cascade. *Infection and Immunity*, Vol. 70, No. 2, pp. 558-68.

Galdiero, M., Galdiero, M., Finamore, E., Rossano, F., Gambuzza, M., Catania, M.R., Teti, G., Midiri, A., & Mancuso, G. (2004). Haemophilus influenzae porin induces Toll-like receptor 2-mediated cytokine production in human monocytes and mouse macrophages. *Infectection and Immunity*, Vol. 72, No. 2, pp. 1204-1209.

Galdiero, M., Vitiello, M., D'Isanto, M., Raieta, K., & Galdiero, E. (2006). STAT1 and STAT3 phosphorylation by porins are independent of JAKs but are dependent on MAPK pathway and plays a role in U937 cells production of interleukin-6. *Cytokine*, Vol. 36, No. 5-6, pp.218-228.

Galdiero, S., Capasso, D., Vitiello, M., D'Isanto, M., Pedone, C. & Galdiero, M. (2003). Role of surface-exposed loops of Haemophilus influenzae protein P2 in the mitogen-activated protein kinase cascade. *Infectection and Immunity*, Vol.71, No. 5, pp.2798-2809.

Galdiero, S., Galdiero, M. & Pedone C. (2007). beta-barrel membrane bacterial proteins: structure, function, assembly and interaction with lipids. *Current Protein & Peptide Science*, Vol.8, No. 1, pp.63-82.

Gupta, S., Kumar, D., Vohra, H., & Ganguly, N.K. (1999). Involvement of signal transduction pathways in *Salmonella* typhimurium porin activated gut macrophages. *Molecular & Cellular Biochemistry*, Vol. 194, No. 1-2, pp. 235-243.

Isibasi, A., Paniagua, J., Rojo, M,P., Martín, N., Ramírez, G., González, C.R., López-Macías, C., Sánchez, J., Kumate, J., & Ortiz-Navarrete, V. (1994). Role of porins from *Salmonella* typhi in the induction of protective immunity. *Proceedings of the National Academy of Sciences USA*, Vol. 7, No. 30, pp. 350-352.

Kisseleva, T., Bhattacharya, S., Braunstein, J., & Schindler, C.W. (2002). Signaling through the JAK/STAT pathway, recent advances and future challenges. *Gene*, Vol. 285, No. 1-2, pp. 1-24.

Lee, J. S., Jung, I. D., Lee, C. M., Park, J. W., Chun, S. H., Jeong, S. K., Ha, T. K., Shin, Y. K., Kim, D. J., & Park, Y. M. (2010). Outer membrane protein a of *Salmonella* enterica serovar Typhimurium activates dendritic cells and enhances Th1 polarisation. BMC *Microbiology*, Vol. 10, pp. 263.

Mangan, D.F., Wahl, S.M., Sultzer, B.M.,& Mergenhagen, S.E. (1992). Stimulation of human monocytes by endotoxin-associated protein: inhibition of programmed cell death (apoptosis) and potential significance in adjuvanticity. *Infection and Immunity*, Vol.60, No. 4, pp. 1684-1686.

Massari, P., Henneke, P., Ho, Y., Latz, E., Golenbock, D.T., & Wetzler, L.M. (2002). Cutting edge: Immune stimulation by neisserial porins is toll-like receptor 2 and MyD88 dependent. *Journal of Immunology.*, Vol. 168, No. 4, pp. 1533-1537.

Massari, P., Ram, S., Macleod, H. & Wetzler, L.M. (2003). The role of porins in neisserial pathogenesis and immunity. *Trends in Microbiology*, Vol. 11, No. 2, pp. 87-93.

Mastroeni, P., Villareal-Ramos, B. & Hormaeche, C.E. (1993). Adoptive transfer of immunity to oral challenge with virulent Sal monella in innately susceptible BALB/c mice requires both immune serum and T cells. *Infectection and Immunity*, Vol. 61, No. 9, pp. 3981-3984.

McCormick, B.A., Miller, S.I., Carnes, D., & Madara, J.L. (1995). Transepithelial signaling to neutrophils by salmonellae: a novel virulence mechanism for gastroenteritis. *Infection and Immunity.*, Vol. 63, No. 6, pp. 2302-2309.

Melancon, J., Murgita, R.A., & Devoe, I.W. (1983). Activation of murine B lymphocytes by Neisseria meningitidis and isolated meningococcal surface antigens. *Infection and Immunity*, Vol. 42, No. 2, pp. 471-479.

Munn, CB.,. Isshiguro, E.E., Kay, W.w., & Trust, T.J. (1982). Role of surface components in serum resistance of virulent Aeromonas salmonicida. *Infection and Immunity*, Vol. 36, pp. 1069-1075.

Muthukkmar, S., & Muthukkaruppan. V.R. (1993). Mechanism of protective immunity induced by porin-lipopolysaccharide against murine salmonellosis. *Infection and Immunity*, Vol. 61, pp. 3017-3025.

Qi, H.L., Tai, J.Y., & Blake, M.S. (1994). Expression of large amounts of neisserial porin proteins in Escherichia coli and refolding of the proteins into native trimers. *Infection and Immunity*, Vol. 62, No. 6, pp. 2432-2439.

Scott, M.J., Godshall, C.J., & Cheadle, W.G. (2002). Jaks, STATs, Cytokines, and Sepsis. *Clinical and Diagnostic Laboratory Immunology*,. Vol. 9, No. 6, pp. 1153-1159.

Shirmer, T., Keller, T. A., Wang, Y. F., & Rosenbusch, J. P. (1995). Structural basis for sugar translocation through maltoporin channels at 3.1 A resolution. *Science*, Vol. 267, No. 5197, pp. 512-514.

Snapper, C.M., Rosas, F.R., Kehry, M.R., Mond, J.J., & Wetzler, L.M. (1997). Neisserial porins mayprovide critical second signals to polysaccharide-activated murine B cells for induction of immunoglobulin secretion. *Infection and Immunity*, Vol. 65, No. 8, pp. 3203-3208.

Song, J., Minetti, C.A., Blake, M.S., & Colombini, M. (1998). Successful recovery of the normal electrophysiological properties of PorB (class 3) porin from Neisseria meningitidis after expression in Escherichia coli and renaturation. *Biochimica et Biophysica Acta*, Vol. 1370, No. 2, pp. 289-298.

Takeuchi, A. (1967). Electron microscope studies of experimental *Salmonella* infection. I. Penetration into the intestinal epithelium by *Salmonella* typhimurium. *American Journal of Pathology.*, Vol. 50, No. 1, pp. 109-136.

Thatte, J., Rath, S. & Bal, V. (1993). Immunization with live vs. killed *Salmonella* typhimurium leads to the generation of an interferon-γ-dominant vs. an IL-4-dominant immune response. *International Immunology*, Vol. 5, No. 11, pp. 1431-1436.

Tufano, M.A., Berlingieri, M.T., Sommese, L., & Galdiero, F. (1984). Immune response in mice and effects on cells by outer membrane porins from *Salmonella* typhimurium. *Microbiology.* Vol. 7, No. 4, pp. 353-366.

Tufano, M.A., Ianniello, R., Galdiero, M., De Martino, L., & Galdiero, F. (1989). Effect of *Salmonella* typhimurium porins on biological activities of human polymorphonuclear leukocytes.. *Microbial Pathogens.*,Vol.7, No. 5, pp. 337-346.

Vitiello, M., D'Isanto, M., Galdiero, M., Raieta, K., Tortora, A., Rotondo, P., Peluso, L., & Galdiero, M. (2004). Interleukin-8 production by THP-1 cells stimulated by *Salmonella* enterica serovar Typhimurium porins is mediated by AP-1, NF-kappaB and MAPK pathways. *Cytokine*, Vol. 27, No. 1, pp. 15-24.

Wetzler, L.M., Ho, Y., & Reiser,H. (1996). Neisserial porins induce B lymphocytes to express costimulatory B7-2 molecules and to proliferate. *Journal of Experimental Medicine*,Vol. 183, No. 3, pp. 1151-1159

Yee S. C., Theam, S. L., Ai Lan, C., Ismail A., & Asma I. (2011). Structural and functional studies of a 50 kDa antigenic protein from *Salmonella enterica* serovar Typhi. *Journal of Molecular Graphics and Modelling*, Vol. 29, No. 6, pp. 834-842.

Part 2

Pathogenesis

8

Insect/Bacteria Association and Nosocomial Infection

Marcos Antônio Pesquero, Lílian Carla Carneiro
and Débora De Jesus Pires
Universidade Estadual de Goiás – UEG – UnU de Morrinhos
Brasil

1. Introduction

The genus *Salmonella* belongs to the family Enterobacteriaceae, a group of bacteria normally found in the intestine of the hosts. More than 2,500 serotypes have been identified in the *Salmonella enterica* complex (Popoff et al., 2004). *Salmonella* species are classified as Gram-negative, rod-shaped, facultatively aerobic bacteria that have mobility in liquid environments and reproduce at temperatures ranging from 5°C to 47°C and pH from 4.5 to 9.0 (Varnan & Evans, 1991). *Salmonella* species produce hydrogen sulfide, are oxidase, indol, and Voges-Proskauer negative, catalase, citrate, lysine-ornithine, decarboxylase, and glucose positive, also presenting other carbohydrates fermentation with acid and gas liberation (Le Minor, 1984). Although the dispersion of these microorganisms is limited due to their incapacity to sporulate and sensitivity within the pasteurization temperature range (Varnan & Evans, 1991; D'Aoust, 2000), they are resistant to desiccation and freezing and are able to survive in the environment for several years (Tortora et al., 2005).

A variety of human foods of plant and animal origin have been identified as vectors of transmission of *Salmonella*. A study with inoculation of *Salmonella* in a tomato plantation soil evidences the risk of human contamination through ingestion of plant foods (Barak & Liang, 2008). According to Hirsh (2003), apparently healthy animals can develop diseases caused by *Salmonella* because of stress factors, such as sudden alterations of the environment temperature, water and food deprivation, overpopulation, gathering animals of different lots, and inappropriate antimicrobial use. As contaminated meat is the most frequent source of human disease caused by *Salmonella*, this type of food is of particular interest concerning epidemiological studies (Gatto et al., 2006). The genus *Salmonella* was found in 20% of the commercialized chicken samples in Malaysia (Rusul et al., 1996). Human disease outbreaks caused by *Salmonella* are generally associated to egg, chicken, and pork consumption (Castagna et al., 2004). Inappropriate food storage also represents an important cause of proliferation and dissemination of these microorganisms (Murmann et al., 2005). *Salmonella enteritidis* and *Salmonella typhimurium* were the most prevalent serotypes involved in foodborne infections registered in the world (Van Der Wolf et al., 2001).

2. Epidemiology

The acute symptoms of the infection caused by *Salmonella* are fever, migraine, nausea, and dysentery. Depending on the patient profile, the extraintestinal chronic form can evolve into sepsis (CDC, 1999; Wilson & Whitehead, 2004; Loureiro et al., 2010). *S. enterica* serotype Blockley is related to up to 29% of the cases of arthritis described in the literature (Dworkin et al., 2001). An aggravating factor in diseases caused by *Salmonella* is the evolution of resistant types due to uncontrolled antibiotic use both in domesticated animals and humans. Antibiotics and chemotherapics are chemical compounds that can inhibit bacterial growth (Cromwell, 1991), but can result in microorganism resistance if used indiscriminately, disregarding bacterial specificity or the minimum inhibitory concentration (Wannmacher, 2004). Therefore, Silva et al. (2004) consider that the use of antimicrobials in bird diets is of great importance to break the cycle of bird disease caused by *Salmonella*. In cases of continuous or prolonged treatment with high-dose antibiotic in humans, instead of curing the patient the infection rates can increase (Barza & Travers, 2002; Wannmacher, 2004). Studies demonstrated that the rates of *Salmonella* multi-drug resistance have increased considerably in recent years (Haneda et al., 2004; Chiu et al., 2006; Carneiro et al., 2008; Pesquero et al., 2008). Bacterial resistance can lead to increased virulence and consequently increased morbidity and mortality of infected people (Mølbak, 2005).

Salmonella resistance to antibiotics has been related to the presence of plasmids. Approximately 30 low-molecular-weight plasmids have been identified in *S. enteritidis* (Rychlik et al., 2001). Plasmids described in *S. enterica* serovar typhi isolates (Boyd et al., 2003) confer pathogenicity to these bacteria (Baker & Dougan, 2007). Some plasmids simultaneously confer resistance and virulence to *Salmonella*. This bacteria-plasmid association presents epidemiological relevance, because a process of recombination with *Salmonella* provides it with advantages to survive in a hostile environment and chances to evolve a new genetic lineage (Majtan et al., 2006).

Antibiotic resistance genes are frequently located within transposons, but they can also be found in the form of gene cassettes captured and clustered in integrons and thereafter mobilized to spread resistance among other organisms (Fluit, 2005). In *S. typhimurium*, antibiotic resistance depends on integrons more frequently associated to a genomic island located on the bacterial chromosome (Tosini et al., 1998). There are two categories of integrons, one represented by repeated direct sequences (IS) and the other represented by inverted sequences (IR). The first integron category, widely distributed in *Salmonella*, consists of two conserved sequences, regions 50 Cs and 30 Cs, which carry the gene int I to the integrase protein and the gene sul I of resistance to the sulfonamide, respectively (Guerra et al., 2002).

Some plasmids are responsible for the phage conversion, which permits bacteria to resist phage infection. For example, the pOG670 plasmid of 54 kb, belonging to the group of incompatibility X (IncX), present in *S. enteritidis*, is capable of converting phages types 1 and 4 into type 6a, and phage 8 into type 13 (Ridley et al., 1996). In *S. abortus-equi*, a plasmid of 85 kb that codifies resistance to toxic heavy metals (chromium, arsenic, cadmium, and mercury) was described. This plasmid was also proven to encode genes that allowed the use of citrate and conferred ß-lactam antibiotic resistance (Ghosh et al., 2000).

Guiney et al. (1995) and Roudier et al. (1992) found serotypes of non-typhoid *Salmonella*, associated with extraintestinal disease, possessing virulent plasmids (spvC) that contain virulence genes (spvC), important for the induction of a systemic and lethal infection. Fierer et al. (1992) reported plasmids in *Salmonella* in 76% of 79 samples of human blood and in 42% of 33 human stool samples. Both in animals and humans, bacteria of the genus *Salmonella* are more frequently found in systemic infections compared with enteric ones. According to Fierer et al. (1993), non-typhoid *Salmonella* pathogenicity is related to the presence of plasmids.

The high susceptibility of hospitalized children to nosocomial infections was attributed to antimicrobial resistance of *S. enterica* (Fonseca et al., 2006). Some disease outbreaks caused by *Salmonella* serotypes resistant to antibiotics have been registered in pediatric settings worldwide and can provoke the death of newborn babies (Pessoa-Silva et al., 2002; Bouallègue-Godet et al., 2005). Diseases caused by *Salmonella* represent 10% to 15% of the acute gastroenteritis cases all over the world (Jay, 2005). Of the 4,012 disease outbreaks of enteric infections that occurred in England and Wales, *Salmonella* was the most frequent microorganism, responsible for 22% of the cases (Guard-Petter, 2001). Estimates indicate that *Salmonella* is responsible for one third of the cases of foodborne illnesses in the US, corresponding to 2-4 million cases a year (Andrews et al., 1992), causing economic losses of about 4 billion dollars annually (Mead et al., 1999). In South American countries, the prevalence of *Salmonella* infections is low (2.5%) compared with the US, although considered one of most important epidemiological illnesses (Franco, 2003).

3. Insect vectors

The transformation of natural ecosystems into urban areas and crop fields results in changes in animal, plant, and microorganism biodiversity and dynamics. In a broad sense, the simplification of the environment reduces biological diversity but, on the other hand, it favors the population growth of other species of bacteria (Fowler, 1983). Opportunistic animals that benefit from human presence are called synanthropic and considered pests if they cause damages to human health and the economy. The current context of worldwide social economic development contributes to environmental deterioration, facilitating horizontal and vertical transmission of illnesses through vectors, mainly insects (Pongsiri & Roman, 2007).

Vectors are organisms that contribute to the dispersion of pathogens by carrying and transmitting them (Purcell & Almeida, 2005), and are known as intermediate or definitive hosts, respectively when the pathogenic organism is carried externally or internally to the vector body. Transmission between hosts can occur indirectly by pathogen spread in the environment, normally through feces and/or secretions, and also due to physical contact, when the pathogen is adhered to the surface of the vector body, or directly by inoculation of the pathogen in the host body through the vector bite. Triatomine species (Hemiptera: Reduviidae) increase human transmission rate of the protozoan *Trypanossoma cruzi* by depositing feces contaminated with infecting metacyclic trypomastigotes on the host face skin near the insect bite.

Insects are among the most diversified, abundant, and widely dispersed animals in the world. They represent more than 50% of all species living on the planet, 71% of animal species, 74% of invertebrates, and 87% of arthropods (Lewinson & Prado, 2002). This success is mainly due to their morphologic characteristics, such as locomotion appendices (legs and wings), exoskeleton, small body size, and metamorphosis. Insects can use specialized types of sexual and asexual reproduction, such as parthenogenesis, pedogenesis, and neoteny, as reproductive strategies (Gullan & Cranston, 2008). The most representative insect orders regarding species richness and abundance are Coleoptera (beetles), Diptera (flies and mosquitoes), Hymenoptera (bees, wasps, and ants), Lepidoptera (butterflies and moths), and Hemiptera (chinch bugs, aphids, whiteflies, mealybugs, and cicadas).

Some species of blood-feeding insects found a direct food source in humans, and many of them are vectors of microorganisms that cause important diseases in tropical and subtropical countries, such as malaria, yellow fever, typhoid fever, dengue fever, filariasis, leishmaniasis, chagas disease, and sleeping sickness. Another group of synanthropic insects indirectly benefits from human foods found in crop fields, or stored, industrialized, and prepared in households, or rejected food waste in landfills and sewage. These insects are known as phytophagous, granivorous, parasites, saprophagous, coprophagous, and generalists, constituting the great majority of insects associated with humans in urban centers and rural areas. Among these insects, the species that are carriers of pathogenic microorganisms from contaminated environments to human food, medical instruments, and kitchen utensils deserve special attention of public health authorities. Therefore, the participation of insects in the transmission of bacteria has been investigated aiming to reduce the occurrence of enteric disease outbreaks in hospitals.

Arthropods are common in hospital environment (Gazeta et al., 2007) and the main vectors of pathogenic microorganisms that infect humans are cockroaches, ants, and flies because of their contact with human feces and other contaminated materials. Sewage and landfills are major sources of pathogenic microorganisms and lack of investments in basic sanitation is a serious public health problem (Andreoli & Bonnet, 2000; Bastos et al., 2003). Cockroaches (Blattodea) are dorsoventrally flattened body insects that eat preferentially decayed vegetable substances. They have specific enzymes and endosymbiont microorganisms that assist them in cellulose digestion and essential amino acid synthesis (Hirose & Panizzi, 2009; Louzada, 2009). *Blattella germanica* and *Periplaneta americana* are cosmopolite cockroach species adapted to urban environments, specially sewage, bathroom, and kitchen, where they find abundant food and absence of predators. Microbiological analyses of cockroaches captured in hospitals and residences have identified these insects as vectors of more than 80 recognized bacteria species belonging to 51 genera, many of them resistant to antimicrobials, posing health risk to the already weakened interns (Table 1). *Salmonella* was found in nine out of twelve studies with cockroaches in hospitals and residences in the world. Although *Salmonella* multiplication might occur in the intestines of the insects (Klowden & Greenberg, 1977), Gram-negative bacteria are more commonly found in their cuticle, facilitating the dispersion of pathogens (Mpuchane et al., 2006).

Ants are eusocial insects that live in colonies with one or more queens and thousands of workers. The latter need to collect food for their own maintenance and to feed both the immature and adult individuals of the colony, including the queen. Through chemical orientation, workers can cover great distances searching for food and return to the nest indicating the way for the other members of the colony. Their feeding habit is varied (Brandão et al., 2009), but the species adapted to urban environments are characterized by the consumption of a generalist diet, mobility of colonies, polygeny, monomorphism, and reduced body size (Passera, 1994). *Monomorium pharaonis, Paratrechina longicornis, Tapinoma melanocephalum,* and species of *Pheidole* and *Solenopsis* are the main ants already found in hospital environments (Fig. 1), and the literature registered 50 bacteria species belonging to 31 genera associated to these and other ant species (Table 2, Fig. 2). Species of the genus *Staphylococcus, Escherichia, Pseudomonas, Enterobacter, Bacillus, Streptococcus,* and *Klebsiella* are the most frequent bacteria associated with ants in Brazilian hospitals (Fig. 3). *Salmonella* was found associated only with *Monomorium pharaonis* (Beatson, 1972) and a species of *Pheidole* (Pesquero et al., 2008) carries bacteria presenting antimicrobial resistance (Carneiro et al., 2008).

Diptera constitutes a great order of insects that present small body size, fly fast, and have a variety of feeding habits, including necrophagy and coprophagy, during the immature phase. When walking over feces and corpses to deposit eggs, female flies acquire and carry pathogenic bacteria to the interior part of residences, restaurants, and hospitals, contaminating mainly foods and kitchen utensils. *Musca domestica* (Muscidae) is a very common cosmopolite and synanthropic species in urban zones, and together with five other species of flies, is associated to 12 bacteria species belonging to 11 genera (Table 3). *Musca domestica* is the most studied fly considered a vector of *Salmonella* that occurs in some countries (Bolaños-Herrera, 1959; Bidawid et al., 1978; Arnold, 1999; Olsen & Hammack, 2000; Mian et al., 2002; Oliveira et al., 2006; Cardozo et al., 2009; Prakash & Tikar, 2009; Butler et al., 2010; Choo et al., 2011).

Cockroaches	Bacteria	Reference
Blattella germanica	Enterobacter Enterococcus Escherichia coli Haemophilus Klebsiela Pseudomonas Shigella Staphylococcus Streptococcus	Salehzadeh et al. (2007)
Periplaneta Blatta	Salmonella bovismorbificans Salmonella oslo Salmonella typhimurium	Devi & Murray (1991)
Blattella germanica	Salmonella enteritidis	Ash & Greenberg (1980).
Periplaneta americana Blattella germânica Supella longipalpa Blatta lateralis Polyphaga aegyptiuca Arenivaga roseni Parcoblatta	Salmonella	Fathpour et al. (2003)

Cockroaches	Bacteria	Reference
cockroaches	*Aeromonas* *Escherichia coli* *Citrobacter freundii* *Enterobacter cloacae* *Klebsiella pneumoniae* *Proteus mirabilis* *Proteus vulgaris* *Pseudomonas aeruginosa* *Salmonella* *Serratia marcescens* *Staphylococcus aureus* *Staphylococcus faecalis* *Staphylococcus epidermidis*	Tatfeng et al. (2005)
Periplaneta americana	*Enterobacter* *Escherichia coli* *Klebsiella* *Proteus vulgaris* *Salmonella* *Serratia* *Shigella* *Staphylococcus aureus* *Staphylocoques epidermidis* *Streptococcus*	Lamiaa et al. (2007)
Blattela germanica	*Bacillus cereus* *Escherichia coli* *Salmonella* *Shigella flexneri* *Staphylococcus aureus*	Tachbele et al. (2006)
Periplaneta americana *Blattella germânica*	*Acinetobacter* *Alcaligenes faecalis* *Arizona* *Bacillus* *Citrobacter freundii* *Enterobacter aerogenes* *Enterobacter agglomerans* *Enterococcus* *Escherichia coli* *Hafnia alvei* *Klebsiella pneumoniae* *Proteus mirabilis* *Proteus vulgaris* *Providencia rettgeri* *Pseudomonas* *Salmonella typhimurium* *Shigella* *Staphylococcus* *Yersinia*	Fakoorziba et al. (1910)

Cockroaches	Bacteria	Reference
Periplaneta americana	*Enterobacter aerogenes* *Escherichia coli* *Citrobacter freundii* *Hafnia alvei* *Salmonella* *Serratia liquefaciens* *Staphylococcus*	Miranda & Silva (2008)
Periplaneta americana *Blatta orientalis*	*Cedecea davisae* *Cedecea lopagei* *Cedecea neteri* *Citrobacter diversus* *Citrobacter freundii* *Edwardsiella ictaluri* *Edwardsiella tarda* *Enterobacter aerogenes* *Enterobacter agglomerans* *Enterobacter asburiae* *Enterobacter cloacae* *Enterobacter gergoviae* *Enterobacter sakasakii* *Escherichia balttae* *Escherichia coli* *Escherichia hermanii* *Escherichia vulneris* *Ewingella americana* *Hafnia alvei* *Klebsiella oxytoca* *Klebsiella ozanae* *Klebsiella pneumoniae* *Klebsiella rhinoscleromatis* *Klebsiella terrigena* *Kluyvera ascorbata* *Moraxella uretharlis* *Morganella morganii* *Obesumbacterium* *Proteus mirabilis* *Proteus myxofeceins* *Proteus penneri* *Proteus vulgaris* *Providencia rustigianii* *Pseudomonas aeruginosa* *Pseudomonas maltophilia* *Pseudomonas pseudoalkaligenes* *Rahnella aquatilis* *Salmonella* *Serratia liquefaciens* *Serratia marcescens* *Serratia odotifera*	Chaichanawongsaroj et al. (2004)

Cockroaches	Bacteria	Reference
Supella supellectilium	*Achromobacter* *Acinetobacter calcoaceticus* *Aeromonas hydrophila* *Alcaligenes faecalis* *Buttiauxella agrestis* *Cedecea* *Citrobacter diversus* *Citrobacter freundii* *Enterobacter aerogenes* *Enterobacter agglomerans* *Enterobacter amnigenus* *Enterobacter cloacae* *Enterobacter sakazakii* *Escherichia adecarboxylata* *Escherichia coli* *Klebsiella oxytoca* *Klebsiella pneunoniae* *Kluyvera* *Proteus mirabilis* *Pseudomonas aeruginosa* *Pseudomonas cepacia* *Pseudonmonas paucimobilis* *Pseudomnonas fluorescens* *Pseudomonas maltophilia* *Pseudomonas stutzeri* *Serratia marcescens* *Serratia liquefaciens* *Staphylococcus aureus*	Le Guyader et al. (1989)
cockroaches	*Actinomyces randingae* *Alcaligenes faecalis* *Arthrobacter cumminnsii* *Aureubacterium* *Bacillus* *Brevibacterium* *Burkholderia vietnamiensis* *Buttiauxella* *Citrobacter* *Corynebacterium* *Enterobacter* *Erwinia* *Escherichia coli* *Hafnia* *Kauri rosea* *Kigali* *Klebsiella* *Kluyvera*	Mpuchane et al., (2006)

Cockroaches	Bacteria	Reference
	Leuconostoc	
	Microbacterium	
	Micrococcus	
	Proteus	
	Providencia ruttier	
	Pseudomonas	
	Rhodococcus australis	
	Rhodococcus rhodochrous	
	Salmonella typhimurium	
	Serratia	
	Shigella	
	Spingobacterium thalpophilum	
	Staphylococcus	
	Stenotrophomonas maltophillia	
	Streptococcus	
	Tsukamurella inchonensis	
	Vibrio metschnikovii	
	Xanthomonas	

Table 1. Registry of occurrences of vector species of the order Blattaria and types of transported bacteria.

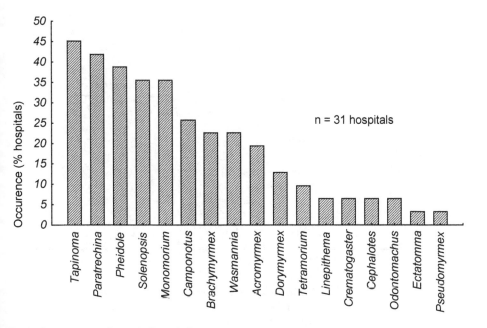

Fig. 1. Occurrence of ants in hospitals.

Ant	Bacteria	Reference
Acromyrmex	*Pseudomonas aeruginosa* *Staphylococcus* *Streptococcus faecalis*	Santos et al. (2009)
Brachymyrmex	*Enterococcus* *Streptococcus agalactiae*	Lise et al. (2006)
Camponotus	*Corynebacterium diphtheria* *Corynebacterium jeikeium* *Streptococcus*	Lise et al. (2006); Santos et al. (2009)
Camponotus vittatus	*Bacillus* *Staphylococcus*	Rodovalho et al. (2007)
Linepithema humile	*Escherichia coli* *Streptococcus*	Santos et al. (2009)
Monomorium pharaonis	*Acinetobacter haemolyticus* *Aeskovia* *Clostridium* *Corynebacterium* *Listeria monocytogenes* *Planococcus* *Pseudomonas aeruginosa* *Pseudomonas luteola* *Salmonella* *Sphingobacterium* *Sphingomonas paucimobilis* *Staphylococcus intermedius* *Stenotrophomonas maltophilia* *Streptococcus bovis* *Enterobacter agglomerans* *Enterococcus faecalis* *Enterococcus faecium* *Gemella haemolysans* *Klebsiella pneumoniae* *Streptococcus acidominimus* *Staphylococcus lugdunensis*	Lise et al. (2006); Moreira et al. (2005); Beatson (1972)
Odontomachus	*Enterococcus* *Escherichia coli* *Pseudomonas aeruginosa* *Staphylococcus* *Streptococcus*	Santos et al. (2009)

Ant	Bacteria	Reference
Paratrechina longicornis	*Acinetobacter haemolyticus* *Alcaligenes faecalis* *Alcaligenes sylosidans* *Bacillus* *Burkholderia cepacia* *Citrobacter diversus* *Comomonas acidoverans* *Corinebacterium* *Enterobacter aerogenes* *Enterobacter agglomerans* *Enterobacter cloacae* *Escherichia coli* *Gemella haemolysans* *Gemella morbillorum* *Klebsiella pneumoniae* *Proteus mirabilis* *Providencia alcalifaciens* *Pseudomonas fluorescens* *Pseudomonas putida* *Pseudomonas stutzieri* *Serratia marcescens* *Serratia rubidae* *Staphylococcus aureus* *Staphylococcus cohnii* *Stenotrophomonas altophilia*	Lise et al. (2006); Moreira et al. (2005); Fontana et al. (2010); Tanaka et al. (2007)
Pheidole	*Aeromonas* *Enterococcus* *Escherichia coli* *Klebsiella* *Pseudomonas aeruginosa* *Salmonella* *Staphylococcus* *Streptococcus*	Santos et al. (2009); Pesquero et al. (2008); Carneiro et al., (2008)
Pheidole megacephala	*Acinetobacter baumannii* *Bacillus* *Escherichia coli* *Pseudomonas aeruginosa* *Serratia liquefaciens* *Shigella sonnei* *Staphylococcus aureus*	Fontana et al. (2010)

Ant	Bacteria	Reference
Solenopsis	*Enterococcus* *Staphylococcus* *Streptococcus*	Santos et al. (2009)
Solenopsis globularia	*Bacillus* *Staphylococcus*	Fontana et al. (2010)
Solenopsis saevissima	*Corynebacterium* *Enterococcus* *Neisseria* *Pseudomonas luteola* *Staphylococcus saprophyticus* *Stenotrophomonas maltophilia*	Lise et al. (2006)
Tapinoma melanocephalum	*Acinetobacter baumanni* *Alcaligenes faecalis* *Bacillus* *Burkholderia cepacia* *Corinebacterium* *Enterobacter aerogenes* *Enterobacter amnigenus* *Enterobacter cloacae* *Enterococcus faecalis* *Escherichia coli* *Gemella morbillorum* *Hafnia alvei* *Klebsiella oxytoca* *Klebsiella pneumoniae* *Pseudomonas aeruginosa* *Pseudomonas fluorescens* *Staphylococcus saprophyticus* *Sphingomonas paucimobilis* *Staphylococcus aureus* *Staphylococcus epidermidis* *Staphylococcus equorum* *Staphylococcus saprophyticus* *Staphylococcus simulans* *Staphylococcus warneri* *Streptococcus viridans*	Lise et al. (2006); Moreira et al. (2005); Fontana et al. (2010); Tanaka et al. (2007); Teixeira et al. (2009); Rodovalho et al. (2007); Santos et al. (2009)
Wasmannia auropunctata	*Pseudomonas aeruginosa* *Staphylococcus* *Streptococcus*	Santos et al. (2009)

Table 2. Registry of occurrences of vector species of the family Formicidae (Hymenoptera) and types of transported bacteria.

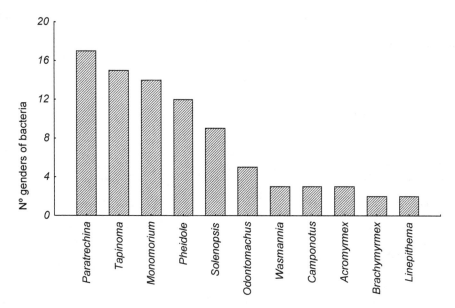

Fig. 2. Occurrence of bacteria genera per ant genus in hospitals.

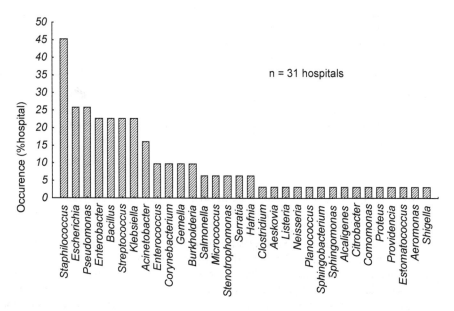

Fig. 3. Occurrence of bacteria genera in hospitals.

Fly	Bacteria	Reference
Hydrotaea aenescens *Musca domestica*	*Salmonella*	Olsen & Hammack (2000)
Chrysomya megacephala *Musca domestica*	*Citrobacter* *Enterobacter* *Escherichia coli* *Klebsiella* *Morganella* *Proteus mirabilis* *Pseudomonas* *Salmonella agona*	Oliveira et al. (2006)
Musca domestica	*Salmonellu* *Shigella*	Bolaños Herrera (1959)
Musca domestica *Fannia caniculares* *Muscina stabulans* *Phaenicia sericata*	*Salmonella* *Shigella*	Bidawid et al. (1978)
Musca domestica	*Salmonella enteritidis*	Mian et al. (2002)
Musca domestica	*Escherichia coli* *Salmonella typhi* *Shigella flexneri* *Yersinia enterocolitica*	Béjar et al. (2006)
Musca domestica	*Campylobacter* *Salmonella*	Choo et al. (2011)

Table 3. Registry of occurrences of vector species of the Order Diptera and types of transported bacteria.

4. Conclusion

The presence of any serotypes of *Salmonella* in any types of food is a reason to classify them as improper for consumption in the international market. Animals contaminated with *Salmonella* destined to human feeding cannot show clinical signals of the illness (Castagna et al., 2004). Therefore, industries that deal with products of animal origin must implement quality control strategies with the purpose of guaranteeing food safety. The main insect vectors of *Salmonella* are cockroaches and flies, but ants also represent a potential risk. Therefore, governments must invest in the construction of modern hospitals equipped with some devices capable of avoiding the entrance and permanence of vectors, particularly these three types of insects.

5. References

Andreoli, C.V., Bonnet, B.R.P. (Coord.). (2000). *Manual de métodos para análises microbiológicas e parasitológicas em reciclagem agrícola de lodo de esgoto*. 2ª Ed. Sanepar, Curitiba.

Andrews, W H., Bruce, V. R., June, G., Satchell, F. & Sherrod, P. (1992). *Salmonella*, In: *Bacteriological analytical manual*, 7th ed., Arlington, Va: Association of Official Analytical Chemists, pp. 51–69, ISBN 0935584498.

Ash, N. & Greenberg, B. (1980). Vector potential of the German cockroach (Dictyoptera: Blattellidae) in dissemination of *Salmonella enteritidis serotype typhimurium*. *J Med Entomol*, Vol. 17, No. 5, pp. 417-423, ISSN 0022-2585.

Baker, S. & Dougan, G. (2007). The genome of *Salmonella enterica* serovar Typhi. *Clin Infect Dis*, Vol. 45, No. 1, pp. S29-S33, ISSN 1058-4838.

Barza, M. & Travers, K. (2002). Excess infection due to antimicrobial resistance: the "attributable fraction". *Clin Infect Dis*, Vol. 34, No. 3, pp.123-125, ISSN 1058-4838.

Bastos, R.K.X., Bevilacqua, P.D., Andrade Neto, C.O. & Von Sperling, M. (2003). Utilização de esgotos tratados em irrigação - aspectos sanitários. In: *Utilização de esgotos tratados em fertirrigação, hidroponia e psicultura*, Bastos, R.K.X. (Coord.). ABES, RiMa, Rio de Janeiro.

Beatson, S.H. (1972). Pharaoh´s ants as pathogens vectors in hospitals. *Lancet*, Vol. 299, No. 1, (feb 1972), pp. 425-427, ISSN 0140-6736.

Béjar, V., Chumpitaz, J., Pareja, E., Valencia, E., Huamán, A, Sevilla, C., Tapia, M. & Saez G. (2006). *Musca domestica* como vector mecánico de bactérias enteropatógenas em mercados Y basurales de Lima Y Callao. Rev Peru Med. Exp. Salud Publica, Vol. 23, No 1, pp. 39-43, ISSN 1726-4634.

Bidawid, S., Edeson, J.F., Ibrahim, J. & Matossian, R.M. (1978). The role of non-biting flies in the transmission of enteric pathogens (*Salmonella* species and *Shigella* species) in Beirut, Lebanon. *Ann Trop Med Parasitol*, Vol. 72, No. 2, (apr 1978), pp. 117-121, ISSN 0003-4983.

Bolaños-Herrera, R. (1959). Frecuencia de *Salmonella y Shigella* em moscas domésticas em la ciudade de San José. *Rev. Biol. Trop.*, Vol. 7, No. 2, pp. 207-210, ISSN 0034-7744.

Bouallègue-Godet, O., Salem, Y.B., Fabre, L., Demartin, M., Grimont, P.A.D., Mzoughi, R. & Weill, F.X. (2005). Nosocomial outbreak caused by *Salmonella enteric* Serotype Livingstone Producing CTX-M-27 Extended-Spectrum ß-Lactamase in a Neonatal Unit in Sousse, Tunisia. *J Clin Microbiol*, Vol. 43, No. 3, (mar 2005), pp. 1037-1044, ISSN 0095-1137.

Boyd, E.F., Porwollik, S., Blackmer, F. & McClelland, M. (2003). Differences in gene content among *Salmonella enterica* serovar Typhi isolates. *J. Clin Microbiol*, Vol. 41, No. 8, (aug 2003), pp. 3823–3828, ISSN 0095-1137.

Brandão, C.R.F., Silva, R.R. & Delabie, J.H.C. (2009). Formigas (Hymenoptera). In: *Bioecologia e nutrição dos insetos*: base para o manejo integrado de pragas, Panizzi, A.R. & Parra, J.R.P. (Eds.), pp. 323-369. Embrapa, ISBN 978-85-7383-452-9, Brasília.

Carneiro, L.C., Carvalhares, T.T., Pesquero, M.A., Quintana, R.C., Feitosa, S.B., Elias Filho, J. & Oliveira, M.A.C. (2008). Identificação de bactérias causadoras de infecção hospitalar e avaliação da tolerância a antibióticos. *Newslab*, Vol. 86, pp. 106-114, ISSN 0104-8384.

Castagna, S.M.F., Schwarz, P., Canal, C.W. & Cardoso, M.R.I. (2004). Prevalência de suínos portadores de *Samonella* sp. ao abate e contaminação de embutidos tipo frescal. *Acta Scientiae Veterinariae*, Vol. 32, No. 2, pp. 141-147, ISSN 1679-9216.

CDC. (2005). *Salmonella* infection-Salmonellose. In: *Center for Disease Control and Prevention*, 10 out 2008, Available from: <http//:http://www.cdc.gov/az/s.html>

Chaichanawongsaroj, N., Vanichayatanarak, K., Pipatkullachat, T., Polrojpanya, M. & Somkiatcharoen, S. (2004). Isolation of gram-negative bacteria from cockroaches trapped from urban environment. *Southeast Asian J Trop Med Public Health*, Vol. 35 No. 3, (sep 2004), pp. 681-684, ISSN 0125-1562.

Chiu, C.H., Su, L.H., Chu, C.H., Wang, M.H., Yeh, C.M., Weill, F.X. & Chu, C. (2006). Detection of multidrug-resistant *Salmonella enterica* serovar typhimurium phage types DT102, DT104, and U302 by multiplex PCR. *J Clin Microbiol*, Vol. 44, No. 7, pp 2354-2358, ISSN 0095-1137.

Choo, L.C., Saleha, A.A., Wai, S.S. & Fauziah, N. (2011). Isolation of *Campylobacter* and *Salmonella* from houseflies (*Musca domestica*) in a university campus and a poultry farm in Selangor, Malaysia. *Tropical Biomedicine*, Vol. 28, No. 1, pp. 16–20, ISSN 0127-5720.

Cromwell, G.L. (1991). Antimicrobial agents. In: *Swine nutricion*, Miller, E.R., Ullrey, D.E. & Lewis, A.J. (Eds.), pp. 297-314, Butterworth-Heinemann, ISBN 0849306965, Boston.

D'Aoust, J.Y. (2000). *Salmonella*. In: *The microbiological safety and quality of food*, Lund, B.M., Baird-Parker, T.C. & Gould, G.W. (Eds.), pp. 1233-1299. Aspen Publishers, ISBN 978-0-8342-1323-4, Maryland.

Devi, S.J.N. & Murray, C.J. (1991). Cockroaches (*Blatta* and *Periplaneta* species) as reservoirs of drug-resistant salmonellas. *Epidemiol Infect*, Vol. 107, pp. 357-361, ISSN 0950-2688.

Dworkin, M.S., Shoemakera, P.C., Goldoft, M.J. & Kobayashi, J.M. (2001). Reactive arthritis and Reiter Õssyndrome following na outbreak of gastroenteritis caused by *Salmonella enteritidis*. *Clin Infect Dis*, Vol. 33, No. 7, pp. 1010-1014, ISSN 1058-4838.

Fakoorziba, M.R., Eghbal, F., Hassanzadeh, J. & Moemenbellah-Fard, M.D. (2010). Cockroaches (*Periplaneta americana* and *Blattella germanica*) as potential vectors of the pathogenic bacteria found in nosocomial infections. *Ann Trop Med Parasitol*, Vol. 104, No. 6, pp. 521–528, ISSN 0003-4983.

Fathpour, H., Emtiazi, G. & Ghasemi, E. (2004). Cockroaches as reservoirs and vectors of drug resistant *Salmonella* spp. In: *Iranian Biomedical Journal*, 23 jan 2011, Available from: < http://www.sid.ir/en/VEWSSID/J_pdf/84920030107.pdf>

Fierer, J., Krause, M., Tauxe, R. & Guiney D. (1992). *Salmonella typhimurium* bacteremia: association with the virulence plasmid. *J Infect Dis*, Vol. 166, No. 3, pp. 639-642, ISSN 0022-1899.

Fierer, J., Eckann, L., Fang, F., Pfeifer, C., Finlay, B.B. & Guiney, D. (1993). Expression of the *Salmonella* virulence plasmid gene spvC in cultured macrophages and nonphagocytic cell. *Infect Immun*, Vol. 61, No. 12, pp. 5231-5236, ISSN 0019-9567.

Fluit, A.C. (2005). Towards more virulent and antibiotic-resistant *Salmonella*? *FEMS Immunol Med Microbiol*, 43: 1-11, ISSN 0928-8244.

Fonseca, E.L., Mykytczuk, O.L., Asensi, M.D., Reis, E.M.F., Ferraz, L.R., Paula, F.L., Ng, L.K. & Rodrigues, D,P. (2006). Clonality and Antimicrobial Resistance Gene Profiles of Multidrug- Resistant *Salmonella enterica* Serovar Infantis Isolates from Four Public

Hospitals in Rio de Janeiro, Brazil. *J Clin Microbiol*, Vol. 44, No. 8, pp. 2767-2772, ISSN 0095-1137.

Fontana, R., Wetler, R.M.C., Aquino, R.S.S., Andrioli, J.L., Queiroz, G.R.G, Ferreira, S.L., Nascimento, I.C. & Delabie, J.H.C. (2010). Disseminação de bactérias patogênicas por formigas (Hymenoptera: Formicidae) em dois hospitais do nordeste do Brasil. *Neotrop Entomol*, Vol. 39, No. 4, (jul-aug 2010), pp. 655-663, ISSN 1519-566X.

Fowler, H.G. (1983). Distribution patterns of paraguayan leaf-cutting ants (*Atta* and *Acromyrmex*) (Formicidae: Attini). *Stud Neotrop Fauna Environ*, Vol. 18, No. 33, pp. 121-128, ISSN 0165-0521.

Franco, B.D.G.M. (2003). Foodborne diseases in Southern South América. In: *International handbook of foodborne pathogens*, Miliotis, M.D. & Bier, J.W. (Eds.), pp. 733-743, Marcell Decker, ISBN 0-8247-0685-4, New York.

Gatto, A.J., Peters, T.M., Green, J., Fisher, I.S., Gill, O.N., O'brien, S.J., Maguire, C., Berghold, C., Lederer, I., Gerner-Smidt, P., Torpdahl, M., Siitonen, A., Lukinmaa, S., Tschäpe, H., Prager, R., Luzzi, I., Dionisi, A.M., Van Der Zwaluw, W.K., Heck, M., Coia, J., Brown, D., Usera, M., Echeita, A. & Threlfall, E.J. (2006). Distribution of molecular subtypes within Salmonella enterica serotype Enteritidis phage type 4 and S. Typhimurium definitive phage type 104 in nine European countries, 2000–2004: results of an international multi-centre study. *Epidemiol Infect*, Vol. 134, No. 4, pp. 729-736, ISSN 0950-2688.

Gazeta, G.S., Freire, M.L. & Ezequiel, O.S. (2007). Artrópodes capturados em ambiente hospitalar do Rio de Janeiro, Brasil. *Rev Patol Trop*, Vol. 36, No. 3, pp. 254-264, ISSN 1980-8178.

Ghosh, A., Singh, A., Ramteke, P.W. & Singh, V.P. (2000). Characterization of large plasmids encoding resistance to toxic heavy metals in Salmonella abortus equi. *Biochem Biophys Res Commun*, Vol. 272, No. 1, pp. 6–11, ISSN 0006-291X.

Guard-Petter, J. (2001). The chicken, the egg and Salmonella enteritidis. Minireview. *Environ Microbiol*, Vol. 3, No. 7, pp. 421-430, ISSN 14622912.

Guerra, B., Soto, S., Helmuth, R. & Mendoza, M.C. (2002). Characterization of a selftransferable plasmid from Salmonella enterica serotype typhimurium clinical isolates carrying two integron-borne gene cassettes together with virulence and drug resistance genes. *Antimicrob Agents Chemother*, Vol. 46, No. 9, pp. 2977-2981, ISSN 0066-4804.

Guiney, D.G., Libby, S., Fang, F.C., Krause, M. & Fierer, J. (1995). Growth-phase regulation of plasmid virulence genes in Salmonella. *Trends in Microbiology*, Vol. 3, No. 7, pp. 275-279, ISSN 0966-842X.

Gullan, P.J. & Cranston, P.S. (2008). Os insetos: um resumo em entomologia. 3ª. ed. Ed. Roca, ISBN 978-85-7241-702-0, São Paulo.

Haneda, T., Okada, N., Miki, T. & Danbara, H. (2004). Sequence analysis and characterization of sulfonamide resistance plasmid pRF-1 from Salmonella enterica serovar choleraesuis. *Plasmid*, Vol. 52, No. 3, pp. 218-224, ISSN 1095-9890.

Hirose, E. & Panizzi, A.R. (2009). Os simbiontes e a nutrição dos insetos. In: *Bioecologia e nutrição dos insetos*: base para o manejo integrado de pragas, Panizzi, A.R. & Parra, J.R.P. (Eds.), pp. 251-276, Embrapa, ISBN 978-85-7383-452-9, Brasília.

Hirsh, D.C. (2003). *Salmonella*. In: *Microbiologia Veterinária*, Hirsh, D.C. & Zee, Y.C. (Eds.), pp. 69-73, Guanabara Koogan, ISBN 8527707845, Rio de Janeiro.

Jay, J.M. (2005). *Microbiologia de alimentos*. 6ª Ed. Artmed, ISBN 853630507X, Porto Alegre.

Klowden, M.J. & Greenberg, B. (1976). *Salmonella* in the American cockroach: evaluation of vector potential through dosed feeding experiments. *J Hyg*, Vol. 77, pp. 105-111, ISSN 0022-1724.

Klowden, M.J. & Greenberg, B. (1977). Effects of antibiotics on the survival of *Salmonella* in the American cockroach. *J Hyg*, Vol. 79, pp. 339-345, ISSN 0022-1724.

Lamiaa, B., Mariam, L. & Ahmed, A. (2007). Bacteriological analysis of *Periplaneta americana* L. (Dictyoptera; Blattidae) and *Musca domestica* L. (Diptera; Muscidae) in ten districts of Tangier, Morocco. Afr J Biotechnol, Vol. 6, No. 17, (sep 2007), pp. 2038-2042, ISSN 1684-5315.

Le Guyadei, A., Rivault, C. & Chaperon, J. (1989). Microbial organisms carried by brown-banded cockroaches in relation to their spatial distribution in a hospital. *Epidem Inf*, Vol. 102, pp. 485-492, ISSN 0950-2688.

Le Minor, L. (1984). Genus III. *Salmonella* Lignières. In: *Bergey's Manual of Systematic Bacteriology*, Krieg, N.R. & Holt, J.G. (Eds.), pp. 427-458, Williams e Wilkins Co., ISBN 978-0-387-98771-2, Baltimore.

Lewinson, TM; Prado, PI. 2002. Biodiversidade brasileira: síntese do estado atual do conhecimento. São Paulo: Contexto.

Lise, F., Garcia, F.R.M. & Lutinski, J.A. (2006). Association of ants (Hymenoptera: Formicidae) with bacteria in hospitals in the State of Santa Catarina. *Rev Soc Bras Med Trop*, Vol. 39, No. 6, (nov-dez 2006), pp. 6523-526, ISSN 0037-8682.

Loureiro, E.C.B., Marques, N.D.B. & Ramos, F.L.P. (2010). *Salmonella* serovars of human origin identified in Pará State, Brazil from 1991 to 2008. Rev. *Pan-Amaz Saude*, Vol. 1, No. 1, pp. 93-100, ISSN 2176-6223.

Louzada, J.N.C. (2009). Insetos detritívoros. In: *Bioecologia e nutrição dos insetos*: base para o manejo integrado de pragas, Panizzi, A.R. & Parra, J.R.P. (Eds.), pp. 637-666, Embrapa, ISBN 978-85-7383-452-9, Brasília.

Majtan, V., Majtan, T., Majtan, J., Monika, S. & Lubica, M. (2006). *Salmonella enterica* serovar Kentucky: antimicrobial resistance and molecular analysis of clinical isolates from the Slovak Republic. *Jpn J Infect Dis*, Vol. 59, pp. 358-62, ISSN 1344-6304.

Mead, P.S., Slutsker, L., Dietz, V., McCaig, L.F., Bresee, J.S., Shapiro, C., Griffin, P.M. & Tauxe, R.V. (1999). Food-related illness and death in the United States. *Emerg Infect Dis*, Vol. 5, No. 5, pp. 607-625, ISSN 1080-6059.

Mian, L.S., Maag, H. & Tacal, J.V. (2002). Isolation of Salmonella from muscoid flies at commercial animal establishments in San Bernardino County, California. *J Vector Ecol*, Vol. 27, No. 1, pp. 82-85, ISSN 1081-1710.

Miranda, R.A. & Silva, J.P. (2008). Enterobactérias isoladas de *Periplaneta americana* capturadas em um ambiente hospitalar. *Ciência et Praxis*, Vol. 1, No. 1, pp. 21-24, ISSN 1984-5782.

Mølbak, K. (2005). Human health consequences of antimicrobial drugs resistant *Salmonella* and other foodborne pathogens. *J. Food Safety*, Vol. 41, No. 11, (dec 2005), pp. 1613-1620, ISSN 0149-6085.

Mpuchane, S., Matsheka, I.M., Gashe, B.A., Allotey, J., Murindamombe, G. & Mrema, N. (2006). Microbiological studies of cockroaches from three localities in Gaborone, Botswana. *AJFAND on line*, Vol. 6, No. 2, pp. 1-17, ISSN 1684-5358.

Murmann, L., Mallmann, C.A. & Dilkin, P. (2005). Temperaturas de armazenamentos de alimentos em estabelecimentos comerciais na cidade de Santa Maria – RS. *Acta Scientiae Veterinariae*, Vol. 33, No. 3, pp. 309-313, ISSN 16799216.

Oliveira, V.C., Almeida, J.M.D., Abalem de Sá, I.V., Mandarino, J.R. & Solari, C.A. (2006). Enterobactérias associadas a adultos de *Musca domestica* (Linnaeus, 1758) (Diptera: Muscidae) e *Chrysomya megacephala* (Fabricius, 1754) (Diptera: Calliphoridae) no Jardim Zoológico, Rio de Janeiro. *Arq Bras Med Vet Zootec*, Vol. 58, No. 4, pp. 556-561, ISSN 0102-0935.

Olsen, A.R. & Hammack, T.S. (2000). Isolation of *Salmonella* spp. from the housefly, *Musca domestica* L., and the dump fly, *Hydrotaea aenescens* (Wiedemann) (Diptera: Muscidae), at caged-layer houses. *J Food Prot*, Vol. 63, pp. 958-960, ISSN 0362-028X.

Passera, L. 1994. Characteristics of tramp species. P.23-43. In: William, DF (ed.), Exotic ants: Biology, impact and control of introduced species. Westview.

Pesquero, M.A., Elias Filho, J., Carneiro, L.C., Feitosa, S.B., Oliveira, M.A.C. & Quintana, R,C. (2008). Formigas em ambientes hospitalar e seu potencial como transmissores de bactérias. *Neotrop Entomol*, Vol. 37, No. 4, pp. 472-477, ISSN 1519-566X.

Pessoa-Silva, C.L., Toscano, C.M., Moreira, B.M., Santos, A.L., Frota, A.C., Solari, C.A., Amorim, E.L., Da Gloria Carvalho, M.S., Teixeira, L.M. & Jarvis, W.R. (2002). Infection due to extended – spectrum beta lactamases producing *Salmonella enterica* subesp. *enterica* sorovar infantis in a neonatal unit. *J. Pediatr*, Vol. 141, No. 3, pp. 381-387, ISSN 0192-8562.

Pongsiri, M.J. & Roman, J. (2007). Examining the links between Biodiversity and human health: an interdisciplinary research initiative at the US Environmental Protection Agency. *EcoHealth*, Vol. 4, (feb 2007), pp. 82-85, ISSN 1612-9210.

Popoff, M.Y., Bockemuhl, J. & Gheesling, L.L. (2004). To the Kauffmann-White scheme. *Research in Microbiology*, Vol. 46, No. 155, pp. 568-570, ISSN 09232508.

Purcell, A.H. & Almeida, R.P.P. (2004). Insects as vectors of disease agents. In: *Encyclopedia of Plant and Crop Science*, Goodman, R.M. (ed.), pp. 5, Taylor & Francis Group, ISBN 978-0-8247-0943-3, London.

Rodovalho, C.M., Santos, A.L., Marcolino, M.T., Bonetti, A.M. & Brandeburgo, M.A.M. (2007). Urban Ants and Transportation of Nosocomial Bacteria. *Neotrop Entomol*, Vol. 36, No. 3, (may-jun 2007), pp. 454-458, ISSN 1519-566X.

Roudier, C., Fierer, J. & Guiney, D.G. (1992). Characterization of translation termination mutations in the spv operon of the *Salmonella* virulence plasmids psdl2. *Journal of Bacteriology*, Vol. 174, No. 20, pp. 6418-6423, ISSN 0021-9193.

Rusul, G., Khair, J., Radu, S., Cheah, C.T. & Yassin, R.M. (1996). Prevalence of *Salmonella* in broilers a tretail outlets, processing plants and farms in Malaysia. *Int J Food Microbiol*, Vol. 33, No. 2-3, pp. 183-194, ISSN 0168-1605.

Rychlik, I., Sebkova, A., Gregorova, D. & Karpiskova, R. (2001). Low-molecular-weight plasmid of *Salmonella enterica* serovar *enteritidis* codes for retron reverse transcriptase and influences phage resistance. *J. Bacteriol.* Vol. 183, No. 9, pp. 2852-2858, ISSN 1098-5530.

Ridley, A.M., Punia, P., Ward, L.R., Rowe, B. & Threlfall, E.J. (1996). Plasmid characterization and pulsed-field electrophoretic analysis demonstrate that ampicillin-resistant strains of *Salmonella enteritidis* phage type 6a are derived from

S. enteritidis phage type 4. *J. Appl. Bacteriol*, Vol. 81, No. 6, (dec 1996), pp. 613–618, ISSN 0021-8847.

Salehzadeha, A., Tavacolb, P. & Mahjubc, H. (2007). Bacterial, fungal and parasitic contamination of cockroaches in public hospitals of Hamadan, Iran. *J Vect Borne Dis*, Vol. 44, (jun 2007), pp. 105 – 110, ISSN 09729062.

Santos, P.F., Fonseca, A.R. & Sanches, N.M. (2009). Formigas (Hymenoptera: Formicidae) como vetores de bactérias em dois hospitais do município de Divinópolis, Estado de Minas Gerais. *Rev Soc Bras Med Trop*, Vol. 42, No. 5, (sep-oct 2009), pp.565-569, ISSN 0037-8682.

Silva, M.C.D., Ramalho, L.S. & Figueiredo, E.T. (2004). *Salmonella* sp. em ovos e carcaças de frangos "in natura" comercializados em Maceió-AL. *Hig Aliment*, Vol. 18, No. 121, pp. 80-84, ISSN 0101-9171.

Tanaka, I.I., Viggiani, A.M.F.S. & Person, O.C. (2007). Bactérias veiculadas por formigas em ambiente hospitalar. *Arq Med ABC*, Vol. 32, No. 2, (aug 2007), pp. 60-3, ISSN 0100-3992.

Tachbele, E., Erku, W., Gebre-Michael, T. & Ashenafi, M. (2006). Cockroach-associated foodborne bacterial pathogens from some hospitals and restaurants in Addis Ababa, Ethiopia: Distribution and antibiograms. *JRTPH*, Vol. 5, pp. 34-41, ISSN 1447-4778.

Tatfenga, Y.M., Usuanleleb, M.U., Orukpeb, A., Digbana, A.K., Okoduac, M., Oviasogied, F. & Turayc, A.A. (2005). Mechanical transmission of pathogenic organisms: the role of cockroaches. *J Vect Borne Dis*, Vol. 42, pp. 129-134, (dec 2005), ISSN 1058-4838.

Teixeira, M.M., Pelli, A., Santos, V.M. & Reis, M.G. (2009). Microbiota Associated with Tramp Ants in a Brazilian University Hospital. *Neotrop Entomol*, Vol. 38, No. 4, (jul-aug 2009), pp. 537-541, ISSN 1519-566X.

Tortora, G.J., Funke, B.R. & Case, C.L. (2005). *Microbiologia*. 8ª ed. Artmed, ISBN 853630488X, Porto Alegre.

Tosini, F., Visca, P., Dionisi, A.M., Pezzella, C., Petrucca, A. & Carattoli, A. (1998). Class 1 integron-borne multiple antibiotic resistance carried by IncFI and IncL/M plasmids in *Salmonella enterica* serotype *typhimurium*. *Antimicrob Agents Chemother*, Vol. 42, no. 12, pp. 3053-3058, ISSN 0066-4804.

Van der Wolf, P.J., Wolbersa, W.B., Elbersa, A.R.W., Van der Heijdenb, H.M.J.F., Koppena, J.M.C.C., Hunnemana, W.A., Van Schiea, F.W. & Tielena, M.J.M. (2001). Herd level husbandry factors associated with the serological *Salmonella* prevalence in nishing pig herds in The Netherlands. *Vet Microbiol*, Vol.78, No. 3, pp. 205-219, ISSN 0378-1135.

Varnan, A.H. & Evans, M.G. (1991). *Foodborne Pathogens*: an illustrated text. Wolfe Publishing Ltda, ISBN 0723415218, England.

Wannmacher, L. (2004). Uso indiscriminado de antibióticos e resistência microbiana: Uma guerra perdida? *Uso Racional de Medicamentos*: Temas Selecionados, Vol. 1, No. 4, (mar 2004), pp. 1-6, ISSN 1810-0791.

Wilson, I.G. & Whitehead, E. (2004). Long termpost *Salmonella* reactive arthritis dueto *Salmonella* Blockley. *Jpn J Infect Dis*, Vol. 57, pp. 210-211, ISSN 1344-6304.

9

The *Salmonella* Pathogenicity Island-1 and -2 Encoded Type III Secretion Systems

Amanda Wisner, Taseen Desin, Aaron White,
Andrew Potter and Wolfgang Köster
Vaccine and Infectious Disease Organization
University of Saskatchewan
Canada

1. Introduction

1.1 General aspects

Salmonellae are motile, facultatively anaerobic, Gram-negative rods measuring 0.3-1.5 by 1.0-2.5 μm in size. The genus Salmonella was named for Dr. Daniel Salmon, a veterinary bacteriologist at the United States Department of Agriculture (USDA) (Gast, 2003, Salyers & Whitt, 2002). The *Salmonella* species are closely related to *Escherichia*, *Yersinia*, and *Shigella*, and contain a circular chromosome approximately 4.7 Mbp in size with an overall GC content of 52% (Marcus, *et al.*, 2000, Salyers & Whitt, 2002, Thomson, *et al.*, 2008). The genus *Salmonella* lies within the kingdom Eubacteria, class Gammaproteobacteria, order Enterobacteriales, and family Enterobacteriaceae. *Salmonella* is divided into two species, *Salmonella bongori* and *Salmonella enterica*. Within *Salmonella enterica* there are 6 subspecies: *salamae, arizonae, diarizonae, houtenae, indica,* and *enterica* (Tindall, *et al.*, 2005). These subspecies can be further classified into approximately 50 serogroups based on their lipopolysaccharide (LPS) O antigen component (Sabbagh, *et al.*, 2010). *Salmonella enterica* subspecies *enterica* finds its niche in warm-blooded animals and is the primary species associated with human infections. *S. bongori* and other *S. enterica* subspecies are more commonly associated with cold-blooded animals, and in some cases can cause disease in these animals (Brenner, *et al.*, 2000).

Salmonella enterica subspecies *enterica* can be further divided into over 2500 serovars based on their flagellar (H) antigen and LPS O antigen structures (Brenner, *et al.*, 2000, Coburn, *et al.*, 2007, Sabbagh, *et al.*, 2010, Tindall, *et al.*, 2005). For the purposes of this document, serovars within *Salmonella enterica* subspecies *enterica* (e.g., Enteritidis, Typhimurium, and Typhi) will be identified by an italicized S, followed by the serovar name (e.g., *S.* Enteritidis, *S.* Typhimurium, and *S.* Typhi). Many serovars are host-adapted, and tend to cause life-threatening systemic disease in their host. For example, *S.* Typhi and *S.* Paratyphi cause systemic disease in humans and some primates, while *S.* Gallinarum and *S.* Pullorum produce systemic disease in chickens, *S.* Dublin causes systemic disease in cattle, and *S.* Choleraesuis in pigs. In contrast, many serovars are non host-adapted and tend to cause gastroenteritis in many different host species. *S.* Typhimurium and *S.* Enteritidis are the most well-known examples and are able to cause different disease outcomes in various host species (Barrow, 2007, Boyle, *et al.*, 2007, Lax, *et al.*, 1995, Spreng, *et al.*, 2006, Zhang & Mosser, 2008). *S.* Typhimurium and *S.* Enteritidis are able to induce a systemic infection in mice, young calves, chicks, and piglets. However, they are also able to colonize poultry and adult cattle without symptoms (Barrow, 2007, Boyle, *et al.*, 2007, Lax, *et al.*, 1995, Spreng, *et al.*, 2006, Zhang & Mosser, 2008). In humans, infection with either of these serovars results in a self-limiting gastroenteritis (salmonellosis) involving fever, diarrhea, and abdominal pain. In rare cases, typically in the very young or immunocompromised, the infection can become systemic and lead to hospitalization and even death. A very small proportion of humans with salmonellosis can develop reactive arthritis (previously referred to as Reiter's syndrome), which is initially characterized by joint pain, eye irritation, and pain during urination (Boyle, *et al.*, 2007, Cogan & Humphrey, 2003, Townes, 2010).

1.2 Human disease, animal reservoirs, and modes of transmission

Infections by *S. enterica* are one of the most common causes of bacterial food-borne gastroenteritis (food poisoning) in the world, along with *E. coli* and *Campylobacter* infections (WHO, 2007). Of the *S. enterica* serovars, *S.* Enteritidis and *S.* Typhimurium are the leading cause of salmonellosis in humans in most countries. *S.* Enteritidis and *S.* Typhimurium are passed to humans primarily via consumption of contaminated poultry meat, water, and eggs. *S.* Enteritidis is more often associated with salmonellosis acquired from eggs, as it has a greater tendency to colonize eggs and reproductive organs of poultry than *S.* Typhimurium (Gantois, *et al.*, 2008). Because chickens mostly do not show symptoms of disease, entire flocks can become colonized quite quickly and shed bacteria in their feces for extended periods of time (Catarame, *et al.*, 2005, Clavijo, *et al.*, 2006, Penha Filho, *et al.*, 2009, Van Immerseel, *et al.*, 2005). Loss of consumer confidence in products because of *Salmonella* contamination can result in substantial economic loss to the poultry industry. Additionally, human cases of salmonellosis place a significant burden on the health care system (Boyle, *et al.*, 2007). There are approximately 1.4 million cases of salmonellosis per year resulting in about 15,000 hospitalizations and 400 deaths per year in the United States of America (USA) (USDA-ERS, 2009). Around 95% of these cases are caused by consumption of contaminated food products, and *S.* Enteritidis is responsible for at least 15% of these cases. *S.* Enteritidis is the second most commonly isolated serovar in North America after *S.* Typhimurium, while *S.* Enteritidis is number one in the European Union (EU) (Barrow, 2007, Callaway, *et al.*, 2008, Cogan & Humphrey, 2003, Foley & Lynne, 2008, Vieira, 2009).

1.3 Virulence factors

1.3.1 Flagella

Flagella are complex motility structures found in members of Prokarya, Archaea, and Eukarya (Gophna, *et al.*, 2003). The presence of flagella has been associated with virulence in many pathogens, including *Salmonella*, which usually expresses between five and ten flagella at random positions on the cell surface (Parker & Guard-Petter, 2001, van Asten & van Dijk, 2005). However, there is conflicting evidence for the contribution of flagella to virulence in *S.* Enteritidis. Flagellar mutants have been shown to be less proficient in colonizing eggs than wild-type *S.* Enteritidis (Cogan, *et al.*, 2004). In 20-day-old chickens, Parker *et al.* (Parker & Guard-Petter, 2001) observed that disruption of flagella (by deletion of transcriptional regulator FlhD) caused enhanced invasiveness upon oral challenge. Other studies have shown that *S.* Enteritidis strains with deletions in major flagellar genes had decreased adherence to chicken intestinal explants and human intestinal epithelial cell lines, suggesting that flagella are important in adherence of *S.* Enteritidis to intestinal epithelial cells prior to invasion (Allen-Vercoe & Woodward, 1999, Dibb-Fuller, *et al.*, 1999). Allen-Vercoe *et al.* (Allen-Vercoe, *et al.*, 1999) also demonstrated that flagella-defective strains were recovered at lower numbers than the wild-type strain from the spleens and livers of 1-day-old chicks after oral challenge, implicating a role for flagella in invasion. This group also showed that flagellar mutants performed similarly to the wild-type strain in colonization of the ceca of 1- and 5-day-old chickens following oral challenge. However, when mutant strains were given in conjunction with wild-type *S.* Enteritidis in a competition experiment, there was much greater shedding of the wild-type strain than the mutants, suggesting that flagella do provide a competitive survival advantage (Allen-Vercoe & Woodward, 1999).

1.3.2 Fimbriae

Fimbriae, or pili, are typically 2-8 nm in width and extend 0.5-10 μm from the cell surface. Fimbriae play an important role in many bacteria, including biofilm formation and the persistence of bacteria in the environment, as well as contribute to colonization and invasion of the host. Many fimbriae are conserved between the *Salmonella* serovars, while some are unique. As each fimbria is typically specific to a given host receptor, the differences in fimbrial distribution among serovars may contribute to host specificity. There are many known and predicted fimbrial operons in *S.* Enteritidis (Gibson, *et al.*, 2007, Sabbagh, *et al.*, 2010, van Asten & van Dijk, 2005). Since fimbriae are not the subject of this review, detailed description of the different fimbrial types and their proposed roles is not included.

1.3.3 *Salmonella* pathogenicity islands

Pathogenicity islands were first identified in uropathogenic *E. coli* (UPEC) in the late 1980s, and have since been described in a wide variety of bacteria (Blum, *et al.*, 1994, Hacker, *et al.*, 1997, Schmidt & Hensel, 2004). Pathogenicity islands have been identified in both Gram-negative and Gram-positive species, and are associated with plant, animal, and human pathogens, as well as non-pathogenic bacteria. They typically harbour large clusters of genes (10 – 200 kb) related to virulence and/or survival and fitness, and have a different GC content in comparison to the rest of the genome. Pathogenicity islands can often be mosaic in structure and are often bordered by transposon insertion sequences and direct repeats, as well as bacteriophage genes, indicating that their insertion into the genome occurred via single or multiple horizontal gene transfer events (Hacker & Kaper, 2000, Schmidt & Hensel, 2004). To date there have been 21 *Salmonella* pathogenicity islands (SPIs) identified; a brief description of each of these islands is listed in Table 1.

SPI	Size (kb)	Function	Reference
SPI-1	40.2	T3SS - Invasion of the intestinal epithelium	(Marcus, *et al.*, 2000, Thomson, *et al.*, 2008, van Asten & van Dijk, 2005)
SPI-2	39.8	T3SS - Systemic infection of mice, survival in intestinal epithelial cells and macrophages	(Marcus, *et al.*, 2000, Thomson, *et al.*, 2008, van Asten & van Dijk, 2005)
SPI-3	16.6	MgtC and B Mg^{2+} transporter, MisL T5SS Implicated in intramactraphage survival Certain components important S. Typhimurium infection of mice, calves and/or chicks	(Blanc-Potard, *et al.*, 1999, Morgan, *et al.*, 2004, Schmidt & Hensel, 2004, Thomson, *et al.*, 2008, van Asten & van Dijk, 2005)
SPI-4	25.0	T1SS (*siiCDF*) and large non-fimbrial adhsin SiiE Co-regulated with SPI-1 Important for membrane ruffling and entry of polarized epithelial cells in conjunction with the SPI-1 T3SS Implicated in *S.* Typhimurium infection of calves	(Gerlach, *et al.*, 2008, Morgan, *et al.*, 2004, Sabbagh, *et al.*, 2010, Thomson, *et al.*, 2008)
SPI-5	6.6	SPI-1 T3SS effector SopB and its chaperone PipC SPI-2 T3SS effectors PipA and PipB; PipD Important for *S.* Dublin induced enteritis in cattle Important for *S.* Typhimurium systemic infection in chicks	(Marcus, *et al.*, 2000, Morgan, *et al.*, 2004, Thomson, *et al.*, 2008, Tsolis, *et al.*, 1999, van Asten & van Dijk, 2005, Wood, *et al.*, 1998)
SPI-6	17.6	The *saf* finbrial operon of chaperone usher class A T6SS and *tcf* fimbrial operon that are absent in *S.* Enteritidis Up to 44 kb in other serovars	(Blondel, *et al.*, 2009, Sabbagh, *et al.*, 2010, Thomson, *et al.*, 2008, van Asten & van Dijk)
SPI-7	Absent	Vi capsule biosynthetic genes; type IV fimbrial operon SopE in *S.* Typhi Only present in *S.* Typhi, *S.* Paratyphi & some *S.* Dublin Largest PI identified so far, size varies between serovars Up to 134 kb	(Sabbagh, *et al.*, 2010, Seth-Smith, 2008, Thomson, *et al.*, 2008, van Asten & van Dijk, 2005)
SPI-8	Absent	Resistance to bacteriocins Absent in *S.* Enteritidis and *S.* Typhimurium 6 – 8 kb in other serovars	(Sabbagh, *et al.*, 2010, Thomson, *et al.*, 2008, van Asten & van Dijk, 2005)

SPI	Size (kb)	Function	Reference
SPI-9	16.3	T1SS, and a RTX-like protein The RTX protein is complete in *S.* Enteritidis, but not *S.* Typhimurium	(Thomson, *et al.*, 2008, van Asten & van Dijk, 2005)
SPI-10	10.0	*Sef* fimbrial operon in *S.* Enteritidis Larger in other serovars (up to 33 kb)	(Sabbagh, *et al.*, 2010, Thomson, *et al.*, 2008, van Asten & van Dijk, 2005)
SPI-11	6.7	PagC, PagD and MsgA important for survival of *S.* Typhimurium in macropahges	(Sabbagh, *et al.*, 2010, Thomson, *et al.*, 2008)
SPI-12	5.8	SPI-2 T3SS effector sspH2 Important for full virulence of *S.* Typhimurium in mice	(Haneda, *et al.*, 2009, Sabbagh, *et al.*, 2010, Thomson, *et al.*, 2008)
SPI-13	25.3	Important for systemic infection in mice by *S.* Typhimurium	(Haneda, *et al.*, 2009, Shi, *et al.*, 2006, Thomson, *et al.*, 2008)
SPI-14	6.8	Electron transfer and putative regulatory genes	(Sabbagh, *et al.*, 2010, Thomson, *et al.*, 2008)
SPI-15	Absent	5 hypothetical proteins Not present in either *S.* Enteritidis or *S.* Typhimurium	(Sabbagh, *et al.*, 2010, Thomson, *et al.*, 2008)
SPI-16	3.3	LPS modification High homology to SPI-17	(Sabbagh, *et al.*, 2010, Thomson, *et al.*, 2008)
SPI-17	3.6	LPS modification; high homology to SPI-16 Present in *S.* Enteritidis and *S.* Typhi, but not *S.* Typhimurium	(Sabbagh, *et al.*, 2010, Thomson, *et al.*, 2008)
SPI-18	Absent	In *S.* Typhi encodes 2 genes for the cytolysin HlyE and the invasion TaiE Not present in either *S.* Enteritidis or *S.* Typhimurium	(Sabbagh, *et al.*, 2010, Thomson, *et al.*, 2008)
SPI-19	14.1	T6SS likely non-functional in *S.* Enteritidis as most of island has been deleted Up to 45 kb in other serovars	(Blondel, *et al.*, 2009, Thomson, *et al.*, 2008)
SPI-20	Absent	T6SS Only identified in *Salmonella enterica* subsp. *arizonae* 34 kb	(Blondel, *et al.*, 2009, Thomson, *et al.*, 2008)
SPI-21	Absent	T6SS Only identified in *Salmonella enterica* subsp. *arizonae* 55 kb	(Blondel, *et al.*, 2009, Thomson, *et al.*, 2008)

Table 1. *Salmonella* pathogenicity islands

2. Type III secretion systems

Type III secretion systems (T3SSs) act as 'injectisomes' and are used by bacteria to deliver effector proteins directly into the cytoplasm of host cells. The first T3SS was isolated in 1998 from *S.* Typhimurium (Kubori, *et al.*, 1998), although it was was initially thought to be an intermediate complex of the flagellar system (Journet, *et al.*, 2005, Moraes, *et al.*, 2008). These structures have since been detected in numerous pathogenic bacterial species. All T3SSs share significant genetic and protein homology and can be divided into five phylogenetic groups: **1)** the Ysc group (such as the plasmid-encoded T3SSs of *Yersinia* species and *Pseudomonas aeruginosa*); **2)** the Hrp1 group (plant pathogens *Pseudomonas syringae* and *Erwinia* species); **3)** the Hrp2 group (such as the mega-plasmid-encoded T3SSs of the plant bacteria *Ralstonia* and *Xanthomonas* species, and one of the T3SSs of *Burkholderia* species); **4)** the Inv/Mxi/Spa group (the SPI-1,T3SS of *Salmonella enterica*, the chromosomally-encoded T3SS of *Shigella*, the non-functional ETT2 T3SS of enterotoxigenic *Escherichia coli* [ETEC], and the second T3SS of *Burkholderia* species); and **5)** the Esa/Ssa group (the locus of enterocyte effacement [LEE] T3SS of ETEC, the SPI-2 T3SS of *Salmonella enterica*, the chromosomally-encoded T3SS of *Yersinia* species, and the plasmid-encoded T3SS of *Shigella* species) (Foultier, *et al.*, 2002, He, *et al.*, 2004). T3SSs are typically encoded on large pathogenicity islands, located either within the chromosome or on a plasmid. T3SSs are mainly found in pathogenic Gram-negative bacteria; however, there are a few exceptions. For instance, T3SSs have been found in the *Chlamydia/Verrucomivrobia* super-phylum that does not resemble either Gram-negative or Gram-positive bacteria. As well, there are a few examples of non-pathogenic symbiotic bacteria of plants having T3SSs, and even a T3SS used for virulence by unicellular Protozoa (Gophna, *et al.*, 2003, Tampakaki, *et al.*, 2004). Flagella are associated with both pathogenic and non-pathogenic bacteria and are most often not involved in direct virulence as T3SSs are. However, there are cases where the flagellar apparatus is responsible for the secretion of virulence factors. For instance, the flagellar apparatus of *Campylobacter jejuni* is essential for virulence and secretes *Campylobacter* invasion antigens (Cia), and the flagellar system of *Bacillus thuringiensis* can secrete the virulence factors hemolysin BL and phosphatidylcholine-preferring phospholipase C (Journet, *et al.*, 2005).

2.1 Flagella

The flagellum of *Salmonella enterica* is made up of 22 structural proteins, six cytoplasmic proteins, four structural chaperones, and three regulatory proteins (Fig. 1). The structure consists of a C ring (FliG, FliM and FliN) and an MS ring (FlgF and FliF) embedded in the cytoplasmic (inner) membrane. An ATPase is located on the cytoplasmic side of the apparatus (FliI). The P ring (FlgI) is located in the peptidoglycan layer and the L ring (FlgH) is in the outer membrane. A rod spanning the two bacterial membranes, made up of FliF, connects the inner membrane and outer membrane rings; other proteins (FliE, FlgB, FlgC, FlgF, and FlgG) are associated with the basal body. A type three secretion (T3S) apparatus is located within the basal body structure (FliO, FliP, FliQ, FliR, FlhA, FlhB, FliH, and the FliI ATPase). The motor-stator (MotA and MotB), which is the driving force for motion, is also located within the basal body. MotA is located within the inner membrane and connects to MotB, which extends into the periplasm. The FlgE hook protein extends from the L/P rings. FliK acts as a 'molecular ruler' to control the length of the hook. The hook is followed by the hook-filament junction (FlgK and FlgL) and the long filament is comprised of either FliC or FljB flagellin; *Salmonella* encodes both proteins, but they are never expressed at the same time. This differential expression may aid *Salmonella* in escaping the host immune defenses by antigenic variation, and/or contribute to host specificity. Finally, the filament is topped off by the FlgD cap (Aizawa, 2001, Liu & Ochman, 2007, Macnab, 2004, McCann & Guttman, 2008, Morgan, *et al.*, 2004, Pallen & Matzke, 2006, van Asten & van Dijk, 2005).

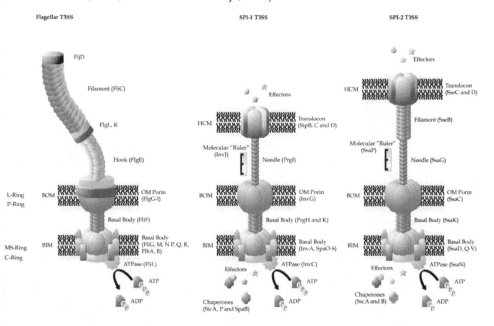

Fig. 1. Schematic representation of the flagellar apparatus, SPI-1, and SPI-2 T3SSs of *Salmonella*. The molecular organization of the flagellar system is depicted above on the left, the SPI-1 T3SS in the middle, and the SPI-2 T3SS on the right. Stoichiometry of proteins was followed where known. Adapted and modified from (Moraes, *et al.*, 2008, Pallen, *et al.*, 2005, Tampakaki, *et al.*, 2004).

2.2 *Salmonella* pathogenicity island-1 type III secretion system

In *S.* Enteritidis, SPI-1 is 40.2 kb in length and has a GC content of 47% (Marcus, *et al.*, 2000, van Asten & van Dijk, 2005). SPI-1 contains 41 genes encoding a T3SS, T3SS regulatory genes, T3SS effectors, and a metal transport system (Fig. 2) (Schmidt & Hensel, 2004, Thomson, *et al.*, 2008). SPI-1 is important for cell invasion of intestinal epithelial cells as well as apoptosis of macrophages (Galán, 2001, Mills, *et al.*, 1995, van der Velden, *et al.*, 2000). *S.* Typhimurium strains defective for InvC (a major structural component of the SPI-1 T3SS) have a 50% higher lethal dose when given orally to Balb/c mice, but perform similarly to wild-type strains when given intraperitoneally, indicating a role for SPI-1 in colonization and invasion during the initial phase of infection, but not during the systemic phase (Galán & Curtiss, 1989).

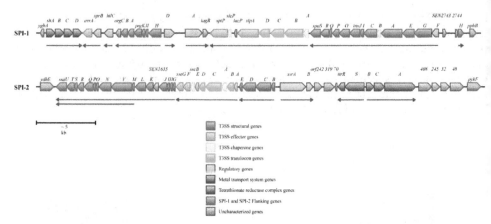

Fig. 2. Genetic organization of *Salmonella* pathogenicity islands 1 and 2. The organization of the ~40kb regions of the *S.* Enteritidis chromosome harbouring SPI-1 and SPI-2 is shown above. Gray arrows represent known or predicted transcriptional units and genes are coloured based on the function of the encoded protein. Based on the published *S.* Enteritidis genome sequence (Thomson, *et al.*, 2008).

2.2.1 Structural components and effectors of the *Salmonella* pathogenicity island-1 type III secretion system

The basal body of the SPI-1 T3SS (Fig. 1) is composed of an inner membrane ring formed by PrgH and PrgK, many inner membrane proteins (SpaP, SpaQ, SpaR, SpaS, and InvA), an ATPase (InvC) and an outer membrane secretin (InvG). Extending from the outer membrane secretin is the needle formed by PrgI, topped by the translocon made up of SipB and SipC (Moraes, *et al.*, 2008). The SPI-1 T3SS is responsible for the secretion of a specific set of effectors. AvrA, SipA, SipB, SipC, SipD, and SptP are all encoded on SPI-1, while the genes encoding GogB, SopE, SopE2, and SspH1 are located on lysogenic bacteriophages in the genome. The gene for SopB is located on SPI-5, and the genes for SopA, SopD, SlrP, SteA, and SteB are located elsewhere within the chromosome. GogB, SlrP, SspH1, SteA, and SteB are also secreted by the SPI-2 T3SS (Abrahams & Hensel, 2006, Bernal-Bayard & Ramos-Morales, 2009, Salyers & Whitt, 2002, Thomson, *et al.*, 2008).

2.2.2 Assembly and regulation of the *Salmonella* pathogenicity island-1 type III secretion system

The assembly of the SPI-1 T3SS proceeds in a similar manner to the assembly of the flagella. The inner membrane and outer membrane rings are formed first in a sec-dependent manner, followed by the association of the rings and formation of the remaining basal body components, including the ATPase. Formation of the needle and translocon is T3S-dependent, and needle length is controlled by InvJ, which acts as a 'molecular ruler' (Deane, *et al.*, 2010, He, *et al.*, 2004, Moraes, *et al.*, 2008).

Expression of the SPI-1 T3SS is regulated by many environmental and genetic signals. Environmental signals include pH, osmolarity, the presence of bile, magnesium concentration, and the presence of short chain fatty acids (Altier, 2005). The preferred invasion site of *Salmonella* is the M-cells of the distal small intestine. When bile is present, indicating the beginning of the small intestine, or when short-chain fatty acids are present, which are produced by microflora of the large intestine, SPI-1 expression is repressed. These environmental signals indicate that the bacterium is not near its preferred site of entry. SPI-1 expression is induced at near neutral pH, and high osmolarity (Altier, 2005, Garmendia, *et al.*, 2003). In the presence of high iron, the ferric uptake regulator (Fur) acts to increase the expression of HilD (a SPI-1 regulator, discussed further in the following text) in an unknown manner. Once in the *Salmonella* containing vacuole, SCV, where there is limited iron, this indirect activation of HilD by Fur is stopped (Altier, 2005, Ellermeier, J. R. & Slauch, 2008). See Fig. 3 for a diagram of the interaction of the regulation pathways outlined below.

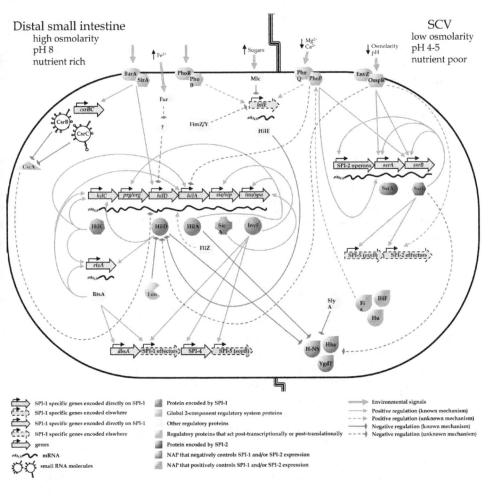

Fig. 3. Regulation of *Salmonella* pathogenicity islands 1 and 2. The major modes of SPI-1 and SPI-2 regulation are depicted above; see text for details (Sections 2.2.2, 2.3.2, and 2.4).

Nucleoid associated proteins (NAPs) affect supercoiling of deoxyribonucleic acid (DNA), and are thus able to alter gene expression. The NAPs Hha and H-NS both repress transcription of many genes, including *rtsA* and the SPI-1 gene *hilA* under conditions of low osmolarity (Altier, 2005, Olekhnovich & Kadner, 2007, Rhen & Dorman, 2005). Hu, IHF, and Fis are also NAPs, and are important for expression of SPI-1 genes (Altier, 2005, Fass & Groisman, 2009).

PhoP/PhoQ and BarA/SirA belong to two-component global regulatory systems that respond to environmental conditions. In low magnesium conditions, for example within the SCV, PhoP can act to negatively regulate HilA, leading to down regulation of the SPI-1 T3SS. SirA positively regulates HilA, by regulating the expression of HilD (Altier, 2005, Ellermeier, J. R. & Slauch, 2007, Hacker & Kaper, 2000, Hueck, 1998). BarA/SirA also controls the *csr* system. CsrA can bind messenger ribonucleic acid (mRNA) at their ribosomal binding site, thus stabilizing, or alternatively, reducing, translation of SPI-1 T3SS proteins, likely at the level of HilD. CsrB and C are small RNA molecules that bind and stop the action of CsrA. BarA/SirA activate CsrB and C, keeping CsrA levels in check. Optimal levels of all three molecules are needed for proper expression of SPI-1. The EnvZ/OmpR system senses osmolarity and has been proposed to regulate HilD post-translationally. The PhoP/PhoQ and PhoR/PhoB systems can activate expression of HilE, which then acts to repress expression of SPI-1 genes, through direct binding to HilD. The type 1 fimbriae regulators FimZ and FimY have also been shown to negatively regulate transcription of SPI-1 genes, likely through activation of *hilE*, while the flagella regulator FliZ positively regulates expression of HilA post-transcriptionally (Altier, 2005, Ellermeier, J. R. & Slauch, 2007). Mlc is a global regulator that detects the presence of sugars such as glucose and mannose, whereby Mlc can repress expression of *hilE* when sugars are readily available, such as in the small intestine (Lim, *et al.*, 2007). The Lon protease (controlled by DnaK and σ^{32}), negatively regulates SPI-1 by degrading HilD in response to the stress of the SCV environment (Matsui, *et al.*, 2008).

HilA belongs to the OmpR/ToxR family of transcriptional regulators, while InvF, HilC, and HilD are in the AraC/XylS family. Each of these genes are located on SPI-1 (Fig. 2) (Hacker & Kaper, 2000, Schmidt & Hensel, 2004). Expression of HilD is likely induced by environmental conditions, and leads to expression of HilC and RtsA. RtsA and HilC can also activate expression of themselves, and each other. RtsA activates *hilA* expression directly, as well as the expression of *slrP*, a SPI-1 T3SS effector, and *dsbA*, which is required for assembly of the T3SS. HilC and D act to derepress transcription of *hilA* and *rtsA* by relieving silencing by H-NS and Hha. HilA is then free to activate transcription of the *prg/org* and *inv/spa* operons (including *invF*). RtsA, HilD, and HilC can also activate transcription of the *inv/spa* operon independently of HilA, but to a lower degree than HilA. InvF activates transcription of the *sic/sip* (including *sicA*) operon of SPI-1, as well as genes within SPI-4 and SPI-5 (Altier, 2005, Ellermeier, J. R. & Slauch, 2007, Hacker & Kaper, 2000, Olekhnovich & Kadner, 2007, Rhen & Dorman, 2005, Schmidt & Hensel, 2004). SicA is the chaperone for the translocator proteins SipB and C. Once the translocon components have been secreted, SicA is free and can activate expression of *invF*, creating a positive feedback loop of secreted effector gene expression once the SPI-1 T3SS is fully formed (He, *et al.*, 2004, Rhen & Dorman, 2005).

2.3 *Salmonella* pathogenicity island-2 type III secretion system

In *S.* Enteritidis, SPI-2 is 39.8 kb in length with a GC content of 43%. SPI-2 is important for survival within the *Salmonella* containing vacuole (SCV) and the systemic phase of infection (Cirillo, *et al.*, 1998, Karasova, *et al.*, 2010, Ochman, *et al.*, 1996, Shea, *et al.*, 1996). There are 44 genes encoded on SPI-2 including a T3SS, T3SS regulatory genes, T3SS effectors, and a tetrathionate reductase system (Fig. 2) (Schmidt & Hensel, 2004, Thomson, *et al.*, 2008).

2.3.1 Structural components and effectors of the *Salmonella* pathogenicity island-2 type III secretion system

The SPI-2 T3SS (Fig. 1) is composed of an inner membrane ring that, in conjunction with many other inner membrane proteins, makes up the basal body. These include SsaD, SsaR, SsaS, SsaT, SsaU, and SsaV, along with the cytoplasmic ATPase SsaN. The outer membrane secretin is made up of SsaC, and is connected to the inner membrane components via SsaJ. A small needle, composed of SsaG, extends from the outer membrane secretin and is extended by a larger filament, made up of SseB. In comparison, many other T3SSs do not have a filament extension. The end of the filament is comprised of the translocon proteins SseC and SseD (Aizawa, 2001, Moraes, *et al.*, 2008, Tampakaki, *et al.*, 2004). SsaP, which acts as a 'molecular ruler', controls the length of the needle (Wilson, 2006).

The SPI-2 T3SS has been shown to secrete many effectors (GogB, PipB, PipB2, SifA, SifB, SopD2, SseF, SlrP, SseG, SseI, SseJ, SseK1, SseK2, SseL, SspH1, SspH2, SteA, SteB, and SteC), although most of their functions are still unknown at this time. Some of the genes encoding these proteins are located directly on the chromosome in the SPI-2 region, but some are located elsewhere on the chromosome, within lysogenic phages (*e.g.* Gifsy-1, -2 and -3) or on the *Salmonella* virulence plasmid. Although GogB, SlrP, SspH1, SteA, and SteB are secreted by the SPI-2 T3SS, these proteins are also known to be secreted by the SPI-1 T3SS (Abrahams & Hensel, 2006, Bernal-Bayard & Ramos-Morales, 2009, Salyers & Whitt, 2002). The functions of these effectors in *Salmonella* pathogenesis are described in section 3.

2.3.2 Assembly and regulation of the *Salmonella* pathogenicity island-2 type III secretion system

Similar to the SPI-1 T3SS and flagellar apparatus, the SPI-2 T3SS is assembled in a step-wise manner involving first the insertion of the inner membrane ring and outer membrane secretin in a sec-dependent manner. Association of the inner membrane and outer membrane rings, placement of further basal body components and recruitment of the ATPase takes place, followed by the subsequent assembly of the rest of the apparatus (Brutinel & Yahr, 2008, Deane, *et al.*, 2010, He, *et al.*, 2004).

Expression of the SPI-2 T3SS is also regulated by many environmental and genetic signals. The preferred replication site of *Salmonella* is within the SCV of macrophages, and environmental signals that mimic the environment of the SCV, such as low magnesium concentration and an acidic pH between 4 and 5, are SPI-2-inducing (Cirillo, *et al.*, 1998, Fass & Groisman, 2009, Rathman, *et al.*, 1996, Rhen & Dorman, 2005). As with the SPI-1 T3SS, expression of SPI-2 genes is affected by the global two-component regulatory systems PhoP/PhoQ and EnvZ/OmpR. Under conditions of low magnesium and calcium PhoP induces SPI-2 gene expression by direct interaction with the *ssrB* gene, and post-transcriptional action on SsrA. In the presence of low osmolarity and acidic pH, OmpR can directly bind both the *ssrA* and *ssrB* promoters, activating transcription. OmpR can also act in conjunction with SsrB to activate transcription of the non-SPI-2-encoded effector SseI (Deiwick, *et al.*, 1998, Fass & Groisman, 2009, Feng, *et al.*, 2003, Garmendia, *et al.*, 2003, Walthers, *et al.*, 2007).

SPI-2 encodes its own two-component regulatory system, SsrA/SsrB. SsrB is able to bind to all SPI-2 promoters, including those of *ssrA*, *ssrB*, and many effectors located outside of SPI-2 (Fass & Groisman, 2009, Walthers, *et al.*, 2007). As with SPI-1, H-NS silences the expression of SPI-2 genes by binding directly to many SPI-2 promoters. This binding can be relieved by the SPI 1 protein HilD under certain conditions, such as stationary phase growth in LB, and may also be relieved by SsrB and/or SlyA (Bustamante, *et al.*, 2008, Fass & Groisman, 2009, Walthers, *et al.*, 2007). The NAPs Hha and YdgT can also repress transcription of SPI 2 genes. Fis, a NAP that is able to bind the promoter regions of *ssr* and *ssa* operons, is also important for expression of SPI-2 as well as SPI-1 genes. Proper levels of Fis are important for activation of *ssrA*, and Fis may also induce SPI-2 gene expression indirectly through controlling expression of PhoP. IHF, another NAP, is also important for expression of both SPI-2 and SPI-1 genes (Fass & Groisman, 2009). Some of the mechanisms controlling regulation of SPI-2 are outlined in Fig. 3.

2.4 Cross-talk between the *Salmonella* flagellar and pathogenicity island-1 and -2 type III secretion systems

The complex regulation of the T3SSs ensures that each system is only expressed under the correct conditions. Expression of multiple versions of each T3SS simultaneously would be energetically expensive, so coordinated expression of the three systems under the specific conditions where each system is required is desirable. Global regulation by two-component regulatory systems that sense divalent cation concentrations, osmolarity, and pH are, in part, responsible for the changes in expression between the flagellar, the SPI-1, and the SPI-2 T3SSs. The SPI-1 T3SS is preferentially expressed within the distal small intestine, which has low oxygen, high osmolarity, a pH of 8, a high concentration of divalent cations, and is rich in nutrients. The environment of the SCV is much different, having low osmolarity, a low concentration of divalent cation concentration, a pH between 4 and 5, and is nutrient poor. In these conditions, the SPI-2 T3SS is preferentially expressed (He, *et al.*, 2004).

The BarA/SirA system positively regulates expression of SPI-1 genes, but negatively regulates expression of flagellar genes. Therefore, in environmental conditions that activate BarA/SirA, the SPI-1 T3SS will be expressed while the flagellar system is downregulated. RtsA and RtsB have also been proposed to be involved in the switch from expression of flagella to expression of the SPI-1 T3SS. RtsA is important for SPI-1 expression, while RtsB represses expression of flagellar genes by interfering with the *flhDC* promoter (Ellermeier, C. D. & Slauch, 2003). In conditions of low divalent cation concentration, PhoP suppresses expression of SPI-1 genes while activating expression of SPI-2 genes. This ensures that once in the SCV, when the SPI-1 T3SS is no longer needed for invasion of non-phagocytic cells, the SPI-2 T3SS expression is induced while the SPI-1 T3SS is downregulated (Rhen & Dorman, 2005).

Interspecies and interkingdom quorum sensing may also be involved in regulating expression of these three systems. In the presence of host norepinephrine there is an upregulation of flagellar genes in *S.* Typhimurium (Bearson & Bearson, 2008). *S.* Typhimurium encodes a putative regulatory protein, YhcS, which has high amino acid similarity to QseA of *E. coli.* QseA activates expression of the LEE T3SS by *E. coli* in response to autoinducer 3 (AI-3) quorum sensing molecules produced by intestinal flora, as well as epinephrine and norepinephrine produced by the host. YhcS may act similarly to QseA in *E. coli* by activating expression of either (or both of) the SPI-1 or SPI-2 T3SSs (Bearson & Bearson, 2008, Choi, *et al.*, 2007, Karavolos, *et al.*, 2008). As mentioned previously, under certain growth conditions HilD can relieve H-NS-mediated repression of SPI-2 genes (Bustamante, *et al.*, 2008). This may account for the fact that SPI-2 is expressed to some extent along with SPI-1 in the intestinal lumen, and that SPI-1 is expressed for a short time in macrophages before the complete switch to SPI-2 expression. The expression of the SPI-2 T3SS before invasion of intestinal epithelial cells would allow the bacteria to ready itself for the SCV environment. Furthermore, the expression of the SPI-1 T3SS is important for inducing macrophage apoptosis during the initial stage of infection while the bacteria is replicating, and before spread to the rest of the body. Some of the interplay between regulation of SPI-1 and SPI-2 can be visualized in Fig. 3.

2.5 Evolution of the type III secretion system

The flagellar systems of Prokarya are completely different from those of Archaea and Eukarya, suggesting that they evolved convergently into structures serving the same function (Gophna, *et al.*, 2003, Liu & Ochman, 2007). However, prokaryotic flagellar systems with a chemotaxis apparatus that controls changes in the direction of motion share their chemotaxis system with archaeal flagellar systems (Liu & Ochman, 2007, Pallen & Matzke, 2006). As some members of Prokarya do not have this chemotaxis system, it may have been acquired by horizontal transfer from a member of Archaea or may have been present for sensing environmental signals before the diversification of Prokarya and Archaea, and has since been lost in some prokaryotic families. While the flagellar systems of Prokarya maintain many of the same genes and proteins among members, they can be quite diverse in their function. For instance, the flagella of *Spirochaetes* are located in the periplasm, while *Vibrio* species express both polar and lateral flagellar systems that share a chemotaxis transduction system but use different motive forces (Na^+ or H^+). Most of the flagellar proteins that serve the same function are homologous, however, not all flagellar system proteins are conserved among all bacterial species. For example, the flagellar structures of Gram-positive bacteria do not have the L and P rings (which would be located in the outer membrane of Gram-negative bacteria). *Spirochaetes* do not have the L and P ring either, as their flagella are located in the periplasm. Some of the structural genes (*flgH, flgI, fliD, fliE,* and *fliH* specifically) are missing in some bacteria; this could indicate a later evolution of these genes combined with limited horizontal transfer, or be an example of sporadic loss of genes from some bacterial families. The latter explanation seems more likely in this case as there are many families of bacteria that contain these genes, and only a few who are lacking (Liu & Ochman, 2007).

The flagella phylogenetic tree is directly related to that of the bacterial speciation genetic tree based on 16S ribosomal RNA. This suggests that flagella have been in existence since before the diversification of bacteria, and have been maintained throughout vertical evolution (McCann & Guttman, 2008). Liu and Ochman propose that the entire flagellar system is actually evolved from a single gene. They suggest, based on sequence similarities, that all of the flagellar genes arose from random duplications and reassortments of a single precursor gene in the ancestor of modern bacteria (Liu & Ochman, 2007). This seems quite unlikely; although there may be sequence similarities between an inner membrane component and an outer membrane component, this does not mean that they are related on an evolutionary scale. Convergent evolution is a more likely explanation for this, in which two different proteins have evolved to serve a similar function, in this case to be embedded in the bacterial membrane.

Unlike flagellar systems, the T3SS phylogenetic tree is not related to that of 16S ribosomal RNA, suggesting that T3SSs were acquired at some point after the diversification of bacteria, and evolved via horizontal transfer events (Foultier, et al., 2002, Gophna, et al., 2003, Liu & Ochman, 2007, Nguyen, et al., 2000). T3SSs are encoded on large pathogenicity islands, while flagellar genes are encoded on the chromosome (Hueck, 1998, Macnab, 2004, van Asten & van Dijk, 2005). It is thought that SPI-2 may have arrived in two separate events, with the *ttr* operon arriving first, followed by the rest of SPI-2 (Marcus, et al., 2000).

The effectors of T3SSs are highly variable between species of bacteria, and are quite often encoded on different regions of the chromosome than the pathogenicity island-encoded T3SSs. The effectors and their evolution will not be discussed here, but information on this topic can be found in a recent review (Stavrinides, et al., 2008). In general, there are about ten core proteins of the flagellar T3S apparatus and the injectisome T3SSs that are highly similar in gene sequence, amino acid sequence, and function (Fig. 1.). For the purposes of this discussion, the flagellar system will be compared only with the two *Salmonella* T3SSs, with homologous proteins given in the order flagella/SPI-1/SPI-2. These homologous proteins are: the cytoplasmic ATPase (FliI/InvC/SsaN), the T3S apparatus (FliH/PrgH/SsaK, FliN/SpaO/SsaQ, FliP/SpaP/SsaR, FliQ/?/SsaS, FliR/SpaR/SsaT, FlhB/SpaS/SsaU and FlhA/InvA/SsaV), part of the connecting rod (FliF/PrgK/SsaJ), and the needle/hook 'molecular ruler' (FliK/InvJ/SsaP) (Blocker, et al., 2003, Desvaux, et al., 2006, He, et al., 2004, Tampakaki, et al., 2004, Wilson, 2006).

The structure of the flagellar apparatus and T3SSs begin to differ more markedly starting at the outer membrane (besides the motor-stator which is only present in the basal body of the flagellar system). The MS ring of the flagellar system is larger than that of the outer membrane secretin of the T3SS (Aizawa, 2001). The secretin of the T3SS belongs to the same family of proteins that make up the T2SS and T4SS secretins, and the pore used by filamentous phages, suggesting that filamentous phages either introduced this type of protein to bacteria, or acquired it from them (Nguyen, et al., 2000). The T3SS needle is straight and thin, as is its filament, although the filament is slightly larger, and notably rigid. The flagellar hook apparatus is larger and curved, and its filament is quite long and flexible. These structures lack significant amino acid and genetic homology, but do share helical symmetry, and overall assembly mechanisms. The flagellum contains approximately 5.6 subunits of flagellin per turn, with an axial rise of 4.7 Å. To compare, the filament of the LEE T3SS in *E. coli* contains 5.5 subunits of EspA per turn, and has an axial rise 4.6 Å (Aizawa, 2001, Journet, et al., 2005, Snyder, et al., 2009, Tampakaki, et al., 2004). The inner diameter of the T3SS filament is between 2 and 3 nm, similar to the inner channel of the flagellum (Blocker, *et al.*, 2003, Journet, *et al.*, 2005, Tampakaki, *et al.*, 2004). The action of the 'molecular rulers' is likely different as well. It has been proposed that the method for measuring hook length in flagella is more 'measuring cup'-like than 'molecular ruler'-like. Journet suggests that the motor-stator switch area of the flagellum acts as a measuring cup, filling with FliK. FliK acts as an accessory to the hook protein, and is secreted at the same time. Once the 'cup' empties of FliK, the apparatus switches its secretion preference from the hook protein (FlgE) to the flagellin protein (FliC or FljB), completing assembly of the flagellum. In contrast, the InvJ and SsaP proteins (similar to the 'molecular rulers' of other bacterial T3SSs) act more like a ruler. It has been suggested that dimers of these proteins are located outside the cell, with one attached to the outer membrane, and the second extending from that. Once the needle complex (PrgI/SsaG) reaches the same height as the InvJ/SsaP dimer, the T3SS switches to secretion and assembly of the translocon of SPI-1 (SipB, C, and D) or filament of SPI-2 (SseB) (Journet, *et al.*, 2005).

Another key area in which the T3SSs and flagellar systems differ is in their chaperones. Although both systems tend to have specific chaperones for specific proteins, the T3SS proteins are recognized by their chaperones at an N-terminal region, while flagellar system chaperones bind at the C-terminal region (Liu & Ochman, 2007). Although the flagellar and T3SS chaperones are different, in some cases the three systems can secrete each other's proteins. For example, both the SPI-1 and SPI-2 T3SSs can secrete the flagellar protein FliC, while the flagellar system can secrete the SPI-1 T3SS effector proteins SptP and SopE, if the SptP and SopE chaperones are absent, and in some instances effectors from T3SSs of other bacterial species (Journet, *et al.*, 2005, Tampakaki, *et al.*, 2004).

3. Pathogenesis of *Salmonella*

Salmonella can enter host cells in at least two ways. The first involves uptake into phagocytic cells (macrophages), while the second is more complicated and involves the action of the SPI-1 T3SS on non-phagocytic cells. After attachment to epithelial cells, the SPI-1 T3SS induces membrane ruffling by secreting effectors into the host cell to trigger cytoskeleton rearrangement. Once inside the epithelial cell, some of these same effectors 'switch off' the membrane ruffling, returning the host cell membrane to its original state (Ibarra & Steele-Mortimer, 2009, Ly & Casanova, 2007, Salyers & Whitt, 2002, Waterman & Holden, 2003). Entry into the host cell (epithelial or macrophage) results in the bacteria being encased within an SCV. While the goal of many intracellular pathogens would be to escape this vacuolar space into the cell cytoplasm, *Salmonella* takes advantage of this space and remains in the SCV (Bhavsar, *et al.*, 2007, Ibarra & Steele-Mortimer, 2009, Salyers & Whitt, 2002).

Once inside the SCV, the SPI-2 T3SS is expressed and begins secreting effector proteins, which are used to manipulate the intracellular environment (Cirillo, *et al.*, 1998, Ibarra & Steele-Mortimer, 2009, Ramsden, *et al.*, 2007). Approximately one hour after entry into the host cell, the SCV switches from early endosomal markers, such as early endosome marker 1 (EE-1), to late endosomal/lysosomal markers, such as lysosomal-associated membrane protein-1 (LAMP-1) and lysosomal glycoproteins (lgps). One important factor that the SCV acquires during this switch is the V-ATPase, which facilitates the acidification of the SCV.

This acidification is an important factor for the induction of *Salmonella* virulence/survival genes (Abrahams & Hensel, 2006, Bhavsar, *et al.*, 2007, Ibarra & Steele-Mortimer, 2009, Kuhle & Hensel, 2004, Ramsden, *et al.*, 2007, Salyers & Whitt, 2002). Another important factor for *Salmonella* survival within host cells is iron acquisition. *Salmonella* releases two siderophores for sequestering Fe^{2+} from the host cell, enterobactin and salmochelin (Ibarra & Steele-Mortimer, 2009). As the SCV matures, it moves along host cell microtubules towards the Golgi apparatus. This process is dependent on many effectors, including SifA, SifB, SopD2, SseF, SseG, SseI, SseJ, SseL, PipB, and PipB2 (Abrahams & Hensel, 2006, Ibarra & Steele-Mortimer, 2009, Ramsden, *et al.*, 2007). SsaB is also important in blocking the fusion of the SCV with lysosomes during this process, which would result in bacterial killing (Kuhle & Hensel, 2004). Movement along the microtubules involves recruitment of a dynein-dynactin motor complex by SifA, SseF, SseG, and PipB2. PipB2 interacts with the motor protein kinesin, while the other three proteins have also been shown to be responsible for keeping the SCV localized to the Golgi apparatus in an unknown manner. These proteins are also very important in *Salmonella*-induced filament (sif) formation, which will be discussed in the following paragraph. SCV membrane integrity is important, and is controlled by a number of SPI-2 T3SS effectors, including SspH2, SseI, SteC, and the *Salmonella* virulence plasmid-encoded protein SpvB. The interaction of these proteins with host filamen and actin causes the formation of an actin-mesh around the SCV (Abrahams & Hensel, 2006, Kuhle & Hensel, 2004, Ramsden, *et al.*, 2007). Another function of the SPI-2 T3SS may be to stop the formation of the NADPH phagocytic oxidase (phox) and inducible nitric oxide synthase (iNOS) on the SCV membrane, ultimately resulting in protection of *Salmonella* from reactive oxygen and nitrogen species (ROS and RNS, respectively) (Abrahams & Hensel, 2006, Coburn, *et al.*, 2005, Salyers & Whitt, 2002). A superoxide dismutase encoded by the Gifsy-2 lysogenic phage helps *Salmonella* survive the oxidative burst, which involves production of ROS and RNS by phagocytic cells that can damage bacteria, and is therefore important for bacterial survival within the SCV (Ibarra & Steele-Mortimer, 2009, Salyers & Whitt, 2002).

The maturation/movement process of the SCV can take around 4 to 6 hours. At this point, when the SCV has been altered to suit the bacteria, *Salmonella* begin to replicate (Abrahams & Hensel, 2006, Finlay & Brumell, 2000). Replication of *Salmonella* is associated with the formation of sifs. Sifs have similar markers to the SCV, and many of the same proteins are responsible for their formation/membrane integrity (SifA, SifB, SseF, SseG, SseJ, SseL, SspH2, SpvB, PipB, and PipB2). The SPI-1 effector SipA has also been shown to be important in sif formation. These sifs extend from the SCV towards the host cell membrane, and other SCVs, if there are multiple SCVs in one cell (Ibarra & Steele-Mortimer, 2009, Kuhle & Hensel, 2004, Ramsden, *et al.*, 2007). The AvrA effector secreted by the SPI-1 T3SS deubiquitinates both IκB-α and β-catenin, which stabilizes the proteins and results in the continued repression of NFκB-mediated gene transcription. This delays apoptosis of intestinal epithelial cells, thereby allowing *Salmonella* to survive within them for longer (Bernal-Bayard & Ramos-Morales, 2009, Bhavsar, *et al.*, 2007, Grassl & Finlay, 2008, Ibarra & Steele-Mortimer, 2009). SlrP also mediates ubiquitination of certain host proteins including Thioredoxin-1 (Trx1). Trx1 can activate the NFκB transcription factor, and has functions among other host cell proteins as well. Binding of SlrP to Trx1 stops its action, which under some conditions can lead to apoptotic cell death, although the exact mechanisms of this need to be studied further (Bernal-Bayard & Ramos-Morales, 2009, Bhavsar, *et al.*, 2007,

Ramsden, *et al.*, 2007). SspH1 can also inhibit NFκB transcription (Ibarra & Steele-Mortimer, 2009, Kuhle & Hensel, 2004, Ramsden, *et al.*, 2007).

3.1 Role of the *Salmonella* pathogenicity island-1 and -2 type III secretion systems in the chicken model of infection

Contaminated poultry and eggs remain a major source of food poisoning caused by *Salmonella*. As the majority of work on *Salmonella* to date has been done in mice, we examined the role of this bacterium in a chicken model of *S*. Enteritidis infection developed using select strains isolated from chickens. We were particularly interested in the role of the SPI-1 and SPI-2 encoded T3SSs of *S*. Enteritidis in the various stages of infection and colonization of birds, as the SPI-1 encoded T3SS has been found to be important for entry into intestinal epithelial cells, and the SPI-2 encoded T3SS has been found to be highly important in the later stages of infection in mice, particularly after *Salmonella* has been vacuolised. To examine this, chickens were challenged orally with 10^6-10^9 *Salmonella* bacteria. In co-challenge trials using 35-day-old chickens we showed that although a *S*. Enteriditis wild-type strain was only slightly more competitive in colonizing the ceca than mutants defective in SPI-1 and SPI-2, the systemic spread of both of these mutant strains to the liver and spleen was significantly less successful than that of the wild-type strain (Desin, *et al.*, 2009, Wisner, *et al.*, 2010). Colonization of the gut was nearly 100% for both wild-type and mutants, whereas *Salmonella* was detected in the liver and spleen in approximately 30% of the birds in these trials. In order to acheive a higher percentage of chickens that were systemically infected, we performed colonization experiments with younger leghorn chickens hatched from specific pathogen-free (SPF) eggs and orally challenged 7 days post-hatch. With respect to colonization of the ceca we found no statistically relevant differences between the wild-type and SPI-1 or SPI-2 mutant strains. However, although 100% of the younger birds developed a systemic infection when challenged with the wild-type strain, challenges with mutant strains devoid of functional SPI-1 and/or SPI-2 T3SSs resulted in a clearly delayed, and less severe, systemic infection. Interestingly, at the end of the test period, 4 days post-challenge, the systemic presence of both the mutant and the wild-type strains was found to be decreasing. From these findings, it is evident that both the SPI-1 and SPI-2 pathogenicity islands are important for the fast and efficient invasion and systemic spread of *Salmonella* in chickens. However, the data obtained 3 and 4 days post-challenge indicates that SPI-1 and SPI-2 are not the only factors needed for systemic dissemination, and that other virulence factors may compensate for the loss of SPI-1 and SPI-2.

4. Concluding remarks

Most of the animal studies in the context of bacterial T3SS-related pathogenicity have been performed with *S*. *Typhimurium* in mice. Based on our findings, however, we see that not all the results obtained from those experiments can be directly transposed to other hosts. Our *in vitro* and *in vivo* results based on a chicken model of *S*. *Enteritidis* infection demonstrated that both SPI-1 and SPI-2 play an important role in invasion and systemic spread, although they do not seem to be essential. Further studies will be necessary to understand the full spectrum of virulence mechanisms of different *Salmonella* strains in various host organisms, including humans.

5. Acknowledgements

Projects in the authors laboratories related to this review were supported through the Industrial Research Chair (IRC) program of the Natural Sciences and Engineering Research Council of Canada (NSERC), the Saskatchewan Health Research Foundation (SHRF) and the Poultry Industry Council (PIC).

6. References

Abrahams, G. L. & Hensel, M. (2006). Manipulating cellular transport and immune responses: dynamic interactions between intracellular *Salmonella enterica* and its host cells. *Cellular Microbiology*, Vol. 8, No. 5, (May), pp. 728-737, ISSN 1462-5814

Aizawa, S. I. (2001). Bacterial flagella and type III secretion systems. *FEMS Microbiology Letters*, Vol. 202, No. 2, (Aug 21), pp. 157-164, ISSN 0378-1097

Allen-Vercoe, E., Sayers, A. R. & Woodward, M. J. (1999). Virulence of *Salmonella enterica* serotype Enteritidis aflagellate and afimbriate mutants in a day-old chick model. *Epidemiology and Infection*, Vol. 122, No. 3, (Jun), pp. 395-402, ISSN 0950-2688

Allen-Vercoe, E. & Woodward, M. J. (1999). Colonisation of the chicken caecum by afimbriate and aflagellate derivatives of *Salmonella enterica* serotype Enteritidis. *Veterinary Microbiology*, Vol. 69, No. 4, (Sep 29), pp. 265-275, ISSN 0378-1135

Allen-Vercoe, E. & Woodward, M. J. (1999). The role of flagella, but not fimbriae, in the adherence of *Salmonella enterica* serotype Enteritidis to chick gut explant. *Journal of Medical Microbiology*, Vol. 48, No. 8, (Aug), pp. 771-780, ISSN 0022-2615

Altier, C. (2005). Genetic and environmental control of *Salmonella* invasion. *Journal of microbiology*, Vol. 43, No. Special, (Feb), pp. 85-92, ISSN 1225-8873

Barrow, P. A. (2007). *Salmonella* infections: immune and non-immune protection with vaccines. *Avian Pathology*, Vol. 36, No. 1, (Feb), pp. 1-13, ISSN 0307-9457

Bearson, B. L. & Bearson, S. M. (2008). The role of the QseC quorum-sensing sensor kinase in colonization and norepinephrine-enhanced motility of *Salmonella enterica* serovar Typhimurium. *Microbial Pathogenesis*, Vol. 44, No. 4, (Apr), pp. 271-278, ISSN 0882-4010

Bernal-Bayard, J. & Ramos-Morales, F. (2009). *Salmonella* type III secretion effector SlrP is an E3 ubiquitin ligase for mammalian thioredoxin. *Journal of Biological Chemistry*, Vol. 284, No. 40, (Oct 2), pp. 27587-27595, ISSN 0021-9258

Bhavsar, A. P., Guttman, J. A. & Finlay, B. B. (2007). Manipulation of host-cell pathways by bacterial pathogens. *Nature*, Vol. 449, No. 7164, (Oct 18), pp. 827-834, ISSN 1476-4687

Blanc-Potard, A. B., Solomon, F., Kayser, J. & Groisman, E. A. (1999). The SPI-3 pathogenicity island of *Salmonella enterica*. *Journal of Bacteriology*, Vol. 181, No. 3, (Feb), pp. 998-1004, ISSN 0021-9193

Blocker, A., Komoriya, K. & Aizawa, S. (2003). Type III secretion systems and bacterial flagella: insights into their function from structural similarities. *Proceedings of the National Acadamy of Sciences*, Vol. 100, No. 6, (Mar 18), pp. 3027-3030, ISSN 0027-8424

Blondel, C. J., Jimenez, J. C., Contreras, I. & Santiviago, C. A. (2009). Comparative genomic analysis uncovers 3 novel loci encoding type six secretion systems differentially distributed in *Salmonella* serotypes. *BMC Genomics*, Vol. 10, No. pp. 354, ISSN 1471-2164

Blum, G., Ott, M., Lischewski, A., Ritter, A., Imrich, H., Tschape, H. & Hacker, J. (1994). Excision of large DNA regions termed pathogenicity islands from tRNA-specific loci in the chromosome of an *Escherichia coli* wild-type pathogen. *Infection and Immunity*, Vol. 62, No. 2, (Feb), pp. 606-614, ISSN 0019-9567

Boyle, E. C., Bishop, J. L., Grassl, G. A. & Finlay, B. B. (2007). *Salmonella*: from pathogenesis to therapeutics. *Journal of Bacteriology*, Vol. 189, No. 5, (Mar), pp. 1489-1495, ISSN 0021-9193

Brenner, F. W., Villar, R. G., Angulo, F. J., Tauxe, R. & Swaminathan, B. (2000). *Salmonella* nomenclature. *Journal of Clinical Microbiology*, Vol. 38, No. 7, (Jul), pp. 2465-2467, ISSN 0095-1137

Brutinel, E. D. & Yahr, T. L. (2008). Control of gene expression by type III secretory activity. *Current Opinion in Microbiology*, Vol. 11, No. 2, (Apr), pp. 128-133, ISSN 1369-5274

Bustamante, V. H., Martinez, L. C., Santana, F. J., Knodler, L. A., Steele-Mortimer, O. & Puente, J. L. (2008). HilD-mediated transcriptional cross-talk between SPI-1 and SPI-2. *Proceedings of the National Acadamy of Sciences*, Vol. 105, No. 38, (Sep 23), pp. 14591-14596, ISSN 0027-8424

Callaway, T. R., Edrington, T. S., Anderson, R. C., Byrd, J. A. & Nisbet, D. J. (2008). Gastrointestinal microbial ecology and the safety of our food supply as related to *Salmonella*. *Journal of Animal Science*, Vol. 86, No. 14 Suppl, (Apr), pp. E163-172, ISSN 1525-3163

Catarame, T. M. G., O'Hanlon, K. A., McDowell, D. A., Blair, I. S. & Duffy, G. (2005). Comparison of a real-time polymerase chain reaction assay with a culture method for the detection of *Salmonella* in retail meat samples. *Journal of Food Safety*, Vol. 26, No. 5, pp. 1 - 15, ISSN 1745-4565

Choi, J., Shin, D. & Ryu, S. (2007). Implication of quorum sensing in *Salmonella enterica* serovar Typhimurium virulence: the luxS gene is necessary for expression of genes in pathogenicity island 1. *Infection and Immunity*, Vol. 75, No. 10, (Oct), pp. 4885-4890, ISSN 0019-9567

Cirillo, D. M., Valdivia, R. H., Monack, D. M. & Falkow, S. (1998). Macrophage-dependent induction of the *Salmonella* pathogenicity island 2 type III secretion system and its role in intracellular survival. *Molecular Microbiology*, Vol. 30, No. 1, (Oct), pp. 175-188, ISSN 0950-382X

Clavijo, R. I., Loui, C., Andersen, G. L., Riley, L. W. & Lu, S. (2006). Identification of genes associated with survival of *Salmonella enterica* serovar Enteritidis in chicken egg albumen. *Applied and Environmental Microbiology*, Vol. 72, No. 2, (Feb), pp. 1055-1064, ISSN 0099-2240

Coburn, B., Grassl, G. A. & Finlay, B. B. (2007). *Salmonella*, the host and disease: a brief review. *Immunology and cell biology*, Vol. 85, No. 2, (Feb-Mar), pp. 112-118, ISSN 0818-9641

Coburn, B., Li, Y., Owen, D., Vallance, B. A. & Finlay, B. B. (2005). *Salmonella enterica* serovar Typhimurium pathogenicity island 2 is necessary for complete virulence in a mouse model of infectious enterocolitis. *Infection and Immunity*, Vol. 73, No. 6, (Jun), pp. 3219-3227, ISSN 0019-9567

Cogan, T. A. & Humphrey, T. J. (2003). The rise and fall of *Salmonella* Enteritidis in the UK. *Journal of Applied Microbiology*, Vol. 94, No. Supplement S1, pp. 114S-119S, ISSN 1364-5072

Cogan, T. A., Jorgensen, F., Lappin-Scott, H. M., Benson, C. E., Woodward, M. J. & Humphrey, T. J. (2004). Flagella and curli fimbriae are important for the growth of *Salmonella enterica* serovars in hen eggs. *Microbiology*, Vol. 150, No. 4, (Apr), pp. 1063-1071, ISSN 1350-0872

Deane, J. E., Abrusci, P., Johnson, S. & Lea, S. M. (2010). Timing is everything: the regulation of type III secretion. *Cellular and Molecular Life Sciences*, Vol. 67, No. 7, (Apr), pp. 1065-1075, ISSN 1420-9071

Deiwick, J., Nikolaus, T., Shea, J. E., Gleeson, C., Holden, D. W. & Hensel, M. (1998). Mutations in *Salmonella* pathogenicity island 2 (SPI2) genes affecting transcription of SPI1 genes and resistance to antimicrobial agents. *Journal of Bacteriology*, Vol. 180, No. 18, (Sep), pp. 4775-4780, ISSN 0021-9193

Desin, T. S., Lam, P. K., Koch, B., Mickael, C., Berberov, E., Wisner, A. L., Townsend, H. G., Potter, A. A. & Köster, W. (2009). *Salmonella enterica* serovar Enteritidis pathogenicity island 1 is not essential for but facilitates rapid systemic spread in chickens. *Infection and Immunity*, Vol. 77, No. 7, (Jul), pp. 2866-2875, ISSN 0019-9567

Desvaux, M., Hebraud, M., Henderson, I. R. & Pallen, M. J. (2006). Type III secretion: what's in a name? *Trends in Microbiology*, Vol. No. (Mar 11), pp. ISSN 0966-842X

Dibb-Fuller, M. P., Allen-Vercoe, E., Thorns, C. J. & Woodward, M. J. (1999). Fimbriae- and flagella-mediated association with and invasion of cultured epithelial cells by *Salmonella enteritidis*. *Microbiology*, Vol. 145 (Pt 5), No. (May), pp. 1023-1031, ISSN 1350-0872

Ellermeier, C. D. & Slauch, J. M. (2003). RtsA and RtsB coordinately regulate expression of the invasion and flagellar genes in *Salmonella enterica* serovar Typhimurium. *Journal of Bacteriology*, Vol. 185, No. 17, (Sep), pp. 5096-5108, ISSN 0021-9193

Ellermeier, J. R. & Slauch, J. M. (2007). Adaptation to the host environment: regulation of the SPI1 type III secretion system in *Salmonella enterica* serovar Typhimurium. *Current Opinion in Microbiology*, Vol. 10, No. 1, (Feb), pp. 24-29, ISSN 1369-5274

Ellermeier, J. R. & Slauch, J. M. (2008). Fur regulates expression of the *Salmonella* pathogenicity island 1 type III secretion system through HilD. *Journal of Bacteriology*, Vol. 190, No. 2, (Jan), pp. 476-486, ISSN 1098-5530

Fass, E. & Groisman, E. A. (2009). Control of *Salmonella* pathogenicity island-2 gene expression. *Current Opinion in Microbiology*, Vol. 12, No. 2, (Apr), pp. 199-204, ISSN 1369-5274

Feng, X., Oropeza, R. & Kenney, L. J. (2003). Dual regulation by phospho-OmpR of *ssrA/B* gene expression in *Salmonella* pathogenicity island 2. *Molecular Microbiology*, Vol. 48, No. 4, (May), pp. 1131-1143, ISSN 0950-382X

Finlay, B. B. & Brumell, J. H. (2000). *Salmonella* interactions with host cells: in vitro to in vivo. *Philosophical Transactions of the Royal Society B: Biological Sciences*, Vol. 355, No. 1397, (May 29), pp. 623-631, ISSN 0962-8436

Foley, S. L. & Lynne, A. M. (2008). Food animal-associated *Salmonella* challenges: pathogenicity and antimicrobial resistance. *Journal of Animal Science*, Vol. 86, No. 14 Supplemental, (Apr), pp. E173-187, ISSN 1525-3163

Foultier, B., Troisfontaines, P., Muller, S., Opperdoes, F. R. & Cornelis, G. R. (2002). Characterization of the ysa pathogenicity locus in the chromosome of Yersinia enterocolitica and phylogeny analysis of type III secretion systems. *Journal of Molecular Evolution*, Vol. 55, No. 1, (Jul), pp. 37-51, ISSN 0022-2844

Galán, J. E. (2001). *Salmonella* interactions with host cells: type III secretion at work. *Annual Review of Cell and Developmental Biology*, Vol. 17, No. pp. 53-86, ISSN 0022-2844

Galán, J. E. & Curtiss, R., 3rd. (1989). Cloning and molecular characterization of genes whose products allow *Salmonella typhimurium* to penetrate tissue culture cells. *Proceedings of the National Acadamy of Sciences*, Vol. 86, No. 16, (Aug), pp. 6383-6387, ISSN 0027-8424

Gantois, I., Eeckhaut, V., Pasmans, F., Haesebrouck, F., Ducatelle, R. & Van Immerseel, F. (2008). A comparative study on the pathogenesis of egg contamination by different serotypes of *Salmonella*. *Avian Pathology*, Vol. 37, No. 4, (Aug), pp. 399-406, ISSN 0307-9457

Garmendia, J., Beuzon, C. R., Ruiz-Albert, J. & Holden, D. W. (2003). The roles of SsrA-SsrB and OmpR-EnvZ in the regulation of genes encoding the *Salmonella typhimurium* SPI-2 type III secretion system. *Microbiology*, Vol. 149, No. 9, (Sep), pp. 2385-2396, ISSN 1350-0872

Gast, R. K. (2003). Chapter 16: *Salmonella* Infections, In: *Diseases of Poultry*, Saif, Y. M., pp. 567-614, Iowa State Press, 0-8138-0423-X, Ames, Iowa

Gerlach, R. G., Claudio, N., Rohde, M., Jackel, D., Wagner, C. & Hensel, M. (2008). Cooperation of Salmonella pathogenicity islands 1 and 4 is required to breach epithelial barriers. *Cellular Microbiology*, Vol. 10, No. 11, (Nov), pp. 2364-2376, ISSN 1462-5814

Gibson, D. L., White, A. P., Rajotte, C. M. & Kay, W. W. (2007). AgfC and AgfE facilitate extracellular thin aggregative fimbriae synthesis in *Salmonella enteritidis*. *Microbiology*, Vol. 153, No. Pt 4, (Apr), pp. 1131-1140, ISSN 1350-0872

Gophna, U., Ron, E. Z. & Graur, D. (2003). Bacterial type III secretion systems are ancient and evolved by multiple horizontal-transfer events. *Gene*, Vol. 312, No. (Jul 17), pp. 151-163, ISSN 0378-1119

Grassl, G. A. & Finlay, B. B. (2008). Pathogenesis of enteric *Salmonella* infections. *Current Opinion in Gastroenterology*, Vol. 24, No. 1, (Jan), pp. 22-26, ISSN 1531-7056

Hacker, J., Blum-Oehler, G., Muhldorfer, I. & Tschape, H. (1997). Pathogenicity islands of virulent bacteria: structure, function and impact on microbial evolution. *Molecular Microbiology*, Vol. 23, No. 6, (Mar), pp. 1089-1097, ISSN 0950-382X

Hacker, J. & Kaper, J. B. (2000). Pathogenicity islands and the evolution of microbes. *Annual Review of Microbiology*, Vol. 54, No. pp. 641-679, ISSN 0066-4227

Haneda, T., Ishii, Y., Danbara, H. & Okada, N. (2009). Genome-wide identification of novel genomic islands that contribute to Salmonella virulence in mouse systemic infection. *FEMS Microbiology Letters*, Vol. 297, No. 2, (Aug), pp. 241-249, ISSN 0378-1097

He, S. Y., Nomura, K. & Whittam, T. S. (2004). Type III protein secretion mechanism in mammalian and plant pathogens. *Biochimica et Biophysica Acta*, Vol. 1694, No. 1-3, (Nov 11), pp. 181-206, ISSN 0006-3002

Hueck, C. J. (1998). Type III protein secretion systems in bacterial pathogens of animals and plants. *Microbiology and Molecular Biology Reviews*, Vol. 62, No. 2, (Jun), pp. 379-433, ISSN 1092-2172

Ibarra, J. A. & Steele-Mortimer, O. (2009). Salmonella--the ultimate insider Salmonella virulence factors that modulate intracellular survival. *Cellular Microbiology*, Vol. 11, No. 11, (Nov), pp. 1579-1586, ISSN 1462-5814

Journet, L., Hughes, K. T. & Cornelis, G. R. (2005). Type III secretion: a secretory pathway serving both motility and virulence (review). *Molecular Membrane Biology*, Vol. 22, No. 1-2, (Jan-Apr), pp. 41-50, ISSN 0968-7688

Karasova, D., Sebkova, A., Havlickova, H., Sisak, F., Volf, J., Faldyna, M., Ondrackova, P., Kummer, V. & Rychlik, I. (2010). Influence of 5 major *Salmonella* pathogenicity islands on NK cell depletion in mice infected with *Salmonella enterica* serovar Enteritidis. *BMC Microbiology*, Vol. 10, No. pp. 75, ISSN 1471-2180

Karavolos, M. H., Bulmer, D. M., Winzer, K., Wilson, M., Mastroeni, P., Williams, P. & Khan, C. M. (2008). LuxS affects flagellar phase variation independently of quorum sensing in Salmonella enterica serovar typhimurium. *Journal of Bacteriology*, Vol. 190, No. 2, (Jan), pp. 769-771, ISSN 1098-5530

Kuhle, V. & Hensel, M. (2004). Cellular microbiology of intracellular *Salmonella enterica*: functions of the type III secretion system encoded by *Salmonella* pathogenicity island 2. *Cellular and Molecular Life Sciences*, Vol. 61, No. 22, (Nov), pp. 2812-2826, ISSN 1420-682X

Lax, A. J., Barrow, P. A., Jones, P. W. & Wallis, T. S. (1995). Current perspectives in salmonellosis. *British Veterinary Journal*, Vol. 151, No. 4, (Jul-Aug), pp. 351-377, ISSN 0007-1935

Lim, S., Yun, J., Yoon, H., Park, C., Kim, B., Jeon, B., Kim, D. & Ryu, S. (2007). Mlc regulation of *Salmonella* pathogenicity island I gene expression via *hilE* repression. *Nucleic acids research*, Vol. 35, No. 6, pp. 1822-1832, ISSN 0305-1048

Liu, R. & Ochman, H. (2007). Origins of flagellar gene operons and secondary flagellar systems. *Journal of Bacteriology*, Vol. 189, No. 19, (Oct), pp. 7098-7104, ISSN 0021-9193

Liu, R. & Ochman, H. (2007). Stepwise formation of the bacterial flagellar system. *Proceedings of the National Acadamy of Sciences*, Vol. 104, No. 17, (Apr 24), pp. 7116-7121, ISSN 0027-8424

Ly, K. T. & Casanova, J. E. (2007). Mechanisms of *Salmonella* entry into host cells. *Cellular Microbiology*, Vol. 9, No. 9, (Sep), pp. 2103-2111, ISSN 1462-5814

Macnab, R. M. (2004). Type III flagellar protein export and flagellar assembly. *Biochimica et Biophysica Acta*, Vol. 1694, No. 1-3, (Nov 11), pp. 207-217, ISSN 0006-3002

Marcus, S. L., Brumell, J. H., Pfeifer, C. G. & Finlay, B. B. (2000). *Salmonella* pathogenicity islands: big virulence in small packages. *Microbes and Infection*, Vol. 2, No. 2, (Feb), pp. 145-156, ISSN 1286-4579

Matsui, M., Takaya, A. & Yamamoto, T. (2008). Sigma32-mediated negative regulation of *Salmonella* pathogenicity island 1 expression. *Journal of Bacteriology*, Vol. 190, No. 20, (Oct), pp. 6636-6645, ISSN 0021-9193

McCann, H. C. & Guttman, D. S. (2008). Evolution of the type III secretion system and its effectors in plant-microbe interactions. *New Phytologist*, Vol. 177, No. 1, pp. 33-47, ISSN 0028-646X

Mills, D. M., Bajaj, V. & Lee, C. A. (1995). A 40 kb chromosomal fragment encoding *Salmonella typhimurium* invasion genes is absent from the corresponding region of the *Escherichia coli* K-12 chromosome. *Molecular Microbiology*, Vol. 15, No. 4, (Feb), pp. 749-759, ISSN 0950-382X

Moraes, T. F., Spreter, T. & Strynadka, N. C. (2008). Piecing together the type III injectisome of bacterial pathogens. *Current Opinion in Structural Biology*, Vol. 18, No. 2, (Apr), pp. 258-266, ISSN 0959-440X

Morgan, E., Campbell, J. D., Rowe, S. C., Bispham, J., Stevens, M. P., Bowen, A. J., Barrow, P. A., Maskell, D. J. & Wallis, T. S. (2004). Identification of host-specific colonization factors of *Salmonella enterica* serovar Typhimurium. *Molecular Microbiology*, Vol. 54, No. 4, (Nov), pp. 994-1010, ISSN 0950-382X

Nguyen, L., Paulsen, I. T., Tchieu, J., Hueck, C. J. & Saier, M. H., Jr. (2000). Phylogenetic analyses of the constituents of Type III protein secretion systems. *Journal of Molecular Microbiology and Biotechnology*, Vol. 2, No. 2, (Apr), pp. 125-144, ISSN 1464-1801

Ochman, H., Soncini, F. C., Solomon, F. & Groisman, E. A. (1996). Identification of a pathogenicity island required for *Salmonella* survival in host cells. *Proceedings of the National Acadamy of Sciences*, Vol. 93, No. 15, (Jul 23), pp. 7800-7804, ISSN 0027-8424

Olekhnovich, I. N. & Kadner, R. J. (2007). Role of nucleoid-associated proteins Hha and H-NS in expression of *Salmonella enterica* activators HilD, HilC, and RtsA required for cell invasion. *Journal of Bacteriology*, Vol. 189, No. 19, (Oct), pp. 6882-6890, ISSN 0021-9193

Pallen, M. J. & Matzke, N. J. (2006). From The Origin of Species to the origin of bacterial flagella. *Nature Reviews Immunology*, Vol. 4, No. 10, (Oct), pp. 784-790, ISSN 1740-1526

Pallen, M. J., Penn, C. W. & Chaudhuri, R. R. (2005). Bacterial flagellar diversity in the post-genomic era. *Trends in Microbiology*, Vol. 13, No. 4, (Apr), pp. 143-149, ISSN 0966-842X

Parker, C. T. & Guard-Petter, J. (2001). Contribution of flagella and invasion proteins to pathogenesis of *Salmonella enterica* serovar enteritidis in chicks. *FEMS Microbiology Letters*, Vol. 204, No. 2, (Nov 13), pp. 287-291, ISSN 1574-6968

Penha Filho, R. A., de Paiva, J. B., Arguello, Y. M., da Silva, M. D., Gardin, Y., Resende, F., Berchieri Junior, A. B. & Sesti, L. (2009). Efficacy of several vaccination programmes in commercial layer and broiler breeder hens against experimental challenge with *Salmonella enterica* serovar Enteritidis. *Avian Pathology*, Vol. 38, No. 5, (Oct), pp. 367-375, ISSN 0307-9457

Ramsden, A. E., Holden, D. W. & Mota, L. J. (2007). Membrane dynamics and spatial distribution of *Salmonella*-containing vacuoles. *Trends in Microbiology*, Vol. 15, No. 11, (Nov), pp. 516-524, ISSN 0966-842X

Rathman, M., Sjaastad, M. D. & Falkow, S. (1996). Acidification of phagosomes containing *Salmonella typhimurium* in murine macrophages. *Infection and Immunity*, Vol. 64, No. 7, (Jul), pp. 2765-2773, ISSN 0019-9567

Rhen, M. & Dorman, C. J. (2005). Hierarchical gene regulators adapt *Salmonella enterica* to its host milieus. *International Journal of Medical Microbiology*, Vol. 294, No. 8, (Mar), pp. 487-502, ISSN 1438-4221

Sabbagh, S. C., Forest, C. G., Lepage, C., Leclerc, J. M. & Daigle, F. (2010). So similar, yet so different: uncovering distinctive features in the genomes of Salmonella enterica serovars Typhimurium and Typhi. *FEMS Microbiology Letters*, Vol. 305, No. 1, (Apr), pp. 1-13, ISSN 0378-109

Salyers, A. & Whitt, D. (2002). *Bacterial Pathogenesis: A Molecular Approach* (2nd), ASM Press, 978-1-55581-418-2, Herndon, Virginia

Schmidt, H. & Hensel, M. (2004). Pathogenicity islands in bacterial pathogenesis. *Clinical Microbiology Reviews*, Vol. 17, No. 1, (Jan), pp. 14-56, ISSN 0893-8512

Seth-Smith, H. M. (2008). SPI-7: *Salmonella*'s Vi-encoding Pathogenicity Island. *Journal of infection in developing countries*, Vol. 2, No. 4, pp. 267-271, ISSN 1972-2680

Shea, J. E., Hensel, M., Gleeson, C. & Holden, D. W. (1996). Identification of a virulence locus encoding a second type III secretion system in *Salmonella typhimurium*. *Proceedings of the National Acadamy of Sciences*, Vol. 93, No. 6, (Mar 19), pp. 2593-2597, ISSN 0027-8424

Shi, L., Adkins, J. N., Coleman, J. R., Schepmoes, A. A., Dohnkova, A., Mottaz, H. M., Norbeck, A. D., Purvine, S. O., Manes, N. P., Smallwood, H. S., Wang, H., Forbes, J., Gros, P., Uzzau, S., Rodland, K. D., Heffron, F., Smith, R. D. & Squier, T. C. (2006). Proteomic analysis of *Salmonella enterica* serovar typhimurium isolated from RAW 264.7 macrophages: identification of a novel protein that contributes to the replication of serovar typhimurium inside macrophages. *Journal of Biological Chemistry*, Vol. 281, No. 39, (Sep 29), pp. 29131-29140, ISSN 0021-9258

Snyder, L. A., Loman, N. J., Futterer, K. & Pallen, M. J. (2009). Bacterial flagellar diversity and evolution: seek simplicity and distrust it? *Trends in Microbiology*, Vol. 17, No. 1, (Jan), pp. 1-5, ISSN 0966-842X

Spreng, S., Dietrich, G. & Weidinger, G. (2006). Rational design of *Salmonella*-based vaccination strategies. *Methods*, Vol. 38, No. 2, (Feb), pp. 133-143, ISSN 1548-7091

Stavrinides, J., McCann, H. C. & Guttman, D. S. (2008). Host-pathogen interplay and the evolution of bacterial effectors. *Cellular Microbiology*, Vol. 10, No. 2, (Feb), pp. 285-292, ISSN 1462-5814

Tampakaki, A. P., Fadouloglou, V. E., Gazi, A. D., Panopoulos, N. J. & Kokkinidis, M. (2004). Conserved features of type III secretion. *Cellular Microbiology*, Vol. 6, No. 9, (Sep), pp. 805-816, ISSN 1462-5822

Thomson, N. R., Clayton, D. J., Windhorst, D., Vernikos, G., Davidson, S., Churcher, C., Quail, M. A., Stevens, M., Jones, M. A., Watson, M., Barron, A., Layton, A., Pickard, D., Kingsley, R. A., Bignell, A., Clark, L., Harris, B., Ormond, D., Abdellah, Z., Brooks, K., Cherevach, I., Chillingworth, T., Woodward, J., Norberczak, H., Lord, A., Arrowsmith, C., Jagels, K., Moule, S., Mungall, K., Sanders, M., Whitehead, S., Chabalgoity, J. A., Maskell, D., Humphrey, T., Roberts, M., Barrow, P. A., Dougan, G. & Parkhill, J. (2008). Comparative genome analysis of *Salmonella* Enteritidis PT4 and *Salmonella* Gallinarum 287/91 provides insights into evolutionary and host adaptation pathways. *Genome Research*, Vol. 18, No. 10, (Oct), pp. 1624-1637, ISSN 1088-9051

Tindall, B. J., Grimont, P. A., Garrity, G. M. & Euzeby, J. P. (2005). Nomenclature and taxonomy of the genus *Salmonella*. *International Journal of Systematic and Evolutionary Microbiology*, Vol. 55, No. 1, (Jan), pp. 521-524, ISSN 1466-5026

Townes, J. M. (2010). Reactive arthritis after enteric infections in the United States: the problem of definition. *Clinical Infectious Diseases*, Vol. 50, No. 2, (Jan 15), pp. 247-254, ISSN 1058-4838

Tsolis, R. M., Adams, L. G., Ficht, T. A. & Baumler, A. J. (1999). Contribution of *Salmonella typhimurium* virulence factors to diarrheal disease in calves. *Infection and Immunity*, Vol. 67, No. 9, (Sep), pp. 4879-4885, ISSN 0019-9567

USDA-ERS. (May 22, 2009). Foodborne Illness Cost Calculator: *Salmonella*, In: *Data Sets*, December 2009, Available from:
http://www.ers.usda.gov/data/foodborneillness/salm_Intro.asp

van Asten, A. J. & van Dijk, J. E. (2005). Distribution of "classic" virulence factors among *Salmonella* spp. *FEMS Immunology and Medical Microbiology*, Vol. 44, No. 3, (Jun 1), pp. 251-259, ISSN 0928-8244

van der Velden, A. W., Lindgren, S. W., Worley, M. J. & Heffron, F. (2000). *Salmonella* pathogenicity island 1-independent induction of apoptosis in infected macrophages by Salmonella enterica serotype typhimurium. *Infection and Immunity*, Vol. 68, No. 10, (Oct), pp. 5702-5709, ISSN 0019-9567

Van Immerseel, F., Methner, U., Rychlik, I., Nagy, B., Velge, P., Martin, G., Foster, N., Ducatelle, R. & Barrow, P. A. (2005). Vaccination and early protection against non-host-specific *Salmonella* serotypes in poultry: exploitation of innate immunity and microbial activity. *Epidemiology and Infection*, Vol. 133, No. 6, (Dec), pp. 959-978, ISSN 0950-2688

Vieira, A. (Year). WHO Global Foodborne Infection Network Country Databank - A resource to link human and non-human sources of *Salmonella*, *ISVEE Conference*, Durban, 2009

Walthers, D., Carroll, R. K., Navarre, W. W., Libby, S. J., Fang, F. C. & Kenney, L. J. (2007). The response regulator SsrB activates expression of diverse *Salmonella* pathogenicity island 2 promoters and counters silencing by the nucleoid-associated protein H-NS. *Molecular Microbiology*, Vol. 65, No. 2, (Jul), pp. 477-493, ISSN 0950-382X

Waterman, S. R. & Holden, D. W. (2003). Functions and effectors of the *Salmonella* pathogenicity island 2 type III secretion system. *Cellular Microbiology*, Vol. 5, No. 8, (Aug), pp. 501-511, ISSN 1462-5822

WHO. (March 2007). Food safety and foodborne illness, In: *World Health Organization: Media Centre*, 2010, Available from:
http://www.who.int/mediacentre/factsheets/fs237/en/

Wilson, J. W. (2006). Chapter 9: Bacterial protein secretion mechanisms, In: *Molecular Paradigms of Infectious Disease: a Bacterial Perspective*, Nickerson, C. A. & Schurr, M. J., pp. 274 - 320, Springer, 978-0-387-30917-0, Detroit, Michigan

Wisner, A. L., Desin, T. S., Koch, B., Lam, P. K., Berberov, E. M., Mickael, C. S., Potter, A. A. & Köster, W. (2010). *Salmonella enterica* subspecies *enterica* serovar Enteritidis SPI-2 T3SS: Role in Intestinal Colonization of Chickens and Systemic Spread. *Microbiology*, Vol. 156, No. 9, (May 20), pp. 2770-2781, ISSN 1350-0872

Wood, M. W., Jones, M. A., Watson, P. R., Hedges, S., Wallis, T. S. & Galyov, E. E. (1998). Identification of a pathogenicity island required for *Salmonella* enteropathogenicity. *Molecular Microbiology*, Vol. 29, No. 3, (Aug), pp. 883-891, ISSN 0950-382X

Zhang, X. & Mosser, D. M. (2008). Macrophage activation by endogenous danger signals. *Journal of Pathology*, Vol. 214, No. 2, (Jan), pp. 161-178, ISSN 0022-3417

10

Animal Models for *Salmonella* Pathogenesis: Studies on the Virulence Properties Using *Caenorhabditis elegans* as a Model Host

Jeong Hoon Cho[1] and Jaya Bandyopadhyay[2]
[1]Department of Biology Education, College of Education
Chosun University, Gwangju
[2]Department of Biotechnology, West Bengal University of Technology, Kolkata
[1]Republic of Korea
[2]India

1. Introduction

Studies associated with host-pathogen interaction and the principal mechanisms of pathogenesis including the systems adopted by the host for its defense has always been a topic of interest to the scientific community that primarily deals with pathogenic microbes causing human diseases. More importantly the post-genomic era has set a milestone in basic and applied science research by proper identification and validation of potential human "disease-causing" or "disease-associated" genes. Although the host-pathogen interaction is a complex biological system (Huffman et. al., 2004) it is equally important to understand the characteristic features of microbes and their respective hosts that always may not culminate into a disease process. So present day researchers are also working with microbes that may exist within hosts without causing any obvious disease, and at the same time trying to explore why some microbes only cause disease in certain hosts. Besides, a wide range of microbes those are pathogenic to the mammals, like bacteria and fungi, have also been known to manifest diseases in simple non-vertebrate hosts as well. In order to understand a pathogen, researchers would preferably screen the microbe's genome at length to identify all its virulence genes. On the other hand, screening in mammalian experimental hosts, such as mice, rats, or other mammals *per se*, sometimes seems unfeasible since they would be required in large numbers and thus quite expensive. In recent days use of simple non-vertebrate hosts, such as the round worm *Caenorhabditis elegans*, the fruit fly *Drosophila melanogaster*, and the plant *Arabidopsis thaliana*, are becoming common for convenience in investigating the virulence strategies adopted by several mammalian pathogenic microbes (bacteria and fungi) (Sifri et al., 2005). Also uses of these model organisms are in great practice to comprehend the conserved molecular pathways that are related to human diseases caused by microbial pathogens.

There is a wide variety of pathogens (both bacteria and fungi) that are known to affect the human health including animals. Amongst them those that have been extensively studied with different laboratory model hosts are, Gram-negative bacteria *Burkholderia, Pseudomonas,*

Salmonella, Serratia and *Yersinia*; Gram-positive bacteria *Enterococcus, Staphylococcus* and *Streptococcus*; and the fungus *Cryptococcus neoformans.* However there are some that are not pathogenic to mammals, but pose as insect pathogens, like *Bacillus thuringiensis* and the nematode-specific *Microbacterium nematophilum.*

Caenorhabditis elegans (commonly known as *C. elegans*) is a free living soil nematode that feeds on bacteria. Under laboratory conditions *C. elegans* feeds on *E. coli* (strain OP50). In order to study the effects of pathogenic strains on animals, researchers have started to use *C. elegans* as an excellent and a convenient model to explore the bacterial pathogenesis on animals by making the worms feed on the pathogens. Amongst the different categories of pathogens that are known to infect the worms in the same manner as in humans / animals, *Salmonella* is one that has been broadly studied in the worm system. The present chapter shall focus the attribute of the nematode *C. elegans* as a convenient model to study host pathogen interaction with special emphasis to *Salmonella.*

2. *Caenorhabditis elegans* as a simple model system to study human diseases

For the past few decades, the free-living bacteriovorous nematode *Caenorhabditis elegans* (Caeno, recent; rhabditis, rod; elegans, elegant) has emerged lately as a powerful model for study of developmental genetics, neurobiology and aging. Ever since it was introduced by Sydney Brenner, this simple multicellular eukaryote has been studied intensively with comprehensive genotypic and phenotypic information now available. This free-living nematode has the following features: small body length (1.5 mm adults), quick generation time (three days), large brood size (approximately 300 progeny per gravid adult), short lifespan (~3 weeks), ease of maintenance, reduced cost, a small genome (one half that of *Drosophila melanogaster*), ability to be stored for long periods by freezing, and the fact that it is a simple and genetically tractable model have made this nematode species an ideal model to study longevity and process of ageing (Ewbank, 2002; Houthoofd et al., 2003; Kurz et al., 2007), and as well as a model organism for molecular and developmental biology. Moreover, under the microscope, the unique transparent body of the worm allows one to observe many biological processes, including organogenesis, behavior, and as well as, pathogenesis. The life cycle is short, temperature dependent and consists of embryogenesis (development from fertilization to hatching) and post-embryonic development that has four larval stages separated by molts followed by the adult stage (Figure 1). Also, at every larval stage a new cuticle of stage-specific composition is secreted and the older one is shed. In L1 larvae, the nervous system, the reproductive system, and the digestive tract begin to develop, and this is completed by the L4 stage. Sometimes these nematodes can adopt another non developmental stage known as the Dauer stage instead of the normal third larval stage (Cassada and Russell, 1975). Entry into dauer is induced by stress like high temperature, starvation or overcrowding at the second molt (Fig. 1).

In 1998, *C. elegans* became the first metazoan to have a completely sequenced genome (The *C. elegans* Sequencing Consortium, 1998). More than 40% of the human disease genes have been predicted to have orthologs in the *C. elegans* genome. Overexpressing human genes in specific cell types in *C. elegans* using tissue- or cell- specific worm promoters, or studying the differential gene expression patterns of worms at the transcriptional level by microarray analysis may well reveal the gene expression profiles (up- or down-regulated) for the worms. This helps the identification of certain members of the signaling cascades that are activated in diseases and may well act as candidates for drug therapies. Double-stranded RNA interference (dsRNAi) has also been another modern approach to study human diseases by treating worms with dsRNAs thus inactivating the function of specific C. elegans gene orthologs (Fire et al., 1998; Tabara et al., 1998; Timmons et al., 2001). Up to date several genomic data on human and animal microbial pathogens that are shown to harm and kill nematodes have been created. With these vast resources of genetic information, there is always a growing need for simple and innovative ways to study microbial virulence strategies and assay the role of individual genes to pathogenesis. Since both the host (i.e. C. elegans) and pathogens are amenable to genetic analysis and high throughput screening, in each of these pathosystems the worm has been successfully utilized both for the identification of microbial virulence factors and as well as the worm's immune-defense mechanisms.

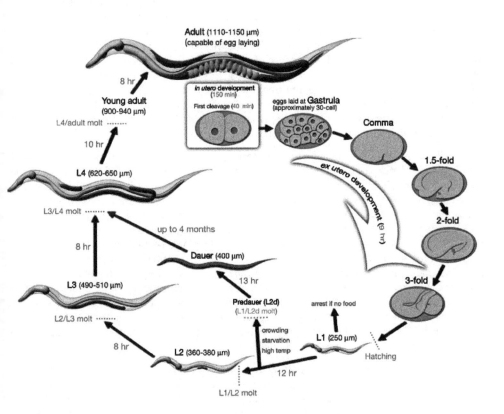

Fig. 1. Life cycle of C. elegans. Life cycle consists of four larval stages. At the first larval stage the life cycle can be interrupted by the "dauer" stage which is triggered by increased pheromone levels that result due to food scarcity, overcrowding and temperature (Courtesy: Introduction to C. elegans anatomy, Handbook-Hermaphrodite; http://www.wormatlas.org)

3. Bacteria as a food source for *C. elegans*: Effect of *Salmonella* and other pathogens as food

C. elegans has a life span of approximately two weeks at room temperature when fed on *Escherichia coli* OP50 bacteria (Brenner, 1974) grown on Nematode Growth Medium (NGM) agar (Garsin et al., 2001). However, when *C. elegans* is fed on other human pathogens, they exhibit a range of significant defects, like shorter life spans. Several human pathogens, including *Pseudomonas aeruginosa, Salmonella typhimurium, Serratia marcescens, Staphylococcus aureus, Vibrio cholerae,* and *Burkholderia pseudomallei,* kill *C. elegans* when supplied as a food source, and a diverse array of bacterial virulence factors have been shown to play a role in both nematode and mammalian pathogenesis (Aballay et al., 2000; Kurz & Ewbank, 2000; Labrousse et al., 2000; Tan et al., 1999). An important feature of *Pseudomonas aeruginosa,* a Gram-negative pathogen, is known to kill *C. elegans* and more particularly, under different media conditions. The *P. aeruginosa* strain PA14 kills *C. elegans* by "slow-killing" (in few days) or even by "fast killing" (few hours) (Tan et al., 1999). *Vibrio cholerae,* another Gram-negative bacterium kills *C. elegans* within few days (~ 5 days) by a "slow-killing" process (Vaitkevicius et al., 2006). A marked decrease in the life span was observed in worms feeding *V. vulnificus* as opposed to the regular food of *E. coli* OP50. In many cases, the intestines of the worms were found to get distended with clumps of pathogenic microorganisms that accumulate within it and probably happen to be primary cause of early deaths (Dhakal et al., 2006). Normally, the pharyngeal grinder of the worm efficiently disrupts the *E. coli* and essentially no intact bacterial cells are found within the intestinal lumen. However, virulent bacterial strains like *V. vulnificus* or *V. cholerae* have been shown to accumulate in both pharynx and the lumen of the worm intestines (Vaitkevicius et al., 2006; Dhakal et al., 2006) as evidenced under fluorescent microscope for the GFP-labeled bacterial strains.

Salmonella is a gram-negative enteric bacterium that represents a major public health problem. *S. enterica* colonizes the *C. elegans* intestine as reported (Aballay et al., 2000; Labrousse et al., 2000). *S. enterica* serovar Typhi causes typhoid fever, a severe systemic infection. *S. enterica* serovar Typhimurium is known to be lethal to mice, causing a typhoid-like disease, and in humans it causes nonfatal infection restricted to the gastrointestinal tract and thus had been studied in the mouse model for systemic infections. When worms were exposed to *S. enterica* for only 3 h, then removed to plates seeded with OP50, there was significant early death. Invasion of host cells is an essential aspect of *Salmonella* sp. pathogenesis in mammalian systems, but *S. enterica* does not appear to invade *C. elegans* cells. Many novel strategies have been devised for understanding its mode of action and its interactions with host cells (Lee & Camilli, 2000; Chiang et al., 1999). The finding that *Salmonella* is capable of infecting *C. elegans,* and that genes important for its full pathogenicity in vertebrates also play a role during infection of *C. elegans,* opens the possibility of taking a new genetic approach to study *Salmonella.*

Salmonella, a gastrointestinal tract pathogen of humans, is responsible for approximately 2 million - 4 million cases of enterocolitis every year in the United States (Tampakakis et al., 2009). During infection, *S. enterica* serovar Typhimurium has the propensity to compete with normal intestinal flora. *Candida albicans* is another opportunistic human fungal pathogen that usually resides in the gastrointestinal tract and on the skin as a commensal and can also cause life-threatening invasive disease. Besides, both these organisms are pathogenic to the nematode *C. elegans,* causing a persistent gut infection and eventually leading to death of the worms (Fig. 2). Mylonakis and his group developed *C. elegans* as a polymicrobial infection model to assess the interactions between *S.* Typhimurium and *C. albicans* (Tampakakis et al., 2009). They reported that when *C. elegans* is infected with *C. albicans* and *S. enterica* serovar Typhimurium, *C. albicans* filamentation is inhibited. They further utilized the host, *C. elegans*, to identify the antagonistic interaction between two human pathogens that reside within the gastrointestinal tract.

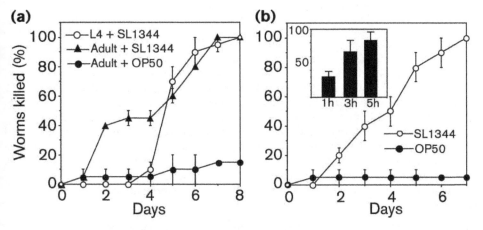

Fig. 2. *S. typhimurium* kills *C. elegans*. (a) L4 stage (open circles) or 1-day-old adult hermaphrodite (solid triangles and solid circles) worms fed either on *S. typhimurium* SL1344 (solid triangles and open circles) or on *E. coli* OP50 (solid circles). (b) *C. elegans* fed on *S. typhimurium* SL1344 (open circles) or *E. coli* OP50 (solid circles) for 5 h, then shifted to *E. coli* OP50. The inset shows the percentages of dead worms after transfer to OP50-containing plates after feeding for 1, 3 or 5 h on SL1344 (Courtesy Aballay et al., 2000).

4. Intestine as the store house of bacterial infection in *C. elegans*

Numerous bacteria infect the intestine of *C. elegans*. In many cases, the intestine becomes inflated; but it is not clear whether this is due to physical pressure exerted by the growing pathogen or as a physiological response of the nematode. The standard laboratory food, i.e. *E. coli* OP50 and *Cryptococcus laurentii* (Tan et al., 1999; Garsin et al., 2001; Mylonakis et al., 2002) does not colonize wild-type *C. elegans*, but various pathogens do. For example, *Enterococcus faecalis*, a gram-positive bacteria, colonizes in *C. elegans* and kills very rapidly (Garsin et al., 2001). It was eventually shown that genes involved directly or indirectly with the quorum-sensing system are involved in killing (Sifri et al., 2002). On the other hand, *Pseudomonas aeruginosa* can kill *C. elegans* rapidly by toxin-mediated mechanisms or slowly in an infectious process. In the "slow killing" model, bacteria colonize the intestine, but within a day exposure to the bacteria, no strong disease symptoms were observed (Tan et al., 1999). With continued exposure, the worms gradually cease pharyngeal pumping, become immobile, and eventually die. Moreover, large quantities of live bacteria like *Salmonella enterica, Burkholderia cepacia, Serratia marcescens, Staphylococcus aureus, Vibrio vulnificus,* V. *cholerae,* and *C. neoformans* are known to kill worms by colonization (Aballay et al., 2000; Garsin et al., 2001; Mylonakis et al., 2002; Kothe et al., 2003; Kurz et al., 2003; Rhee et al., 2006; Vaitkevicius et al., 2006). A screen of 960 transposon insertions in *S. enterica* produced 15 mutations with reduced killing of *C. elegans*, of which only some were virulent

(Tenor et al., 2004). Although bacterial colonization is greatly correlated with worm killing, it is not adequate for killing. For instance, aerobically grown *Enterococcus faecium* although accumulates to high levels, it does not kill (Garsin et al., 2001). *S. enterica*, *S. marcescens*, and *E. faecalis* are pathogens also known to cause persistent infections (Aballay et al., 2000; Labrousse et al., 2000; Garsin et al., 2001; Kurz et al., 2003) in *C. elegans* in contrast to *Pseudomonas aeruginosa* and *S. aureas*.

Different strains of *Salmonella*, such as *S. typhimurium* as well as other *Salmonella enteric* serovars including *S. enteritidis* and *S. dublin* are all effective in killing *C. elegans* (Aballay et al., 2000). When worms are placed on a lawn of *S. typhimurium*, the bacteria have been shown to accumulate in the intestinal lumen and the nematodes die over the course of several days (Fig. 3). This killing in particular requires direct contact with live bacterial cells. Interestingly, the worms die in the same manner even when placed on a lawn of *S. typhimurium* for a relatively short period of time (3–5 hours) before transfer to a lawn of *E. coli*, their natural food. A high titer of *S. typhimurium* still persists in the *C. elegans* intestinal lumen for the rest of the worms' life even after their transfer to an *E. coli* lawn. Killing is directly correlated with an increase in the titer of *S. typhimurium* in the *C. elegans* lumen. Even a small inoculum of *S. typhimurium* has been shown to be enough to establish a persistent infection *C. elegans* which is probably due to the presence of *C. elegans* intestinal receptors to which bacteria might adhere (Fig. 4).

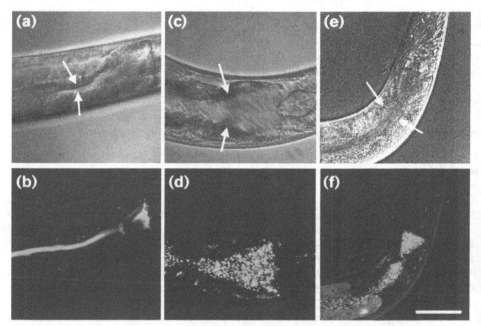

Fig. 3. Bacterial colonization of the *C. elegans* intestine. Confocal images showing young adult hermaphrodite worms fed on (a,b) *E. coli* DH5α–GFP for 72 h, (c,d) *S. typhimurium* SL1344–GFP for 72 h, or (e,f) *P. aeruginosa* PA14–GFP for 24 h. (a,c,e) Transmission images showing the intestinal margins (indicated with arrows). (b,d,f) Merged images showing bacterial fluorescence (green channel) and the gut autofluorescence (red channel). Scale bar, 50 μm (Courtesy Aballay et al., 2000).

Animal Models for Salmonella Pathogenesis: Studies on the Virulence Properties Using Caenorhabditis
elegans as a Model Host
209

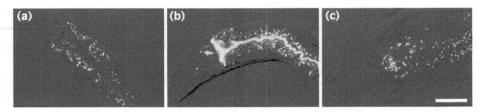

Fig. 4. *S. typhimurium* colonizes the worm intestine. Young adult worms were fed on (a) *E. coli* DH5α–GFP or (b, c) *S. typhimurium* SL1344–GFP for 5 h and then transferred to *E. coli* OP50 for (a, b) 24 h or (c) 96 h. Scale bar, 50 μm (Courtesy Aballay et al., 2000).

Bacterial proliferation and persistence can be easily determined by monitoring the worms in due course under microscope for the presence of GFP-labeled bacteria. In particular, pathogenic strains expressing green fluorescent protein (GFP) are therefore extremely useful in examining the fate of such microbes upon ingestion by the worms (Fig. 5). A virulent strain of

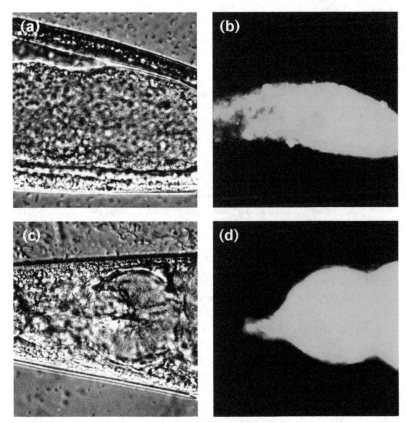

Fig. 5. Accumulation of *S. typhimurium* within the intestine and pharynx of *C. elegans*. (a,c) Nomarski and (b,d) fluorescence photomicrographs of the (a,b) posterior and (c,d) anterior of a worm after contact for 5 days with GFP-expressing *S. typhimurium*. The intestine and terminal bulb of the pharynx are seen to be full of intact bacteria (Courtesy Labrousse et al., 2000).

S. typhimurium expressing GFP (12023 ssaV–GFP) is known to kill *C. elegans* as the wild-type strain. The grinder which is located in the terminal bulb of the pharynx of the wormsnormally breaks bacteria (Albertson & Thomson, 1976). However, with increasing infection, the number of *S. typhimurium* significantly increases beyond the terminal bulb and gradually starts to mount up within the intestinal lumen (Fig. 5). Increase in the intestinal lumen of the worms is accompanied by the decrease in the volume of the intestinal cells. Nonetheless, the cells of the terminal bulb of the pharynx get progressively destroyed and their place is taken up by bacteria. Also worms with defective grinders have been found to be more susceptible to *Salmonella* infection and therefore less resistant to the pathogenic effects. For example, *phm-2* worm mutants possess abnormal terminal bulb and therefore are more susceptible to bacterial attack than the N2 worms (Fig. 6) (Labrousse et al., 2000).

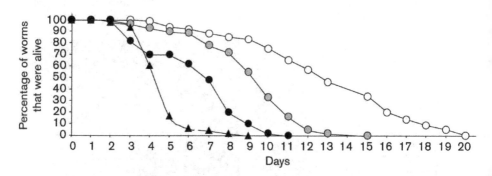

Fig. 6. Survival of *C. elegans* fed on *E. coli* and *S. typhimurium*. Wild-type worms (circles) *or phm-2* mutants (triangles) fed on *E. coli* strain OP50 until the larval L4 stage and then kept on OP50 (open circles), or transferred *to S. typhimurium* strain 12023 (black symbols), or after 8 h Thimerosal sterilization and returned to OP50 (grey circles). Dead worms were scored accordingly. (Courtesy Labrousse et al., 2000).

5. Assessment of pathogenicity of microbes to *C. elegans*

Both genetic and environmental factors play an important role in determining the virulence of a pathogen. Host mortality assays are generally performed to assess the pathogenicity of the microbes. This is generally done by measuring the time (TD50: time to death for 50% of the host) required by the microbe to kill a fixed percentage of host (Mahajan-Miklos et al., 1999; Garsin et al., 2001). As already mentioned earlier, *S. enterica* serovar Typhimurium colonizes the nematode intestine (Aballay et al., 2000; Labrousse et al.,2000). Adult worms transferred plates seeded with *S. enterica* and incubated at 25° C, the TD50 was shown to be was 5.1 days, compared to 9.9 days for control animals fed on *E. coli* OP50 (Aballay et al., 2000). When worms were exposed to *S. enterica* for merely 3 h, then removed to OP50, there was a significant early death in the worm population suggesting the pathogenic effect of *S. enterica* on *C. elegans*. Although invasion of host cells is an essential aspect of *Salmonella* sp. pathogenesis in higher animal systems, yet it has been demonstrated that *S. enterica* does not appear to invade *C. elegans* cells.

Animal Models for Salmonella Pathogenesis: Studies on the Virulence Properties Using Caenorhabditis
elegans as a Model Host
211

Salmonella pathogenicity islands -1 and -2 (SPI-1 and SPI-2), PhoP and a virulence plasmid are required for the establishment of a persistent infection (Alegado & Tan, 2008). It was observed that the PhoP regulon, SPI-1, SPI-2 and spvR are induced in *C. elegans* and isogenic strains lacking these virulence factors exhibited significant defects in the ability to persist in the worm intestine. *Salmonella* infection also led to induction of two *C. elegans* antimicrobial genes, *abf-2* and *spp-1*, which operate to limit bacterial proliferation. Thus resistance to host antimicrobials in the intestinal lumen has been found to be a key mechanism for *Salmonella* persistence. Apart from genetic factors there are environmental factors, such as, the composition of the media on which the pathogen is grown that has been shown to have influence on the host's mortality rate. For example, *Escherichia coli* OP50, which is non pathogenic otherwise can be rendered pathogenic almost as pathogenic as *Enterococcus faecalis* when it is grown on brain heart infusion (BHI) agar (Garsin et al., 2001). *Salmonella enterica* strains grown on NGM are rendered infectious depending on their serotypes (Table 1).

Strain	Growth media	Pathogenicity status	References
S. enterica ser. Paratyphi	NGM	Non-pathogenic	Aballay et al., 2000
S. enterica ser. Typhi	NGM	Non-pathogenic	Aballay et al., 2000
S. enterica ser. Dublin	NGM	Infectious	Aballay et al., 2000
S. enterica ser. Enteritidis	NGM	Infectious	Aballay et al., 2000
S. enterica ser. Typhimurium	NGM	Infectious	Aballay et al., 2000, Labrousse et al. (2000)

Table 1. Effect of media on *C. elegans* exposed to *Salmonella* (Adapted from Alegado et al., 2003).

6. *C. elegans* inherent immune response to *Salmonella* infection

Innate immunity consists of a variety of defense machinery used by metazoans to avert microbial infections. These nonspecific defense responses used by the innate immune system in animals are governed by interacting and intersecting pathways that not only directs the immune responses but also governs the longevity and responses to different stresses. Even though ample research on *C. elegans* immune response is still ongoing, yet there has not been enough information on the worms' innate immune response towards bacterial pathogens in contrast to the fruit fly, *Drosophila,* and mammals where a fundamental feature like Toll signaling pathway exists. For example, isolation of a strain carrying a mutation in *nol-6*, which encodes a nucleolar RNA-associated protein in *C. elegans* or RNAi-mediated depletion of *nol-6* as well as other nucleolar genes led to an enhanced resistance to *S. enterica* mediated killing that was associated with a reduction of pathogen accumulation. These results also demonstrated that animals deficient in *nol-6* are more resistant to infections by Gram-negative and Gram-positive pathogens signifying that nucleolar disruption activates immunity against different bacterial pathogens (Fuhrman et al., 2009). Studies also indicated that nucleolar disruption through RNAi ablation of ribosomal genes resulted in an increased resistance to pathogen that requires P53/CEP-1. Thus from the reports it is quite evident that *C. elegans* activates innate immunity against bacterial infection in a *p53/cep-1*-dependent manner (Fig. 7). Furthermore, *C. elegans* mutants which exhibited reduced pathogen accumulation (Rpa), displayed enhanced resistance to *S. enterica*-mediated killing (Fig. 8).

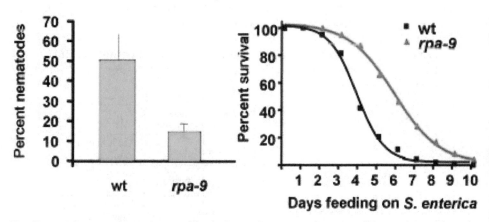

Fig. 7. *rpa-9* mutants are resistant to both *S. enterica* accumulation and *S. enterica*-mediated killing (Courtesy Fuhrman et al., 2009).

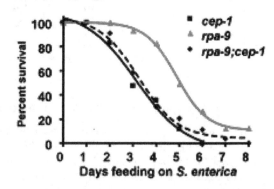

Fig. 8. *rpa-9* mutation activates immunity against *S. enterica* in a p53/cep-1- dependent manner. (Courtesy Fuhrman et al., 2009).

To date different molecular approaches, including forward genetics screens and RNAi have facilitated the identification of certain signaling pathways involved in the response of *C. elegans* to infection. For example, *Salmonella enterica* serovars is also known to trigger programmed cell death (PCD), and *C. elegans* cell death (ced) mutants have been shown to be more susceptible to *Salmonella*-mediated killing (3) (Aballay et al., 2003). *Salmonella*-elicited PCD was shown to require p38 mitogen-activated protein kinase (MAPK) encoded by the *pmk-1* gene. On the other hand inactivation of *pmk-1* by RNAi blocked *Salmonella*-induced cell death. *C. elegans* innate immune response triggered by *S. enterica* was thus shown to require intact lipopolysaccharide (LPS) and is mediated by a MAPK signaling pathway. Besides innate immunity in *C. elegans* is known to be regulated by neurons expressing NPR-1/GPCR, a G-protein-coupled receptor related to mammalian neuropeptide Y receptors that functions to suppress innate immune responses (Styer et al., 2008).

With regard to the conserved Toll signaling, *C. elegans* too possesses a toll-signaling pathway comparable to the innate immunity found in *Drosophila* or mammals. As opposed to the fly and mammalian tolls, *C. elegans* *tol-1* (the *C. elegans* homolog of Toll) was previously stated to be required for the worm development and recognition of pathogens but not important for resistance to the pathogens (Pujol et al., 2001). However, later evidences subsequently support that TOL-1 is required to prevent *Salmonella enterica* invasion of the pharynx, which comprise one of the first barriers against pathogens in *C. elegans*. It was also illustrated that TOL-1 is required for the correct expression of ABF-2, which is a defensin-like molecule expressed in the pharynx, and heat-shock protein 16.41 (HSP-16.41), which is also expressed in the pharynx, and is part of a HSP superfamily of proteins required for *C. elegans* immunity. Thus, TOL-1 has been shown to have a direct role in *C. elegans* defence against pathogens (Tenor & Aballay, 2008).

7. Influence of probiotic bacteria on *Salmonella*-infected *C. elegans*

Probiotic bacteria have been defined as living microorganisms that exert useful effects on human health when ingested in sufficient numbers. Lactic acid bacteria (LAB) are the most frequently used probiotic microorganisms. LAB have been found to have a wide range of physiological influences on their hosts, including antimicrobial effects, microbial interference, supplementary effects on nutrition, antitumor effects, reduction of serum cholesterol and lipids, and immunomodulatory effects. Lactobacilli and bifidobacteria fed worms were shown to display increased life span and resistance to *Salmonella* clearly showing that LAB can enhance the host defense of *C. elegans* by prolonging the life span (Ikeda et al., 2007). Hence the nematode may once again emerge out as an appropriate model for screening useful probiotic strains or dietetic antiaging substances.

8. Role of NRAMPs and autophagy in bacterial infection

The *C. elegans* intestine also presents many advantages because this system can mimic the host–pathogen interactions that occur specially during phagocytosis. Macrophages play a pivotal role in the resolution of microbial infections via the process of phagocytosis. Nramp1 (Natural resistance-associated macrophage protein-1) is a functionally conserved iron-manganese transporter in macrophages and manganese, a superoxide scavenger, which is required in trace amounts and functions as a cofactor for most antioxidants. Nramp homologues, *smfs*, have been identified in the nematode *C. elegans* (Bandyopadhyay et al., 2009). We have demonstrated that hypersensitivity to the pathogen *Staphylococcus aureus*, an effect that was rescued by manganese feeding or knockdown of the Golgi calcium/manganese ATPase, *pmr-1*, indicating that manganese uptake is essential for the innate immune system. Reversal of pathogen sensitivity by manganese feeding suggested a protective and therapeutic role of manganese in pathogen evasion systems thus proposing that the *C. elegans* intestinal lumen may mimic the mammalian macrophage phagosome and thus could be a simple model for studying manganese-mediated innate immunity. Similar experiments with *Salmonella enterica* in the near future may open more possibilities in favor of utilizing the nematode intestine as a model for manganese-mediated innate immunity.

Autophagy, a lysosomal degradation pathway, plays a crucial role in controlling intracellular bacterial pathogen infections. Jia et al., (2009) showed the outcome of autophagy gene inactivation by feeding RNAi techniques on *Salmonella enterica* serovar Typhimurium infection in *C. elegans*. Genetic inactivation of the autophagy pathway increased bacterial intracellular replication, decreased animal lifespan, and resulted in apoptotic independent death. In *C. elegans*, genetic knockdown of autophagy genes abrogates pathogen resistance conferred by a loss-of-function mutation, *daf- 2(e1370)*, in the insulin-like tyrosine kinase receptor or by overexpression of the DAF-16 FOXO transcription factor. Therefore, autophagy genes play an essential role in host defense *in vivo* against an intracellular bacterial pathogen and mediate pathogen resistance in long-lived mutant nematodes.

9. *C. elegans* as a target for drug discovery

By means of genomics technologies, *C. elegans* is growing into a prominent model organism for functional characterization of novel drugs in biomedical research. In fact many biomedical discoveries, for example diabetes type 2 diseases, depression (relating to serotonergic signaling) or the neurodegenerative Alzheimer's disease have been made for the first time using the worms. The simple body plan of the worms has always made it an appropriate model for the fastest and most amenable to cost-effective medium/high-throughput drug screening technologies. Besides, *C. elegans* has always been a better choice over *in vitro* or cellular models to study drug-reporter interaction and in doing so monitoring the actual behavioral responses of the animals. Conventionally, antimicrobial drug discovery has brought about screening candidate compounds directly on target microorganisms (Johnson & Liu, 2000). In order to discover such novel antimicrobials, a series of antibiotics are therefore being screened to identify those that help in the survival of the worms or markedly reduce the number of bacteria colonizing the nematode intestine. For such high throughput screening of compound libraries, conventional agar-based infection experiments in *C. elegans* are later assessed in liquid media contained in standard 96-well microtiter plates for carrying out the curing assays. Interestingly, these simple infection systems may allow one to screen nearly 6,000 synthetic compounds and more than 1000 natural extracts. Moreover, the *in vivo* effective dose of many of these compounds was significantly lower than the minimum inhibitory concentration (MIC) needed to prevent the growth of the pathogens *in vitro*. More importantly, many of the compounds and extracts had not as much of affect on in bacterial growth *in vitro*. Screening synthetic compound libraries and as well as extracts of natural products for substances that cure worms from bacterial persistent infection allows one to identify compounds that not only blocks pathogen replication *in vitro* but in addition identifies virulence of the pathogen, may kill it, or may augment the host's immune response. Nevertheless, activities of some these compounds or extracts are considerably high only in whole animal assay *in vivo*, and hence the rationale for using a whole-animal screen in a drug discovery program.

10. Closing remark

Attention must be given to the *C. elegans* natural bacterial food, pathogens and their virulence factors. A better understanding about the dietary behavior and the natural pathogenic organisms of the *C. elegans* shall open the gates for more information about this worm. Besides, the introduction of genomics and combinatorial chemistry has firmly enabled one to make use of defined targets to identify new antibiotics. The nematode *C. elegans* has undoubtedly proven to be a simple model for studying the interaction between microbial pathogens and host factors, and further examining the roles of specific gene products to virulence and immunity. It is apparent that there are conserved pathogenic genes involved in *C. elegans* killing and mammalian pathogenesis. An important experimental advantage of *C. elegans* as a model to study bacterial pathogenesis is that genetic analysis may as well be carried out in both the pathogens and in the host, simultaneously, a process termed as "interactive genetic analysis." It would undoubtedly be more useful to further focus on the characterization of chemical suppressors of virulent factor expressions or secretions as candidate novel antibiotics, taking *C. elegans* as the model. Additionally the worm model would also be useful to address questions with regard to the pathophysiology of worm death in case of lethal infections and further extend to identify the groups of virulent factors that are important in *C. elegans* killing.

The various categories of experiments so far carried out has provided a proof-of-principle that screening experiments may be useful in identifying new bacterial virulence factors, not only in *Salmonella*, but perhaps other pathogens that are able to cause a persistent infection in *C. elegans*, such as *S. aureus* (Sifri et al., 2003). Until date several loci have been identified from screens not having direct implication in *Salmonella* virulence. Thus, a saturating genome-wide screen would be extremely fruitful in identifying the predominance of *Salmonella* genes that are required for persistent infection in *C. elegans*, some of which could also be important for pathogenesis in other hosts.

11. Acknowledgment

All publications and figures referred in this chapter have been cited to justify the theme of the present review article. We are deeply indebted to all the authors of the original papers. At the same time we sincerely regret for those references that have been left out unintentionally.

12. References

Aballay, A.; Drenkard, E.; Hilbun, L.R. & Ausubel, F.M. (2003). *Caenorhabditis elegans* innate immune response triggered by *Salmonella enterica* requires intact LPS and is mediated by a MAPK signaling pathway. *Current Biology*, Vol. 13, pp 47–52, ISSN 0960-9822.

Aballay, A.; Yorgey, P. & Ausubel, F.M. (2000). *Salmonella typhimurium* proliferates and establishes a persistent infection in the intestine of *Caenorhabditis elegans*. *Current Biology*, Vol.10, pp 1539–1542, ISSN 0960-9822.

Albertson, D.G. & Thomson, J.N. (1976). The pharynx of *Caenorhabditis elegans*. *Philos Trans R Soc Lond B Biol Sci*, Vol. 275, pp 299-325, ISSN 0962-8436.

Alegado, R.A.; Campbell, M.C.; Chen, W.C.; Slutz, S.S. & Tan, M.W. (2003). Characterization of mediators of microbial virulence and innate immunity using the *Caenorhabditis elegans* host-pathogen model. *Cellular Microbiology*, Vol. 5, No. 7, pp 435-444, ISSN 1462-5814.

Alegado, R. A. & Tan M.W. (2008). Resistance to antimicrobial peptides contributes to persistence of *Salmonella typhimurium* in the *C. elegans* intestine. *Cellular Microbiology*, Vol.10, No. 6, pp 1259–1273, ISSN 1462-5814.

Bandyopadhyay, J.; Song, H. O.; Singaravelu, G.; Sun, J.L. ; Park, B.J.; Ahnn, J. & Cho, J.H. (2009). Functional assessment of Nramp-like Metal Transporters and Manganese *in C. elegans*. *Biochem Biophys Res Commun*, Vol. 390, No.1. pp 136-141, ISSN 0006-291X.

Brenner, S. (1974). The genetics of *Caenorhabditis elegans*. *Genetics*, Vol. 77, pp 71-94, ISSN 0016-6731.

Cassada, R. C. & Russell, R. L. (1975). The dauer larva, a post-embryonic developmental variant of the nematode *Caenorhabditis elegans*. *Dev Biol*, Vol.46, pp 326-342, ISSN 0012-1606.

Chiang, S.L.; Mekalanos, J.J. & Holden, D.W. (1999). *In vivo* genetic analysis of bacterial virulence. *Annu Rev Microbiol*, Vol. 53, pp 129-154, ISSN 0066-4227.

Dhakal, B.K.; Lee, W.; Kim, Y.P.; Choy, H.E.; Ahnn. J. & Rhee, J.H. (2006). *Caenorhabditis elegans* as a simple model host for *Vibrio vulnificus* infection. *Biochem Biophys Res Comm*, Vol. 346, pp 751-757, ISSN 0006-291X.

Ewbank, J. J. (2002). Tackling both sides of the host-pathogen equation with *Caenorhabditis elegans*. *Microbes Infect*, Vol. 4, pp 247-256, ISSN 1286-4579.

Fire, A.; Xu, S.; Montgomery, M.K.; Kostas, S.A.; Driver, S.E. & Mello, C.C. (1998). Potent and specific genetic interference by double-strand RNA in *Caenorhabditis elegans*. *Nature* Vol. 391, pp 806-811, ISSN 0028-0836.

Fuhrman, L.E.; Goel, A.K.; Smith, J. ; Shianna, K.V. & Aballay, A. (2009). Nucleolar proteins suppress *Caenorhabditis elegans* innate immunity by inhibiting p53/CEP-1. *Plos Genetics*, Vol.5, No. 9, pp 1-14, ISSN 1553-7390.

Garsin, D.A.; Sifri, C.D.; Mylonakis, E.; Qin, X.; Singh, K.V.; Murray, B.E.; Calderwood, S.B. & Ausubel, F.M. (2001). A simple model host for identifying Gram-positive virulence factors. *Proc Natl Acad Sci USA*, Vol. 98, pp 10892–10897, ISSN 0027-8424.

Houthoofd, K.; Braeckman, B. P.; Johnson, T. E. & Vanfleteren, J. R. (2003). Life extension via dietary restriction is independent of the Ins/IGF-1 signalling pathway in *Caenorhabditis elegans*. *Exp Gerontol*, Vol. 38, pp 947-954, ISSN 0531-5565.

Huffman, D.L.; Bischof, L.J.; Griffitts, J.S. & Aroian, R.V. (2004). Pore worms: using *Caenorhabditis elegans* to study how bacterial toxins interact with their target host. *Int J Med Microbiol* , Vol. 293, pp 599-607, ISSN 1438-4221.

Ikeda, T.; Yasui, C.; Hoshino, K.; Arikawa, K. & Nishikawa, Y. (2007). Influence of lactic acid bacteria on longevity of *Caenorhabditis elegans* and host defense against *Salmonella enterica* serovar Enteritidis. *Applied and Environmental Microbiology*, Vol. 73, No.20, pp 6404–6409, ISSN 0099-2240.

Jia, K.; Thomas, C.; Akbar, M.; Sun, Q.; Adams-Huet, B.; Gilpin, C. & Beth Levine. (2009). Autophagy genes protect against *Salmonella typhimurium* infection and mediate insulin signaling-regulated pathogen resistance. *Proc Natl Acad Sci USA*, Vol.106, No.34, pp 14564–14569, ISSN 0027-8424.

Johnson, C.D. & Liu, L.X. (2000). Novel antimicrobial targets from combined pathogen and host genetics. *Proc Natl Acad Sci USA*, Vol. 97, pp 958-959, ISSN 0027-8424.

Kothe, M.; Antl, M.; Huber, B.; Stoecker, K.; Ebrecht, D.; Steinmetz, I. & Eberl. L. (2003). Killing of *Caenorhabditis elegans* by *Burkholderia cepacia* is controlled by *cep* quorum-sensing system. *Cellular Microbiology*, Vol. 5, pp 343-351, ISSN 1462-5814.

Kurz, C.L.; Chauvet, S.; Andrès, E.; Aurouze, M; Vallet, I; Michel, G.P.; Uh, M.; Celli, J.; Filloux, A.; De Bentzmann, S.; Steinmetz, I.; Hoffmann, J.A.; Finlay, B.B.; Gorvel, J.P.; Ferrandon, D. & Ewbank, J.J. (2003). Virulence factors of the human opportunistic pathogen *Serratia marcescens* identified by *in vivo* screening. *EMBO J*, Vol. 22, pp 1451-1460, ISSN 0261-4189.

Kurz, C. L. & Ewbank, J. J. (2000). *Caenorhabditis elegans* for the study of host-pathogen interactions. *Trends Microbiol*, Vol. 8, pp 142–144, ISSN 0966-842X.

Kurz, C. L.; Shapira, M.; Chen, K.; Baillie, D. L. & Tan, M.-W. (2007). *Caenorhabditis elegans pgp-5* is involved in resistance to bacterial infection and heavy metal and its regulation requires TIR-1 and a p38 Map Kinase cascade. *Biochem Biophys Res Commun*, Vol. 363, pp 438-443, ISSN 0006-291X.

Labrousse, A.; Chauvet, S.; Couillault, C.L. & Ewbank, J.J. (2000). *Caenorhabditis elegans* is a model host for *Salmonella typhimurium*. *Curr Biol*, Vol.10, pp 1543–1545, ISSN 0960-9822.

Lee, S.H. & Camilli, A. (2000). Novel approaches to monitor bacterial gene expression in infected tissue and host. *Curr Opin Microbiol*, Vol.3, pp 97-101, ISSN 1369-5274.

Mahajan-Miklos, S.; Tan, M. W.; Rahme, L.G. & Ausubel, F.M. (1999). Molecular mechanisms of bacterial virulence elucidated using *Pseudomonas aeruginosa-Caenorhabditis elegans* pathogenesis model. *Cell*, Vol.96, pp 47-56, ISSN 0092-8674.

Mylonakis, E.; Ausubel, F.M.; Perfect, J.R.; Heitman, J. & Calderwood, S.B. (2002). Killing of *Caenorhabditis elegans* by *Cryptococcus neoformans* as a model of yeast pathogenesis. *Proc Natl Acad Sci USA*, Vol. 99, pp 15675–15680, ISSN 0027-8424.

Pujol, N.; Link, E.M.;Liu, L.X.; Kurz, C.L.; Alloing, G., Tan, M.W.; Ray, K.P.; Solari, R.; Johnson, C.D. & Ewbank, J.J. (2001). A reverse genetic analysis of components of the Toll signaling pathway in *Caenorhabditis elegans*. *Curr Biol*, Vol.11, pp 809-821, ISSN 0960-9822.

Rhee, J.E.; Jeong, H.G.; Lee, J.H. & Choi, S.H. (2006). AphB influences acid tolerance of *Vibrio vulnificus* by activating expression of the positive regulator CadC. *J Bacteriol*, Vol.188, pp 6490-6497, ISSN 0021-9193.

Sifri, C.D.; Begun, J.; Ausubel, F.M. & Calderwood, S.B. (2003). *Caenorhabditis elegans* as a model host for *Staphylococcus aureus* pathogenesis. *Infect Immun*, Vol. 71, No. 4, pp 2208-2217, ISSN 0019-9567.

Sifri, C.D.; Begun, J. & Ausubel, F.M. (2005). The worm has turned--microbial virulence modeled in *Caenorhabditis elegans*. *Trends Microbiol*, Vol.13, No. 3, pp 119-127, ISSN 0966-842X.

Sifri, C.D.; Mylonakis, E.; Singh, K.V.; Qin, X.; Garsin, D.A.; Murray, B.E.; Ausubel, F.M. & Calderwood, S.B. (2002). Virulence effect of *Enterococcus faecalis* protease genes and the quorum-sensing locus *fsr* in *Caenorhabditis elegans* and mice. *Infect Immun*, Vol. 70, pp 5647–5650, ISSN 0019-9567.

Styer, K. L.; Singh, V.; Macosko, E.; Steele, S.E.; Bargmann, C. I. & Aballay, A. (2008). Innate immunity in *Caenorhabditis elegans* is regulated by neurons expressing NPR-1/GPCR. *Science*, Vol. 322, No. 5900, pp 460–464, ISSN 0036-8075.

Tabara, H.; Grishok, A. and Mello, C.C. (1998). RNAi in *C. elegans*: soaking in the genome sequence. *Science*, Vol. 282, pp 430-431, ISSN 0036-8075.

Tampakakis, E.; Peleg, A. Y. & Mylonakis, E. (2009). Interaction of *Candida albicans* with an intestinal pathogen, *Salmonella enterica* serovar Typhimurium. *Eukaryotic Cell*, Vol. 8, No. 5, pp 732–737, ISSN 1535-9778.

Tan, M.W.; Rahme, L.G.; Sternberg, J.A.; Tompkins, R.G. & Ausubel, F.M. (1999). *Pseudomonas aeruginosa* killing of *Caenorhabditis elegans* used to identify *P. aeruginosa* virulence factors. *Proc Natl Acad Sci USA* Vol. 96, pp 2408–2413, ISSN 0027-8424.

Tenor, J. L. & Aballay, A. (2008). A conserved Toll-like receptor is required for *Caenorhabditis elegans* innate immunity. *EMBO Reports*, Vol. 9, No. 1, pp 103-109, ISSN 1469-221X.

Tenor, J.L.; McCormic, B.A.; Ausubel, F.M. & Aballay, A. (2004). *Caenorhabditis elegans*-based screen identifies *Salmonella* virulence factors required for conserved host–pathogen interactions. *Curr Biol*, Vol. 14, pp 1018–1024, ISSN 0960-9822.

The *C. elegans* Sequencing Consortium. (1998). Genome sequence of the nematode *C. elegans*: a platform for investigating biology. *Science*, Vol. 282, pp 2012-2018, ISSN 0036-8075.

Timmons, L; Court, D.L. & Fire, A. (2001). Ingestion of bacterially expressed dsRNAs can produce specific and potent genetic interference in *Caenorhabditis elegans*. *Gene*, Vol. 263, pp 103-112, ISSN 0378-1119.

Vaitkevicius, K.; Lindmark, B; Ou, G.; Song, T; Toma, C.; Iwanaga, M; Zhu, J.; Andersson, A; Hammarstrom, M.L.; Tuck, S. & Wai, S.N. (2006). A *Vibrio cholerae* protease needed for killing of *Caenorhabditis elegans* has a role in protection from natural predator grazing . *Proc Natl Acad Sci USA*, Vol. 103, pp 9280-9285, ISSN 0027-8424.

Part 3

Novel Techniques

11

Immunoimmobilization of Living *Salmonella* for Fundamental Studies and Biosensor Applications

Zhiyong Suo[1], Muhammedin Deliorman[1], Sukriye Celikkol[1,3],
Xinghong Yang[2] and Recep Avci[1]
[1]Department of Physics, Montana State University, Bozeman
[2]Department of Immunology and Infectious Diseases
Montana State University, Bozeman
[3]Institute of Environmental Sciences, Bogazici University, Istanbul
Turkey

1. Introduction

Currently, there is no technique available to probe an individual bacterium in its physiological environment for a prolonged period of time to determine its response to environmental stimuli or to conduct measurements on it. Controlling and manipulating individual bacteria will facilitate fundamental studies of such bacterial characteristics as morphology, adhesion, biomineralization and mechanical properties under physiological conditions. Access to specific individual cells will open new research frontiers in areas such as differentiation among the individual offspring of a predetermined bacterium (Arnoldi et al. 1998; Chao et al. 2011; Gao et al. 2011). In many applications, it is necessary to immobilize bacteria on flat substrates or particles. For example, there have been many reports on using living bacteria as sensors for environmental monitoring because of their low cost, fast growth, easy genetic modification and handling, and sensitivity to a wide variety of environmental stimuli (Kuang et al. 2004; Mbeunkui et al. 2002; Premkumar et al. 2002). It is often necessary to immobilize living bacteria on a designated area of a detecting surface to build a practical sensor. Reliable, controllable and efficient immobilization of bacteria is crucial for the success of pathogen detection. Typically, a captured bacterium triggers an event that converts the capturing process into a signal which is detectable by optical, electrochemical, mechanical or other means (O'Kennedy et al. 2009). The technique developed by our group appears to be very promising for these applications. Before we describe our methodology here, we will give a brief review of the previous approaches and methods for the immobilization of bacteria on material surfaces.

1.1 Nonspecific immobilization of bacteria through physical adsorption or entrapment

The majority of reported immobilization approaches utilize either the nonspecific adsorption of bacterial cells on charged surfaces by means of electrostatic forces or the physical entrapment of cells in gel or micro-holes.

1.1.1 Physical absorption

Typically, bacterial surfaces are negatively charged under physiological conditions for most Gram-positive and Gram-negative species. Thus, it is possible for bacteria to adhere to a positively charged substrate prepared by modifying the surface using positively charged polymers or silanes. Various bacteria have been attached to substrates decorated with polylysine (Rozhok et al. 2006; Rozhok et al. 2005), polyethyleneimmine (Razatos et al. 1998), amino-terminated silanes (Arnoldi et al. 1998), gelatin (Doktycz et al. 2003) and alginate (Polyak et al. 2001). However, unlike eukaryotic cells, bacterial cells are still very challenging to immobilize reliably and reproducibly under their physiological conditions using positively charged polymers. For example, none of the experiments we report here could be done using this approach. This is mostly because, in contrast to a eukaryotic cell, only a very small fraction of a bacterial cell surface can come into close enough contact with a charged substrate surface to adhere, preventing the bacterium from attaching to the surface effectively. Additionally, many bacterial species, including *Salmonella*, have a layer of capsular extracellular polymeric substances (EPS) covering their outer surface, as shown in Fig. 1A (Suo et al. 2007), which further weakens interactions with and adhesions to the substrate surface.

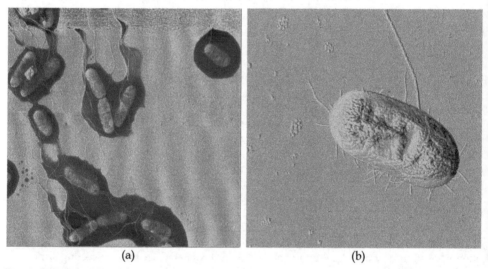

(a)   (b)

Fig. 1. AFM phase images of *S.* Typhimurium showing the detailed structures of (A) flagella and capsular EPS (scan size: 20 μm), and (B) a single bacterium with its fimbriae and part of its flagellum (scan size: 4 μm).

1.1.2 Physical entrapment

Because bacterial motion can be significantly slowed in viscous media or micro-cavities, living bacteria can be physically entrapped in hydrogels or inside microwells. Micro- (Xu et al. 2007) and macro-contact printing (Weibel et al. 2005) have been employed to transfer live bacteria onto the surface of a nutrient-rich matrix such as agarose or hydrogel. Bacterial microarrays have also been prepared by loading individual bacterial cells into microwells (2.5 μm wide, ~3 μm deep) at the distal end of an optical fiber bundle by centrifuge (Brogan and Walt 2005). A bacterial array printed onto porous nylon has also been reported

(Heitkamp and Stewart 1996), in which cells were physically entrapped in the pores of a special nylon substrate in close contact with a nutrient medium. Akselrod et al. reported three-dimensional heterotypic arrays of living cells in hydrogels created by means of high-precision (submicron accuracy) time-multiplexed holographic laser trapping (Akselrod et al. 2006). However, this technique has limited applications in practice, as, besides the need for a trapping laser, excessive exposure to laser light may cause photodamage to the cells; furthermore, arrays are expected to merge in a few hours because of cell division. The entrapment methods suffer from slow response times, low loading rates into microwells and easy detachment from surfaces.

2. Antibody-mediated immobilization of bacteria

Another approach to bacterial immobilization takes advantage of the interaction between a receptor and an appropriate ligand on a bacterial surface. Many different types of receptors can be used for this purpose, including enzyme receptors, nucleic acid receptors, polysaccharide receptors (lectins) (Gao et al. 2010) and antibodies against bacterial surface antigens. Since receptors often recognize specific types of ligands on a bacterial surface, such immobilization could achieve a high degree of specificity and efficiency. In this chapter, we focus on antibody-mediated immobilization, referred to as immunoimmobilization. Readers can refer to a recent review for the current status of the field of immobilization using a broad spectrum of receptors (Velusamy et al. 2010b).

The large variety of bacterial surface antigens and corresponding antibodies offers a number of choices for immunoimmobilization, which could be highly specific for a given species. This approach has been used to detect *Salmonella* (Table 1) (Mantzila et al. 2008; Oh et al. 2004) and other bacterial pathogens (Byrne et al. 2009; Skottrup et al. 2008; Velusamy et al. 2010a). Efficient capturing is always desired for bacterial detection, since it will facilitate converting captured pathogens into a detectable signal and, most importantly, a higher capture efficiency will result in a higher sensitivity (lower detection limit). Extensive research has been reported on the development of new detection methods that involve converting an already captured pathogen into an output signal by optical, electrochemical, mechanical or other means (O'Kennedy et al. 2009). However, there has been little study of how to enhance the capture efficiency. In fact, poor immobilization of bacterial cells is often observed. For example, only 2% surface coverage of the bacteria was achieved for a sensor using *E. coli* to monitor environmental toxicity (Premkumar et al. 2001).

In order to achieve reliable and efficient immunoimmobilization, the substrate should be decorated with a dense layer of an antibody which targets the most abundant antigen on the bacterial surface. This requirement draws attention to the two most critical aspects in immunoimmobilization: optimization of the surface chemistry to maximize the antibody density on the substrate surface and selection of an antibody which targets the appropriate bacterial surface antigen such as fimbriae (Fig. 1B).

2.1 Surface chemistry

The surface chemistry for linking antibody molecules is similar to that which has been popularly used to prepare protein microarrays and protein-modified resin (Hermanson et al. 1992). However, in order to achieve a high immobilization efficiency, a substrate for bacterial immobilization should have a larger number of antibody molecules on the surface, with the paratope of each antibody molecule pointing away from the substrate. It would also be desirable for the antibody molecules to have sufficient freedom of movement to orient themselves in a proper binding direction towards the bacterial antigens.

As shown in Fig. 2A, thiolated tethers including 16-mercaptohexadecanoic acid (MHA) and 11-mercapto-undecanoic acid (MUA) have been commonly used to activate gold and silver surfaces. The carboxyl terminal of MHA and MUA can link to amino groups after activation. Silanes with an active terminal are widely used for silicon oxide, silicon nitride, glass, indium tin oxide (ITO), aluminum, titanium and steel surfaces. In Fig. 2, two popular silanes are shown, aminopropyltriethoxylsilane (APTES) (Fig. 2B) and (3-glycidoxypropyl)-trimethoxysilane (GOPTMS) (Fig. 2C). APTES is further modified with N-(3-maleimidopropionyloxy)succinimide (BMPS), a short cross-linker, to provide an active maleimido terminal to link to the cysteine residue of antibodies. The glycidyl terminal can react with amino or hydroxyl groups of an antibody. Another popularly used linkage is the biotin-avidin (streptavidin/neutravidin) system (Fig. 2E), which involves the covalent linking of biotin to antibodies followed by the binding of biotin-labeled antibodies to an avidin layer on the substrate (Taitt et al. 2004). Antibodies can also be linked to substrate surfaces through protein A/G/L (Fig. 2D), which specifically binds to the Fc region of IgG (Choi et al. 2008; Gao et al. 2006).

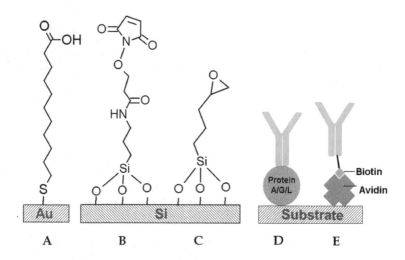

Fig. 2. Chemistry of antibody linkage

Many short cross-linkers are commercially available for linking antibodies; some examples are shown in Fig. 2. An antibody attached to a substrate through one of these cross-linkers forms a dense two-dimensional (2-D) monolayer on the substrate in which the paratopes of the individual antibody molecules are randomly oriented and the individual antibody molecules have very limited freedom of movement for reorientation. A short cross-linker works well with a high-purity antibody against a bacterial surface antigen: this gives a fairly high expression level, and a satisfactory immobilization of bacteria can be achieved. However, when the antibody is not pure or has a low affinity, or when the bacterial surface antigen has a low expression level, it becomes necessary to boost the antibody binding probability by aligning the antibody molecules so that their paratopes point away from the substrate and using long cross-linkers to obtain the necessary degree of freedom of movement.

When linked through flexible tether molecules with lengths varying from tens to hundreds of nanometers, the antibody molecules will form a three-dimensional (3-D) network on the substrate surface. Such tethers provide the necessary degree of freedom of movement for an antibody to access a larger fraction of the bacterial surface and increase the immobilization efficiency. These tethers can be constructed from brush polymers, dendrimers (Han et al. 2010), certain peptides and block copolymers. Because of their flexibility and polydispersity, these highly branched tethers will maximize the loading of antibody molecules onto the substrate surface in a 3-D network, hence increasing the number of antibody molecules per unit area to more than can be linked by short tethers forming a 2-D network.

Poly(ethylene glycol) (PEG) has been widely used for surface modification since the early 1990's, and a variety of PEG-based cross-linkers are now commercially available. The aqueous solubility and flexibility of PEG make these linkers ideal for antibody molecules. It should be noted that most available PEG cross-linkers can only link one antibody molecule to a terminal, so it is expected that surfaces modified by such PEG cross-linkers will be covered by only one monolayer of antibody.

2.2 Antibody-antigen selection

Antibody-mediated immobilization works in a complex environment, such as growth medium, blood or a food sample, which simplifies the sample preparation. However, previous work using antibodies against whole bacterial cells often resulted in low immobilization efficiency (Premkumar et al. 2001; Rozhok et al. 2005). A general guideline for antibody selection is to use antibodies targeting antigens on the surface of the bacterium (the outer membrane of Gram-negative and the peptidoglycan layer of Gram-positive bacteria). Antigens inside bacterial cells or embedded in cell wall components usually should be avoided since it is impossible for antibody molecules to reach them in living bacteria. A systematic evaluation and comparison of the immobilization efficiencies of selected antibody-antigen pairs associated with common bacterial surface antigens is still needed.

We have evaluated the immobilization efficiencies of IgG antibodies against four different types of surface antigens of *S.* Typhimurium and *E. coli*: lipopolysaccharides (LPS), flagella, fimbriae, and a capsular protein (Suo et al. 2009a). The results show that, with the exception of the capsular protein, all the surface antigens tested can in principle be targeted to achieve some degree of immobilization and that the immobilization efficiency is correlated to multiple factors, especially to the choice of antibody-antigen pairs.

2.2.1 Method

The immobilization efficiency is defined by the number of immunoimmobilized bacteria per unit area within a specific time period for a specific concentration of bacteria in the medium. The antibody solution is deposited onto an activated silicon substrate as small droplets. Because of the specificity of the antibody-antigen interaction, bacterial cells are immobilized only inside the antibody-modified areas. Therefore, a sharp separation of the bacteria-covered areas from those which are not covered is expected. An optical image focused on the antibody-coated areas is used to determine the immobilization efficiency.

2.2.2 Anti-fimbria antibodies

Antibodies against various fimbriae have been tested in our lab, and they usually result in very efficient immobilization. We have studied the immobilization efficiency of both engineered and wild-type strains of *S.* Typhimurium and *E. coli* using antibodies against various types of fimbriae, including K88ab (F4), K88ac (F4), K99 (F5), 987P (F6), F41 and CFA/I. An example of two CFA/I-expressing strains (*S.* Typhimurium Δ*asd::kan*R H71-pHC and *E. coli* H681-pBBScfa) being immobilized on a silicon substrate modified with anti-CFA/I is given in Fig. 3A,B. A sharp boundary can be observed separating the bacteria-covered area from the area that is not covered, indicating the high specificity and efficiency of the antibody binding. The cell coverage within the antibody-modified area approached a dense monolayer. The cell density of *E. coli* H681-pBBScfa was slightly lower than that of *S.* Typhimurium Δ*asd::kan*R H71-pHC, because of the lower CFA/I expression level of the *E. coli* strain relative to that of the *S.* Typhimurium strain. The areas outside the antibody-modified regions (upper right-hand sides of the panels in Fig. 3) serve as a negative control for evaluaing the immobilization efficiency. Usually no cells or only sparsely attached cells were observed in these control areas. We conducted additional control experiments on similar substrates using no antibody and using an irrelevant antibody (anti-cytochrome *c*), for which no immobilization was observed.

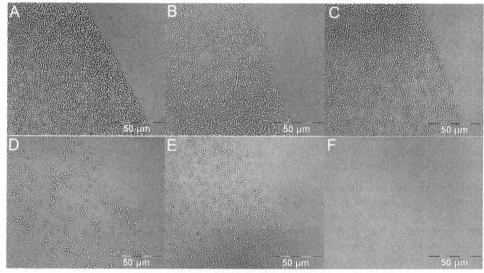

Fig. 3. Immobilization of bacteria using antibodies targeting various bacterial surface antigens: (A) *S.* Typhimurium Δ*asd::kan*R H71-pHC on substrate modified with anti-CFA/I, (B) *E. coli* H681-pBBScfa on substrate modified with anti-CFA/I, (C) *S.* Typhimurium *motA3::cat* H683-pTP2fliC (with flagella motion paralyzed) on substrate modified with anti-flagellin (notice that all the bacteria are lying down), (D) *S.* Typhimurium H647 (with active flagella) on substrate modified with anti-flagellin (notice that almost all the bacteria are standing up) (E) H647 on substrate modified with anti-O4 antigen (sc52224), and (F) *S.* Typhimurium Δ*asd::kan*R H71-pF1 on substrate modified with anti-F1.

2.2.3 Anti-flagellin

The efficiency of immobilization using anti-flagellin and anti-LPS was tested on *S.* Typhimurium H647, which expresses flagella but lacks CFA/I fimbriae. As expected, this strain could not be immobilized on substrate modified with anti-CFA/I antibody. However, H647 could be immobilized on substrates modified with anti-flagellin (Fig. 3C,D), although with a relatively low cell density as compared with CFA/I fimbriae. In spite of the fact that anti-flagellin shows a reasonably strong affinity to purified *S.* Typhimurium flagella (data not shown), the coverage density of H647 could not be improved by increasing the incubation time (Suo et al. 2009a). We speculate that the low cell coverage density of H647 is a result of the high-speed rotary motion of the flagella, which can be as high as 10,000 rpm at 35°C (Magariyama et al. 2001), hindering antibody-antigen interactions. Preliminary results showed that the immobilization efficiency could be enhanced while using the same anti-flagellin by paralyzing the flagella motion, as shown in Fig. 3C for *motA3::cat* H683-pTP2fliC. Notice in Fig. 3D that almost all the bacteria are standing up, most likely because of the flagella motion, as opposed to lying down when their flagella are paralyzed (Fig. 3C).

2.2.4 Anti-LPS

The antibodies against *S.* Typhimurium LPS showed the anticipated results, in that only one of the four antibodies tested demonstrated efficient immobilization of H647. LPS is an important amphiphilic molecule extending out from the bacterial outer membrane. It is composed of three covalently linked domains: lipid A, core antigen (oligosaccharide) and O-antigen (polysaccharide) (Raetz 1996). The lipid A is embedded in the outer membrane lipid bilayer, and hence it is expected that it would be difficult for an antibody to recognize it in a living bacterium. Our observations support this expectation, as the antibody against lipid A failed to provide any immobilization of *S.* Typhimurium. The saccharides, including both the core antigen and O-antigen, protrude from the phospholipid bilayer of the outer membrane and, therefore, can serve as potential targets for immunoimmobilization. Our experiments also showed no successful immobilization for the antibody against the core antigen. This implies that the core antigen is shielded by the O-antigen, preventing antibody-antigen interactions.

About 67 types of O-antigens have been identified for *Salmonella* serovars (Grimont and Weill 2007). These O-antigens are long chains of polysaccharides with a total length of up to 40 repeating units, typically with three to six sugars in each repeating unit (Raetz 1996). The sugar composition and the alteration of linkages among the sugars determine the serogroup to which a specific strain belongs (Selander et al. 1996). We tested three commercially available antibodies targeting *S.* Typhimurium O-antigen (Santa Cruz Biotech. Inc., Santa Cruz, CA): sc52221, sc52223 and sc52224. The latter two antibodies (sc52223 and sc52224) are specific to the O4-antigen. Only one antibody (sc52224) showed successful immobilization (Fig. 3E). *S.* Typhymurium usually contains O-antigens 4, 5 and 12 (Grimont and Weill 2007), all of which share a common tetrasaccharide repeating unit (with different sizes) given by α-D-mannose-1$\rightarrow$2-α-L-rhamnose-1$\rightarrow$3-α-D-galactose trisaccharide, to which an abequose is α1,3 linked (Curd et al. 1998; Weintraub et al. 1992). So far there is not enough information on the O-antigen structure of H647, and we infer that H647 may not contain O-4 antigen. H647 is a recombinant strain constructed by complementing an *asd* mutation strain H683 with an *asd*$^+$ plasmid (Ascon et al. 1998). There are results indicating the *asd*$^+$ plasmid may interfere with the expression of cell wall components including LPS in our experiments. Based on the fact that only one of the two monoclonal O4-specific antibodies showed positive results in immobilization experiments, we hypothesize that H647 altered its O-antigen structure to a different group, O-9 (D1), which has a repeating unit of α-D-mannose-1→2-α-L-rhamnose-1→3-α-D-galactose trisaccharide to which a tyvelose is αl,3 linked (Curd et al. 1998). This is a reasonable hypothesis considering that this structure is very similar to serogroup O-4 in that tyvelose differs from abequose only in the 3-D orientation of their OH groups at the 2 and 4 positions. Although more experiments are needed to verify this hypothesis, the results demonstrate that the antibodies targeting the O-antigen can be used for bacterial immobilization. The high specificity of anti-O-antigens can be used for rapid serotyping of *Salmonella* (Cai et al. 2005).

2.2.5 Anti-capsular protein

A polyclonal antibody raised against the proteinaceous capsular antigen (F1-antigen) was tested for immobilizing S. Typhimurium. F1-antigen, originally discovered for *Yersinia pestis*, can form a dense amorphous capsule that covers the bacterium (Friedlander et al. 1995; Titball and Williamson 2001). S. Typhimurium strain Δ*asd::kan*R H71-pF1 was constructed to express F1-antigen as a model bacterium. The expression of F1 antigen was confirmed by immunofluorescence and Western blot analysis (Yang et al. 2007). It was expected that such a capsular proteinaceous antigen would have a relatively strong interaction with the corresponding antibody. However, Δ*asd::kan*R H71-pF1 cells could not be immobilized on the substrate premodified with anti-F1 antigen (Fig. 3F). Most likely this is because the F1 capsule (Fig. 1), unlike CFA/I fimbriae, is not tightly bound to the bacterial cell wall and, therefore, can easily slough off from the surface of the bacterium, leading to the failure of bacterial immobilization.

2.3 Other factors

Other factors, such as the clonal type, the purity of the antibody, the antigen expression level, and the incubation time, also contribute to immobilization efficiency. The optimization of all these factors will lead to a very low detection limit. For example, we can readily detect S. Typhimurium or *E. coli* at a concentration of 10^3 to 10^4 CFU/ml.

3. A platform for studying an individual living bacterium

Immunoimmobilization offers access to controlling and manipulating living bacteria at the single-cell level. With this technique, it is possible to deploy a single living bacterium at a designated location, and then conduct measurements on the living bacterium with optical, mechanical, electrochemical and other tools. Combining these with PCR techniques, it is also possible to investigate the bacterial genomes of individual cells. Below, we present the preparation of bacterial micropatterns and the study of the mechanical properties of living S. Typhimurium.

3.1 Preparation of bacterial micropatterns

During the past decades there have been great advances in micro-electromechanical systems (MEMS) and nano-electromechanical systems (NEMS) technologies. Most of the techniques, such as microcontact printing, focused ion beam (FIB) etching, micro-plotting, e-beam lithography and dip-pen nanolithography (DPN), can be used to prepare micro- and nano-scale patterns on a flat substrate (Salaita et al. 2007).

The successful preparation of bacterial patterns relies mainly on the preparation of high-quality antibody micropatterns, which can be made using two general approaches: (1) modifying the substrate surface chemically to form a chemical micropattern in which the antibody will bind only inside or outside the modified areas and (2) depositing the antibody directly onto designated locations on an activated substrate surface. Most patterning techniques, such as FIB etching, DPN, e-beam lithography and microcontact printing, fall into one of these categories. The advantage of the first approach is the convenience of preparing patterns at nanoscale resolution (except for microcontact printing, which has a practical resolution of around 1 μm (Huck 2007)), which makes it possible to prepare single-cell arrays. However, it is difficult to represent multiple antibodies on the substrate using this approach since there are fairly limited chemical linkages available for antibody differentiation. The second approach (direct deposit approach) often refers to the microplotting method, which uses antibody solutions as ink for preparing antibody microspots on various substrates. We have tried two different microplotters whose antibody pattern sizes reached down to 25 μm in diameter. The antibody spot size is related to many factors, including the tip size, the viscosity of the protein solution, the hydrophobicity of the substrate surface and the moving speed of the tip of the microplotter (Larson et al. 2004), but generally it is very difficult to obtain spots smaller than 25 μm. The microplotting approach has a great advantage in that there is, in principle, no upper limit on the number of antibodies that can be represented in the microarray. For both approaches, it is necessary to passivate the substrate areas to prevent the nonspecific absorption of bacterial cells. In the first approach, the passivation is done by modifying the substrates using PEGlyated tethers (Lahiri et al. 1999) before patterning the surface with antibodies. In the microplotting methods, the passivation is done by post-exposing the protein micropatterns to BSA or milk proteins before exposing them to bacterial cultures.

Once prepared, an antibody micropattern is incubated with bacterial cultures. The bacterial cells are immobilized only on the antibody-patterned areas and thus form bacterial micropatterns. In Fig. 4, micropatterns of living S. Typhimurium prepared using FIB, microplotting and DPN are presented as examples. As can be seen in Fig. 4A, a cellular resolution of bacterial patterning is achieved in FIB patterning. Some of the line thicknesses in the patterns, ~1 μm, are comparable to the dimensions of the bacteria, and the bacteria are concentrated along these narrow lines; very few cells are observed outside the lines. Fig. 4B shows well-defined circular patterns obtained by microplotting. Single-cell resolution is best

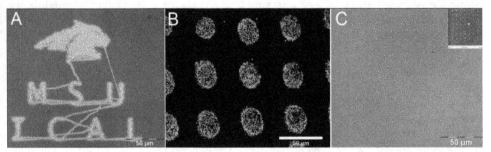

Fig. 4. Micropatterns of S. Typhimurium prepared by incubating the bacterial culture with antibody micropatterns fabricated using (A) FIB etching, (B) microplotting, and (C) DPN. Inset of panel C: A DPN pattern on gold substrate before antibody linking (scan size, 20 μm).

demonstrated by the microarray in Fig. 4C, which was prepared using the DPN method. An array of 16×16 antibody spots (each submicron in diameter) was placed on a chip using a sharp AFM tip. Notice that almost all the antibody spots captured at least one bacterium, most of the time multiple bacteria. This shows that immunoimmobilization, applied properly, is an extremely efficient technique.

3.2 Bacterial cells remain bioactive while immunoimmobilized

The question of whether or not immunoimmobilized microorganisms will maintain bioactivities such as cell division is highly relevant. An excessive number of antibodies surrounding a bacterium will eventually alter its behavior and perhaps even kill it. The technique used in immunoimmobilization is a fairly mild treatment of the bacterium that is immobilized. All the antibodies are bound to the surface; hence, only those bacteria that are near the surface of the substrate will interact with the antibodies. Of the many antibody molecules (~thousands per μm^2) evidence points to only a handful (tens per bacteria) interacting with the organism and keeping it bound to the surface. The evidence for this is that an immobilized bacterium still moves around with a considerable degree of freedom; in fact, when necessary it stands up, as mentioned above and also shown below, indicating that it is held to the surface by a small number of antibody-antigen pairs. This fact enables organisms to maintain their usual bioactivities, while giving rise to secondary activities triggered by the process of immobilization. We have seen evidence of such activity in the excessive production of flagella by bound bacteria as compared to that of the planktonic variety of the same bacteria (Suo et al. 2009a). Below we give some examples of the bioactivities of immobilized bacteria.

3.2.1 Immobilized bacteria are capable of division

Our results, as well as those of previous studies, indicate that immunoimmobilization does not hinder such physiological activities as the cell division (Suo et al. 2008), gene expression or bioluminescence of bacteria at the locations of their immobilization (Premkumar et al. 2001).

Immunoactivated areas exposed to low concentrations of bacteria (10^3-10^4 CFU/ml) for long periods of times (~20 hr) become populated by living bacteria for two reasons: (1) As time passes, some bacteria, even at low concentrations, are eventually captured by the antibodies within the activated areas. (2) The offspring of the captured bacteria start populating the antibody-activated areas (Fig. 5) because they too are captured at these locations following cell division, adding to the density of immunoimmobilized bacteria. Experiments with low cell concentrations are currently in progress, and the results will be published elsewhere.

It should be noted that once the surface is covered by a monolayer of live cells, the bacterial patterns on the antibody-modified areas do not change physical dimensions as a result of cell division. After the activated area is fully covered with bacteria, excess bacteria are released into the medium and become planktonic. This feature is not available for patterns prepared by embedding (Mbeunkui et al. 2002; Polyak et al. 2001; Premkumar et al. 2002; Weibel et al. 2005; Xu et al. 2007) or physical entrapment (Kuang et al. 2004). For these patterns, bacteria are held on the substrates or inside the microwells with weak forces, and the cells do not stay fixed to the patterned areas for long periods of time, causing the eventual disintegration of the pattern.

Fig. 5. Division of immobilized bacterial cells: (A) Four cells were captured inside the antibody-modified area (the bright square) when the substrate was incubated with a *S.* Typhimurium culture of 10^4 CFU/ml and (B) more cells were observed after the sample was incubated in cell-free medium for 20 hr at room temperature.

3.2.2 Crowded bacteria stand up

A surprising result is that bacteria initially immobilized in a lying-down orientation take a standing-up orientation as their density increases (Fig. 6A,B). This standing-up orientation of crowded cells has also been confirmed by laser scanning confocal microscopy (LSCM) images (Fig. 6C,D). Images (A) and (B) correspond to a sample incubated in growth medium at 37°C for 3 and 15 h, respectively. Notice that all the cells that were lying down in panel (A) appear to stand up in panel (B), corresponding to the increased bacterial density (see also panel (D)). Fig. 6C shows an LSCM image of a sample incubated in growth medium for

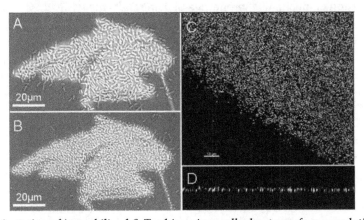

Fig. 6. Reorientation of immobilized *S.* Typhimurium cells due to surface population density. (A) Sample incubated in growth medium at 37°C for 3 h. (B) Same sample, incubated in growth medium at 37°C for 15 h. (C) LSCM image of a sample incubated in growth medium for 15 hr and then stored in PBS buffer at 4°C for 6 h and stained with viability stains. (D) Reconstituted Z-section image of the immobilized cells in (C).

15 hr and then stored in PBS buffer at 4°C for 6 h and stained with viability stains, which indicated that a majority of the cells were alive. The reconstituted Z-section image of the immobilized cells (Fig. 6D) further confirms that a majority of the cells took a standing-up orientation when crowded. Although at this time this behavior is not well understood, it might be related to the depletion of nutrients at the crowded bacterial positions and to the struggle of the bacteria to move away from their immobilized positions. This hypothesis is supported by the observation of an excess number of flagella produced by the immobilized bacteria (Suo et al, *Langmuir* 2008), presumably in an effort to free themselves from their positions.

3.2.3 Biofilm formation

The immobilized bacteria retained their cellular functions, including division, and continued to divide until a dense monolayer of bacteria filled the patterned area of the surface (Suo et al. 2008). We observed that biofilm started to form on the substrate after the initial two weeks. We incubated fifteen identical antibody-modified silicon chips with *S.* Typhimurium Δ*asd::kan*[R] H71-pHC to form the initial bacterial patterns, and then all of the chips were rinsed with PBS and transferred to non-flow cell-free growth medium at room temperature. These silicon chips were prepared such that the patterned area was modified with anti-CFA/I fimbriae and the rest of the area was passivated using PEG molecules to prevent nonspecific attachment (Suo et al. 2008). After a predetermined incubation period, one silicon chip was taken out and rinsed with PBS to remove the unattached and loosely attached cells, then imaged under the optical microscope to determine the bacterial content on the substrate surface and its viability. The bacterial patterns were maintained for about two weeks, after which a bacterial biofilm started to form on the substrate, both inside the antibody-modified area and in the PEG-covered areas (Suo et al. 2008).

3.3 Investigation of mechanical properties of living *Salmonella*

We measured the physical properties of living *S.* Typhimurium using an atomic force microscope (AFM) at room temperature (Suo et al. 2009b). In a previous study of bacterial turgor pressure, a fairly gentle pressure, not enough to break the bacterial cell wall, was applied to bacterial cells (Yao et al. 2002). A fairly common belief is that a bacterium will lyse and die if its cell wall suffers severe mechanical damage, such as being punctured by an AFM tip. Our results show not only that *S.* Typhimurium cells can survive such damage, but also that the cells are still capable of self-replication.

3.3.1 Puncturing curves

A layer of *S.* Typhimurium cells in their exponential growth phase was immobilized in well-defined square patterns on a silicon wafer with ~20 cells within each square. Such immobilization lends itself to the continuous observation of the same group of cells under a light microscope before and after they are subjected to AFM probing. A sharp AFM tip was brought into contact with a live *S.* Typhimurium cell, and the loading force on the tip was increased until the tip punctured the cell wall (Fig. 7A). This was marked by a sharp decrease in the cantilever deflection value from ~150 nm to ~35 nm (inset in Fig. 7B). The tip was pushed continuously until it stopped penetrating the cell, which was interpreted as the tip contacting the hard silicon surface underneath the organism. When the loading force reached a preset value (i.e. ~4 nN) the tip was lifted to ~2 μm above the cell surface in order to start another puncture cycle at an adjacent location on the organism. This puncturing process was repeated at a rate of 1 Hz or 2 Hz until each cell within a 7 × 7 μm^2 or 10 × 10 μm^2 area had undergone ~20 or ~40 puncturing events per μm^2. Each puncturing event generated a pair of force vs. displacement curves, or "puncturing curves" (Fig. 7B). These curves reveal a variety of information, including the true height of a live bacterium, its mechanical properties under physiological conditions, the pressure required to puncture the cell wall for a given tip geometry, and amount of cell deformation (indentation) before the cell wall is punctured. The inset shows the raw cantilever deflection versus displacement curves from which the puncturing curves were obtained by subtracting the cantilever deflection from the z-piezo displacement using a MatLab code developed by our group.

When a sharp tip punctures the cell wall of a Gram-negative bacterium it tears the three-layer cell wall structure before entering the bacterial cytoplasm. The pressure required to tear, or puncture, the cell wall is determined from the puncturing curve. The distance (~800 nm) between where the tip makes contact with the bacterial surface and where it touches the substratum is the measure of the true height of the bacterium in its physiological medium. The maximum penetration of F ≈ 2 nN at ~100 nm suggests that the bacterium deformed under the pressure exerted by the sharp tip and that the bacterial surface was indented by ~100 nm just before the tip punctured the cell wall. The modulus of elasticity of the living cell at the initial contact and the turgor pressure of the organism can be determined from the early part of the loading vs. tip penetration curve, using simple models such as the Hertzian model (Yao et al. 2002) or more complicated ones (Arnoldi et al. 2000).

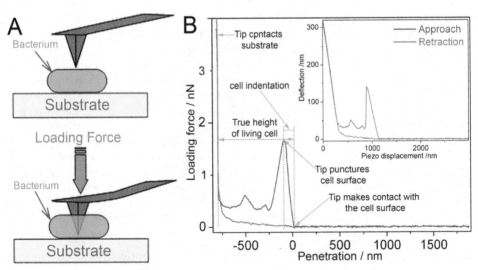

Fig. 7. (A) Schematics of puncturing experiments and (B) a typical puncturing curve

The pressure, P, required to penetrate the bacterial cell wall can be determined from P= $F/\pi r^2 \approx 5.0\pm0.8$ atm where r refers to the radius of the AFM tip. In our experiments, the average modulus of elasticity was calculated to be $0.4\pm0.2\times10^6$ Pa for living S. Typhimurium cells in growth medium or in PBS buffer. This is comparable to results reported for *P. aeruginosa* and *E. coli* (Ingraham and Marr 1996; Yao et al. 2002).

The fine structures in the penetration range from ~250 nm through ~600 nm are ascribed to the resistance the tip experienced as it penetrated the cytoplasm of the organism. These forces are related to the tip geometry and, more importantly, to the tip aspect ratio. The force the tip experiences when it is pushed into the cytoplasm can be traced to two sources: (1) vertical resistance from the cell membranes, particularly from the peptidoglycan layer, during the initial 50 to 200 nm indentation (depending on the tip radius), until the tip overcomes this resistance and punctures the cell wall, and (2) lateral resistance as the tip tears the peptidoglycan layer. A tip with a high aspect ratio, such as a "spike tip" with a conical geometry, will give virtually no fine structures after the initial puncturing. A lower aspect ratio tip (pyramid tip) will tear the cell wall much more as it pushes into the cytoplasm and will introduce more severe cell damage than a spike tip (Suo et al. 2009b).

The common feature among all the puncturing curves is that there was very little or no resistance as the tip was pulled back from the cytoplasm of the bacteria (red line in Fig. 7B). This lack of resistance during tip retraction implies that there are only weak tip/cell wall interactions once the tip breaks the cell wall. The lipids, which are in continuous contact with the tip, offer no resistance to the motion of the tip as it is withdrawn from the cytoplasm.

3.3.2 Bacteria maintain their viability after being punctured multiple times

The size of the puncture hole while a tip penetrates a bacterium depends on the tip radius, the aspect ratio of the tip and the depth of penetration. However, there is no evidence of damage due to a puncture hole left behind from a puncturing event. It appears that puncture holes self-repair so that bacterial integrity and functionality are maintained, which is supported by the observation that punctured cells are capable of cell division, as is shown in Fig. 8. In this experiment, the colony of bacteria encircled by the blue dashed square in Fig. 8B was imaged before and after the puncturing. The data show that there was a one-to-one correspondence between the bacteria inside the blue square and the bright features in the AFM force-volume image. The light pixels in the dark background mark the locations of the punctured bacteria with a density of 20 puncturing events per square micron. Fig. 8B corresponds to the optical image acquired immediately after the puncturing experiment. We then kept the bacteria in their growth medium while taking images of the same location at one-minute intervals for 100 minutes. The effect of puncturing on the viability of *S.* Typhimurium was also studied using viability dyes. The majority of cells that had been punctured multiple times were able to divide, and there was no statistical difference found in survival rate between the punctured and the un-punctured cells, regardless of tip geometry or puncturing density varying between 20 and 40 puncturing events per μm^2.

It is also notable that the puncturing curves for dead *S.* Typhimurium cells were markedly different from those for living cells (Suo et al. 2009b). *S.* Typhimurium cells killed using glutaraldehyde did not show the force maximum associated with the puncturing event for live bacteria. Further analysis of the puncturing curves of dead cells revealed that dead bacteria shrink by about 40% and also appear to be softer, with an elastic modulus of ~0.13 ± 0.07 MPa in PBS as opposed to an elastic modulus of ~0.50 ± 0.10 MPa for living bacteria.

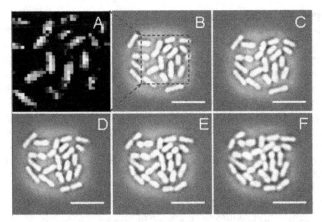

Fig. 8. Viability test: *S.* Typhimurium cells divide after being punctured multiple times. (A) An AFM force-volume image with bright spots showing the locations of the punctured bacteria. These spots have a one-to-one correlation with the cells shown in the blue dashed square of the optical image in panel B. Panels B-F show time-lapse images of the same group of bacteria taken 0, 25, 50, 75, and 100 min after the puncturing experiment, respectively. Cell division is highlighted by red, olive, and black ovals. Scale bar: 5 µm. Reprinted from Suo et al, *Langmuir* 2009 with permission from ACS

A model has been proposed to explain how bacteria survive the harsh damage done to their cell walls by an AFM tip. A detailed description of this model can be found elsewhere (Suo et al. 2009b). Briefly, the survival of punctured cells is owed to the remarkable features of the phospholipid bilayer that surrounds the cytoplasm of each bacterium. Phospholipid bilayers, at temperatures above the glass transition temperature, are composed of highly dynamic, fluid-like phospholipid molecules decorated with cross-membrane proteins (Osborn et al. 1972; Voet and Voet 1995). We believe that phospholipid molecules, including cross-membrane proteins with their hydrophobic regions intersecting the bilayers, undergo a reconfiguration (in a ns time frame) in response to AFM tip penetration (in a ms time frame). As the tip moves into or out of the cytoplasm, the fast-responding lipid membrane and membrane proteins are always in close contact with the tip surface and continuously seal against leakage into or out of the cytoplasmic and perhaps even the periplasmic regions regardless of the tip geometry. Apparently, a torn peptidoglycan layer is not a serious threat to cell viability.

4. Immunosensors for rapid detection of bacterial species

The rapid detection of microbial pathogens at a contamination site is critical for preventing the spread of the disease-causing microorganisms before the disease becomes epidemic. Diseases caused by water- and food-borne pathogens have been a serious threat to public health. A comprehensive examination of all the pathogen detection studies conducted over a period of 20 years showed that 38% were related to the food industry, 18% to clinical analysis, 16% to water and the environment and 27% to other areas (Lazcka et al. 2007), which emphasizes the importance of biosensor applications in these fields. *Salmonella* and *E. coli* were the most commonly detected pathogens in the aforementioned samples, with percentages of 33% and 27%, respectively (Turner 2011).

Bacterial pathogens are generally detected and identified using either a polymerase chain reaction (PCR) or antibody-based techniques (Velusamy et al. 2010a). The PCR approach offers an accurate determination of pathogens at the genomic level, but requires a proper design of primers targeting specific genes (Alexa et al. 2001; Holoda et al. 2005; Malorny et al. 2009). Antibody-based assays focus on the detection of bacterial antigens, and it is possible to detect multiple pathogens in a single assay using microarray techniques (Cai et al. 2005; Choi et al. 2008). Antibody-based techniques usually involve two events: capturing of the targeted pathogen on the sensor surface and follow-up signal generation. Efficient capturing is always desired, since it will facilitate converting captured pathogens into a detectable signal and, most importantly, a higher capture efficiency will result in a higher sensitivity (lower detection limit). Extensive research has been reported on the development of new detection methods that involve converting an already captured pathogen into an output signal by optical, electrochemical, mechanical or other means (O'Kennedy et al. 2009). During the past decade considerable advances were made in detecting pathogens by coupling immunological techniques with chemical and electronic actuators and techniques based on chemoluminescence (Wolter et al. 2008), electrochemical impedance (Geng et al. 2008), surface plasmon resonance (SPR) (Zordan et al. 2009), quartz crystal microbalance (QCM) (Adanyi et al. 2006; Boujday et al. 2008; Hirst et al. 2008) and wave guides (Adanyi et al. 2006). However, there has been little study of how to enhance the capture efficiency.

There have been many reports on the detection of *Salmonella* using antibody-based methods in the past decade, and in Table 1 we summarize these reports. The traditional detection of *Salmonella* spp. in food industries using differential growth media usually takes days to weeks (Amaguafia and Andrews 2000; Eaton et al. 2005). Detection based on immunoimmobilization or other antibody-based methods requires much less time, as evidenced by Table 1, showing typical assay times for these methods between 20 min and 6 hr. Such assays significantly reduce the time and effort needed for *Salmonella* detection and can possibly be used as on-site analysis in the critical initial stage of food poisoning.

As shown in Table 1, there have been many efforts to sense bacteria captured on an antibody-modified surface, ranging from the rather "traditional" method (ELISA) to new techniques such as microcantilevers. Most reports claim a detection limit close to 10^3 cells/ml, which could produce meaningful measurements in real applications, considering the infectious dose for human salmonellosis is around 10^3 CFU (Blaser and Newman 1982). Many different antibodies have been used, including commercial antibodies and antibodies prepared by the authors. It is desired for the readers to know the source, specificity, clonal type and purity of the antibodies used. However, it is not uncommon for inadequate information to be provided in this respect. For example, the term "anti-*Salmonella*" is often used without further specification.

Microarrays prepared using multiple antibodies are particularly efficient in the detection and identification of *Salmonella*. Cai et al. constructed a 8×15 model array for the identification of 20 common *Salmonella* serovars and evaluated the use of 117 target and 73 nontarget *Salmonella* strains (Cai et al. 2005). A total of 35 polyclonal antibodies against O-antigens or H-antigens (flagella) were used for serotyping *Salmonella* strains. Microarrays have been prepared using antibodies targeting multiple bacterial species. Choi et al. prepared a microarray containing four monoclonal antibodies against *E. coli* O157:H7, *S.* Typhimurium, *Yersinia enterocolitica*, and *Legionella pneumophila* (Choi et al. 2008). One problem with an antibody microarray is that generally the signal intensity of each antibody spot varies as a function of both the antibody concentration and the strain. This is actually

	Sensing element	Method	Detection limit	Medium	Assay time	Ref.
S. Infantis, S. Berta, S. enteritidis, S. Thompson, S. Typhimurium, S. Agona, S. Braenderup, S. Heidelberg, S. Dublin	paramagnetic beads coated with specific antibodies	immunomagnetic separation and enzyme immunoassay	10^4–10^6 CFU/ml	poultry environ mental samples	48 h	(Leon-Velarde et al. 2009)
S. Typhimurium (ATCC 53648)	anti-S. Typhimurium mAb; physical absorption	sandwich ELISA; electrochemical detection (chronoamperometry)	5×10^3 cells/ml	spiked chicken meat	n.a.	(Salam and Tothill 2009)
	anti-S. Typhimurium mAb; covalent linking		20 CFU/ml			
S. Typhimurium LT 2	anti-Salmonella magnetic beads; anti-Salmonella-HRP (rabbit pAb)	electrochemical magneto-immunosensing	5×10^3 CFU/ml	pure culture	50 min	(Liebana et al. 2009)
			7.5×10^3 CFU/ml	milk diluted 1/10 in LB		
			$0.108 \times$ CFU/ml	spiked milk, pre-enriched		
S. Typhymurium	mAb	ellipsometry	10^3 CFU/ml.	pure culture	n.a.	(Bae et al. 2005)
S. Typhimurium	goat IgG against Salmonella CSA-1	fluorescence resonance energy transfer based method	10^3 cells/ml.	pure culture	n.a.	(Ko and Grant 2006)
			10^5 CFU/g	spiked pork	5 min	
S. Typhimurium (ATCC 14028)	AP labeled anti-Salmonella, Ab-coated magnetic microbeads	immunomagnetic separation and immuno-optical absorption	2.2×10^4 CFU/ml	chicken carcass rinse	2 h	(Liu et al. 2001)
			2×10^4 CFU/ml	pure culture		
S. enteritidis isolated from food samples	HRP-labeled pAb	microcantilever	10^5 CFU/ml	pure culture	40 min	(Ricciardi et al. 2010)
S. Typhimurium	biotinylated anti-Salmonella CSA-1	surface plasmon resonance	1×10^6 CFU/ml	spiked chicken carcass rinse	n.a.	(Lan et al. 2008)
S. enteritidis	mAb MO9; anti-S. enteridis	piezoelectric immunosensor	1×10^5 cells/ml	pure culture	35 min	(Si et al. 2001)
S, Typhimurium (ATCC 14028)	anti-Salmonella CSA-1	quartz crystal microbalance	10^5 and 10^7 cells/ml	pure culture	hours	(Su and Li 2005)

	Sensing element	Method	Detection limit	Medium	Assay time	Ref.
S. Typhymurium	pAb, polyvalent somatic O antibody	quartz crystal acoustic wave device	10^2 to 10^{10} cells/ml	pure culture	3 h 40 min	(Olsen et al. 2003)
S. paratyphi A S. enteriditis S. Typhimurium	mAb MO2, MO4 and MO9	quartz crystal microbalance	6×10^4 cells/ml 6×10^4 cells/ml 8×10^4 cells/ml	pure culture	40 min	(Wong et al. 2002)
S. Typhi (strain SKST), a clinical isolate	pAb flagellar and mAb	sandwich ELISA	10^4-10^5 CFU/ml 10^2 CFU/ml	pure culture spiked milk, vegetable /meat/ chicken rinse	6 h	(Kumar et al. 2008)
S. Typhimurium (KCCM 11806)	mAb against S. Typhimurium	surface plasmon resonance	10^2 CFU/ml	pure culture	n.a.	(Oh et al. 2004)
S. Typhimurium	rabbit pAb against *Salmonella*	Faradic impedimetric immunosensor	10^5 CFU/mL	spiked milk	2 h	(Mantzila et al. 2008)
S. Typhimurium	mAb against S. Typhimurium	microarray, fluorescence staining	n.a.	pure culture	n.a.	(Choi et al. 2008)
>20 *Salmonella* serovars	35 antisera against O- and H- antigens	microarray, fluorescence staining	n.a.	pure culture	n.a.	(Cai et al. 2005)
S. Typhimurium (ATCC 14028), heat killed	anti-*Salmonella* goat pAb (capture Ab); pAb HRP-anti-*Salmonella* (detection Ab)	microarray, microfluidics, chemiluminescence	2×10^7 cells/ml 3×10^6 cells/ml	pure culture	18 min 13 min	(Karsunke et al. 2009) (Wolter et al. 2008)
S. Typhimurium (ATCC 14028), heat killed	Antibody-labeled microspheres	flow cytometry	2.5-500	pure culture	180 min	(Dunbar et al. 2003)
S. Typhimurium (ATCC 14028 and wild type)	anti-S. Typhimurium mouse mAb, rabbit pAb	multiplexed assay, chemiluminescence	10^4 - 10^5 cells/ml	spiked human fecal and beef samples	60 min	(Magliulo et al. 2007)
S. Typhimurium, heat killed	rabbit pAb anti-*Salmonella* sp. (capture Ab), anti-S. Typhimurium LPS, mAb (detection Ab)	flow-through fluorescence assay	8×10^4 8×10^3	spiked food and fecal samples	15 min 60 min	(Taitt et al. 2004)

pAb: poyclonal antibody; mAb: monoclonal antibody; n.a.: not available; CSA: common structural antigens; HRP: horseradish peroxidase; AP: alkaline phosphatase.

Table 1. Antibody-based detection of *Salmonella*

an expected result, considering the immobilization efficiency is affected by multiple factors, including the substrate surface chemistry; the purity, clonal type and affinity of the antibody; the type and expression level of the bacterial antigen; the incubation duration; and the medium. Since microarrays employ multiple antibodies it is important to evaluate their cross-reactivities to avoid false positive results. One example is the work by Rivas et al. evaluating the binding capacities and cross-reactivities of 200 different antibodies for the detection of environmental toxins (Rivas et al. 2008). Our work focuses on determining how these factors affect the efficiency of capturing pathogenic bacteria. When coupled with microfluidics techniques, antibody-based detection can lead to miniaturized and automated detectors, which are in great demand for field applications (Karsunke et al. 2009; Wolter et al. 2008). Our work on using immunoimmobilization for biosensor applications has just been submitted for publication. Besides the direct capturing of bacterial cells, microarrays prepared using antibodies, proteins or carbohydrates have been used to detect pathogenic bacteria including *Salmonella*. These microarrays do not target the bacterial cell, and further reviews can be found in Bacarese-Hamilton's work (Bacarese-Hamilton et al. 2002).

5. Summary and outlook

We have presented the immobilization and manipulation of *Salmonella* cells on flat substrates in order to conduct tests on individual bacteria and to develop a biosensor technology based on capturing living organisms. Such immobilization is achieved using antibodies covalently tethered to flat substrates such as silicon wafers in order to capture living bacteria in their physiological environment. The immobilization process takes advantage of the specific interaction between an antibody and the corresponding antigen on a bacterial surface. Thus, the method offers a high immobilization efficiency with a rate of initial attachment of over 100 microorganisms per minute per 100×100 μm^2 area for a bulk concentration of microorganisms of $\sim 2 \times 10^6$ CFU/ml. Because the surfaces to which the targeted organisms are tethered are highly polished and flat, the detection of these organisms is highly efficient: a single bacterium in a field with an area of 200×200 μm^2 can easily be imaged under optical microscope. The key to the success of immunoimmobilization is the combination of proper selection of antibody-antigen pairs and optimization of the surface chemistry for antibody linkage. If the surface chemistry is designed carefully, living bacterial cells can be immobilized on a variety of substrates, including silicon wafer, glass, gold and steel. Our work of the last 5-6 years on *Salmonella* and *E. coli* suggests that the most efficient and reliable immunoimmobilization involves a limited number of specific surface antigens such as pili, flagella or O-antigens and the corresponding antibodies. Bacterial cells immobilized in this way are linked robustly enough to be tethered to their locations but still maintain their viability and functionality without any noticeable hindrance. The technique provides a promising platform for the *in situ* investigation of individual or small groups of localized bacterial cells in their natural physiological environments, which offers substantial promise for the future. For example, our work has proven that multiple puncturings of the cell wall of a bacterium by means of an AFM tip does not kill the organism. Until this technique was published, many prominent microbiologists had believed that the puncturing process would undoubtedly kill the organism. This new phenomenon opens up the possibility of introducing macromolecules or nanoparticles into the cytoplasm of an individual living bacterium and observing the response of the bacterium to the intrusion. This possibility is a fertile ground for new science, and only time will show how fruitful.

The high efficiency and specificity of immunoimmobilization can also be utilized in biosensor technology for the rapid detection and identification of pathogenic species in field applications and the sorting of specific species from mixed consortia. To sum up, immunoimmobilization as described in this chapter has great potential both in fundamental and in practical applications.

6. Acknowledgements

The content of this work covers many years of research. Our research is currently supported by funds from various sources including ONR Multidisciplinary University Research Initiative (MURI) grant N00014-10-1-0946, Montana State University funds supporting the Imaging and Chemical Analysis Laboratory (ICAL), National Institutes of Health grants P20 RR020185 and R21 AI080960, an equipment grant from the M.J. Murdock Charitable Trust, and the Montana State University Agricultural Experimental Station. In the past, this work also was supported in part by funds from NASA-EPSCOR under grant NCC5-579, in part by U.S. Public Service Grant AI-41123, and in part by Montana Agricultural Station and USDA Formula funds. We thank Dr. Jean Starkey and Dr. Kate McInnerney for their help on the Bio-Rad system. We thank Betsey Pitts of the Center for Biofilm Engineering for her help with LSCM studies. The help of the ICAL staff, particularly Ms. Linda Loetterle, Ms. Laura Kellerman and Nancy Equall on handling bacteria and on AFM and SEM analysis, is greatly appreciated. We also would like to acknowledge TUBITAK for providing a scholarship to Ms. S. Celikkol during her PhD research in ICAL. Finally, we acknowledge and appreciate greatly the editing work by Mrs. Lois Avci and her meticulous attention to detail.

7. References

Adanyi, N., Varadi, M., Kim, N., Szendro, I., 2006. Development of new immunosensors for determination of contaminants in food. Curr. App. Phys. 6(2), 279-286.

Akselrod, G.M., Timp, W., Mirsaidov, U., Zhao, Q., Li, C., Timp, R., Timp, K., Matsudaira, P., Timp, G., 2006. Laser-guided assembly of heterotypic three-dimensional living cell microarrays. Biophys. J. 91(9), 3465-3473.

Alexa, P., Stouracova, K., Hamrik, J., Rychlik, I., 2001. Gene typing of the colonisation factors K88 (F4) in enterotoxigenic Escherichia coli strains isolated from diarrhoeic piglets. Veterinarni Medicina 46(2), 46-49.

Amaguafia, R.M., Andrews, W.H., 2000. Detection by classical cultural techniques. In: Robinson, R.K., Batt, C.A., Patel, P.D. (Eds.), Encyclopedia of food microbiology, pp. 1948-1951. Academic Press, San Diego.

Arnoldi, M., Fritz, M., Bauerlein, E., Radmacher, M., Sackmann, E., Boulbitch, A., 2000. Bacterial turgor pressure can be measured by atomic force microscopy. Phys Rev E 62(1), 1034-1044.

Arnoldi, M., Kacher, C.M., Bauerlein, E., Radmacher, M., Fritz, M., 1998. Elastic properties of the cell wall of Magnetospirillum gryphiswaldense investigated by atomic force microscopy. Applied Physics a-Materials Science & Processing 66, S613-S617.

Ascon, M.A., Hone, D.M., Walters, N., Pascual, D.W., 1998. Oral immunization with a Salmonella typhimurium vaccine vector expressing recombinant enterotoxigenic

Escherichia coli K99 fimbriae elicits elevated antibody titers for protective immunity. Infect. Immun. 66(11), 5470-5476.

Bacarese-Hamilton, T., Bistoni, F., Crisanti, A., 2002. Protein microarrays: From serodiagnosis to whole proteome scale analysis of the immune response against pathogenic microorganisms. BIOTECHNIQUES, 24-29.

Bae, Y.M., Park, K.W., Oh, B.K., Lee, W.H., Choi, J.W., 2005. Immunosensor for detection of Salmonella typhimurium based on imaging ellipsometry. Colloids and Surfaces a-Physicochemical and Engineering Aspects 257-58, 19-23.

Blaser, M.J., Newman, L.S., 1982. A Review of Human Salmonellosis .1. Infective Dose. Reviews of Infectious Diseases 4(6), 1096-1106.

Boujday, S., Briandet, R., Salmain, M., Herry, J.M., Marnet, P.G., Gautier, M., Pradier, C.M., 2008. Detection of pathogenic Staphylococcus aureus bacteria by gold based immunosensors. Microchim. Acta 163(3-4), 203-209.

Brogan, K.L., Walt, D.R., 2005. Optical fiber-based sensors: application to chemical biology. Curr. Opin. Chem. Biol. 9(5), 494-500.

Byrne, B., Stack, E., Gilmartin, N., O'Kennedy, R., 2009. Antibody-Based Sensors: Principles, Problems and Potential for Detection of Pathogens and Associated Toxins. Sensors 9(6), 4407-4445.

Cai, H.Y., Lu, L., Muckle, C.A., Prescott, J.F., Chen, S., 2005. Development of a novel protein microarray method for Serotyping Salmonella enterica strains. J Clin Microbiol 43(7), 3427-3430.

Chao, S.H., Shi, X., Lin, L.I., Chen, S.Y., Zhang, W.W., Meldrum, D.R., 2011. Real-time PCR of single bacterial cells on an array of adhering droplets. Lab on a Chip 11(13), 2276-2281.

Choi, J.W., Kim, Y.K., Oh, B.K., 2008. The development of protein chip using protein G for the simultaneous detection of various pathogens. Ultramicroscopy 108(10), 1396-1400.

Curd, H., Liu, D., Reeves, P.R., 1998. Relationships among the O-antigen gene clusters of Salmonella enterica groups B, D1, D2, and D3. J. Bacteriol. 180(4), 1002-1007.

Doktycz, M.J., Sullivan, C.J., Hoyt, P.R., Pelletier, D.A., Wu, S., Allison, D.P., 2003. AFM imaging of bacteria in liquid media immobilized on gelatin coated mica surfaces. Ultramicroscopy 97(1-4), 209-216.

Dunbar, S.A., Vander Zee, C.A., Oliver, K.G., Karem, K.L., Jacobson, J.W., 2003. Quantitative, multiplexed detection of bacterial pathogens: DNA and protein applications of the Luminex LabMAP (TM) system. Journal of Microbiological Methods 53(2), 245-252.

Eaton, A.D., Clesceri, L.S., Rice, E.W., Greenberg, A.E., 2005. Standard methods for the examination of water and wastewater, 21st Edition ed. American Public Health Association, Washington D.C.

Friedlander, A.M., Welkos, S.L., Worsham, P.L., Andrews, G.P., Heath, D.G., Anderson, G.W., Pitt, M.L.M., Estep, J., Davis, K., 1995. Relationship between Virulence and Immunity as Revealed in Recent Studies of the F1 Capsule of Yersinia-Pestis. Clinical Infectious Diseases 21, S178-S181.

Gao, D., McBean, N., Schultz, J.S., Yan, Y.S., Mulchandani, A., Chen, W.F., 2006. Fabrication of antibody arrays using thermally responsive elastin fusion proteins. Journal of the American Chemical Society 128(3), 676-677.

Gao, J.Q., Liu, D.J., Wang, Z.X., 2010. Screening Lectin-Binding Specificity of Bacterium by Lectin Microarray with Gold Nanoparticle Probes. Analytical Chemistry 82(22), 9240-9247.

Gao, W.M., Zhang, W.W., Meldrum, D.R., 2011. RT-qPCR based quantitative analysis of gene expression in single bacterial cells. Journal of Microbiological Methods 85(3), 221-227.

Geng, P., Zhang, X.N., Meng, W.W., Wang, Q.J., Zhang, W., Jin, L.T., Feng, Z., Wu, Z.R., 2008. Self-assembled monolayers-based immunosensor for detection of Escherichia coli using electrochemical impedance spectroscopy. Electrochim. Acta 53(14), 4663-4668.

Grimont, P.A.D., Weill, F.-X., 2007. Antigenic Formulae of the *Salmonella* Serovars. 9 ed. Institute Pasteur.

Han, H.J., Kannan, R.M., Wang, S.X., Mao, G.Z., Kusanovic, J.P., Romero, R., 2010. Multifunctional Dendrimer-Templated Antibody Presentation on Biosensor Surfaces for Improved Biomarker Detection. Advanced Functional Materials 20(3), 409-421.

Heitkamp, M.A., Stewart, W.P., 1996. A novel porous nylon biocarrier for immobilized bacteria. Appl. Environ. Microbiol. 62(12), 4659-4662.

Hermanson, G.T., Mallia, A.K., Smith, P.K., 1992. Immobilized affinity ligand techniques. Academic Press, Inc., San Diego.

Hirst, E.R., Yuan, Y.J., Xu, W.L., Bronlund, J.E., 2008. Bond-rupture immunosensors - A review. Biosens Bioelectron 23(12), 1759-1768.

Holoda, E., Vu-Khac, H., Andraskova, S., Chomova, Z., Wantrubova, A., Krajnak, M.K., Pilipcinec, E., 2005. PCR assay for detection and differentiation of K88ab(1), K88ab(2), K88ac, and K88ad fimbrial adhesins in E-coli strains isolated from diarrheic piglets. Folia Microbiol. 50(2), 107-112.

Huck, W.T.S., 2007. Self-assembly meets nanofabrication: Recent developments in microcontact printing and dip-pen nanolithography. Angewandte Chemie-International Edition 46(16), 2754-2757.

Ingraham, J.L., Marr, A.G., 1996. Effect of temperatrure, pressure, pH and osmotic stress on growth. In: Neidhardt, F.C., Curtiss III, R., Ingraham, J.L., Lin, E.C.C., Low, K.B., Magasanik, B., Reznikoff, W.S., Riley, M., Schaechter, M., Umbarger, H.E. (Eds.), *Escherichia Coli* and *Salmonella*: Cellular and Molecular Biology, p. 1575, 2nd ed. ASM Press, Washington, D.C.

Karsunke, X.Y.Z., Niessner, R., Seidel, M., 2009. Development of a multichannel flow-through chemiluminescence microarray chip for parallel calibration and detection of pathogenic bacteria. Anal. Bioanal. Chem. 395(6), 1623-1630.

Ko, S.H., Grant, S.A., 2006. A novel FRET-based optical fiber biosensor for rapid detection of Salmonella typhimurium. Biosens Bioelectron 21(7), 1283-1290.

Kuang, Y., Biran, I., Walt, D.R., 2004. Living bacterial cell array for genotoxin monitoring. Anal. Chem. 76(10), 2902-2909.

Kumar, S., Balakrishna, K., Batra, H., 2008. Enrichment-ELISA for detection of Salmonella typhi from food and water samples. Biomedical and Environmental Sciences 21(2), 137-143.

Lahiri, J., Isaacs, L., Tien, J., Whitesides, G.M., 1999. A strategy for the generation of surfaces presenting ligands for studies of binding based on an active ester as a common reactive intermediate: A surface plasmon resonance study. Analytical Chemistry 71(4), 777-790.

Lan, Y.B., Wang, S.Z., Yin, Y.G., Hoffmann, W.C., Zheng, X.Z., 2008. Using a Surface Plasmon Resonance Biosensor for Rapid Detection of Salmonella Typhimurium in Chicken Carcass. Journal of Bionic Engineering 5(3), 239-246.

Larson, B.J., Gillmor, S.D., Lagally, M.G., 2004. Controlled deposition of picoliter amounts of fluid using an ultrasonically driven micropipette. Rev Sci Instrum 75(4), 832-836.

Lazcka, O., Del Campo, F.J., Munoz, F.X., 2007. Pathogen detection: A perspective of traditional methods and biosensors. Biosens Bioelectron 22(7), 1205-1217.

Leon-Velarde, C.G., Zosherafatein, L., Odumeru, J.A., 2009. Application of an automated immunomagnetic separation-enzyme immunoassay for the detection of Salmonella enterica subspecies enterica from poultry environmental swabs. Journal of Microbiological Methods 79(1), 13-17.

Liebana, S., Lermo, A., Campoy, S., Cortes, M.P., Alegret, S., Pividori, M.I., 2009. Rapid detection of Salmonella in milk by electrochemical magneto-immunosensing. Biosens Bioelectron 25(2), 510-513.

Liu, Y.C., Che, Y.H., Li, Y.B., 2001. Rapid detection of Salmonella typhimurium using immunomagnetic separation and immune-optical sensing method. Sensors and Actuators B-Chemical 72(3), 214-218.

Magariyama, Y., Sugiyama, S., Kudo, S., 2001. Bacterial swimming speed and rotation rate of bundled flagella. FEMS Microbiol. Lett. 199(1), 125-129.

Magliulo, M., Simoni, P., Guardigli, M., Michelini, E., Luciani, M., Lelli, R., Roda, A., 2007. A rapid multiplexed chemiluminescent immunoassay for the detection of Escherichia coli O157 : H7, Yersinia enterocolitica, salmonella typhimurium, and Listeria monocytogenes pathogen bacteria. Journal of Agricultural and Food Chemistry 55(13), 4933-4939.

Malorny, B., Huehn, S., Dieckmann, R., Kramer, N., Helmuth, R., 2009. Polymerase Chain Reaction for the Rapid Detection and Serovar Identification of Salmonella in Food and Feeding Stuff. Food Anal. Met. 2(2), 81-95.

Mantzila, A.G., Maipa, V., Prodromidis, M.I., 2008. Development of a faradic impedimetric immunosensor for the detection of Salmonella typhimurium in milk. Anal. Chem. 80(4), 1169-1175.

Mbeunkui, F., Richaud, C., Etienne, A.L., Schmid, R.D., Bachmann, T.T., 2002. Bioavailable nitrate detection in water by an immobilized luminescent cyanobacterial reporter strain. Appl. Microbiol. Biotech. 60(3), 306-312.

O'Kennedy, R., Byrne, B., Stack, E., Gilmartin, N., 2009. Antibody-Based Sensors: Principles, Problems and Potential for Detection of Pathogens and Associated Toxins. Sensors 9(6), 4407-4445.

Oh, B.K., Kim, Y.K., Park, K.W., Lee, W.H., Choi, J.W., 2004. Surface plasmon resonance immunosensor for the detection of Salmonella typhimurium. Biosens Bioelectron 19(11), 1497-1504.

Olsen, E.V., Pathirana, S.T., Samoylov, A.M., Barbaree, J.M., Chin, B.A., Neely, W.C., Vodyanoy, V., 2003. Specific and selective biosensor for Salmonella and its detection in the environment. Journal of Microbiological Methods 53(2), 273-285.

Osborn, M.J., Gander, J.E., Parisi, E., Carson, J., 1972. Mechanism of Assembly of Outer Membrane of Salmonella-Typhimurium - Isolation and Characterization of Cytoplasmic and Outer Membrane. J. Biol. Chem. 247(12), 3962-3972.

Polyak, B., Bassis, E., Novodvorets, A., Belkin, S., Marks, R.S., 2001. Bioluminescent whole cell optical fiber sensor to genotoxicants' system optimization. Sensor. Actuat. B 74(1-3), 18-26.

Premkumar, J.R., Lev, O., Marks, R.S., Polyak, B., Rosen, R., Belkin, S., 2001. Antibody-based immobilization of bioluminescent bacterial sensor cells. Talanta 55(5), 1029-1038.

Premkumar, J.R., Rosen, R., Belkin, S., Lev, O., 2002. Sol-gel luminescence biosensors: Encapsulation of recombinant E-coli reporters in thick silicate films. Anal. Chim. Acta 462(1), 11-23.

Raetz, C.R.H., 1996. Bacterial Lipopolysaccharides: a Remarkable Family of Bioactive Macroamphiphiles. In: Neidhardt, F.C. (Ed.), Escherichia coli and Salmonella: Cellular and Molecular Biology, pp. 1035-1063, 2th ed. ASM Press, Washington D.C.

Razatos, A., Ong, Y.L., Sharma, M.M., Georgiou, G., 1998. Molecular determinants of bacterial adhesion monitored by atomic force microscopy. Proc. Natl. Acad. Sci. USA 95(19), 11059-11064.

Ricciardi, C., Canavese, G., Castagna, R., Digregorio, G., Ferrante, I., Marasso, S.L., Ricci, A., Alessandria, V., Rantsiou, K., Cocolin, L.S., 2010. Online Portable Microcantilever Biosensors for Salmonella enterica Serotype Enteritidis Detection. Food and Bioprocess Technology 3(6), 956-960.

Rivas, L.A., Garcia-Villadangos, M., Moreno-Paz, M., Cruz-Gil, P., Gomez-Eivira, J., Parro, V., 2008. A 200-Antibody Microarray Biochip for Environmental Monitoring: Searching for Universal Microbial Biomarkers through Immunoprofiling. Anal. Chem. 80(21), 7970-7979.

Rozhok, S., Fan, Z.F., Nyamjav, D., Liu, C., Mirkin, C.A., Holz, R.C., 2006. Attachment of motile bacterial cells to prealigned holed microarrays. Langmuir 22(26), 11251-11254.

Rozhok, S., Shen, C.K.F., Littler, P.L.H., Fan, Z.F., Liu, C., Mirkin, C.A., Holz, R.C., 2005. Methods for fabricating microarrays of motile bacteria. Small 1(4), 445-451.

Salaita, K., Wang, Y.H., Mirkin, C.A., 2007. Applications of dip-pen nanolithography. Nature Nanotechnology 2(3), 145-155.

Salam, F., Tothill, I.E., 2009. Detection of Salmonella typhimurium using an electrochemical immunosensor. Biosens Bioelectron 24(8), 2630-2636.

Selander, R.K., Li, J., Nelson, K., 1996. Evolutionary Genetics of Salmonella enterica. In: Neidhardt, F.C. (Ed.), Escherichia coli and Salmonella: Cellular and Molecular Biology, pp. 2691-2720, 2nd ed. ASM Press, Washington D.C.

Si, S.H., Li, X., Fung, Y.S., Zhu, D.R., 2001. Rapid detection of Salmonella enteritidis by piezoelectric immunosensor. Microchemical Journal 68(1), 21-27.

Skottrup, P.D., Nicolaisen, M., Justesen, A.F., 2008. Towards on-site pathogen detection using antibody-based sensors. Biosens. Bioelectron. 24(3), 339-348.

Su, X.L., Li, Y.B., 2005. A QCM immunosensor for Salmonella detection with simultaneous measurements of resonant frequency and motional resistance. Biosens Bioelectron 21(6), 840-848.

Suo, Z., Yang, X.H., Avci, R., Deliorman, M., Rugheimer, P., Pascual, D.W., Idzerda, Y., 2009a. Antibody Selection for Immobilizing Living Bacteria. Anal. Chem. 81(18), 7571-7578.

Suo, Z.Y., Avci, R., Deliorman, M., Yang, X.H., Pascual, D.W., 2009b. Bacteria Survive Multiple Puncturings of Their Cell Walls. Langmuir 25(8), 4588-4594.

Suo, Z.Y., Avci, R., Yang, X.H., Pascual, D.W., 2008. Efficient immobilization and patterning of live bacterial cells. Langmuir 24(8), 4161-4167.

Suo, Z.Y., Yang, X.H., Avci, R., Kellerman, L., Pascual, D.W., Fries, M., Steele, A., 2007. HEPES-stabilized encapsulation of Salmonella typhimurium. Langmuir 23(3), 1365-1374.

Taitt, C.R., Shubin, Y.S., Angel, R., Ligler, F.S., 2004. Detection of Salmonella enterica serovar typhimurium by using a rapid, array-based immunosensor. Applied and Environmental Microbiology 70(1), 152-158.

Titball, R.W., Williamson, E.D., 2001. Vaccination against bubonic and pneumonic plague. Vaccine 19(30), 4175-4184.

Turner, M., 2011. Microbe outbreak panics Europe. Nature 474, 137-137.

Velusamy, V., Arshak, K., Korostynska, O., Oliwa, K., Adley, C., 2010a. An overview of foodborne pathogen detection: In the perspective of biosensors. Biotech. Adv. 28(2), 232-254.

Velusamy, V., Arshak, K., Korostynska, O., Oliwa, K., Adley, C., 2010b. An overview of foodborne pathogen detection: In the perspective of biosensors. Biotechnology Advances 28(2), 232-254.

Voet, D., Voet, J.G., 1995. Biochemistry, 2nd ed. John Wiley and Sons, Inc., New York.

Weibel, D.B., Lee, A., Mayer, M., Brady, S.F., Bruzewicz, D., Yang, J., DiLuzio, W.R., Clardy, J., Whitesides, G.M., 2005. Bacterial printing press that regenerates its ink: Contact-printing bacteria using hydrogel stamps. Langmuir 21(14), 6436-6442.

Weintraub, A., Johnson, B.N., Stocker, B.A.D., Lindberg, A.A., 1992. Structural and Immunochemical Studies of the Lipopolysaccharides of Salmonella Strains with Both Antigen-O4 and Antigen-O9. J. Bacteriol. 174(6), 1916-1922.

Wolter, A., Niessner, R., Seidel, M., 2008. Detection of Escherichia coli O157 : H7, Salmonella typhimurium, and Legionella pneumophila in water using a flow-through chemiluminescence microarray readout system. Anal. Chem. 80(15), 5854-5863.

Wong, Y.Y., Ng, S.P., Ng, M.H., Si, S.H., Yao, S.Z., Fung, Y.S., 2002. Immunosensor for the differentiation and detection of Salmonella species based on a quartz crystal microbalance. Biosens Bioelectron 17(8), 676-684.

Xu, L., Robert, L., Ouyang, Q., Taddei, F., Chen, Y., Lindner, A.B., Baigl, D., 2007. Microcontact printing of living bacteria arrays with cellular resolution. Nano lett. 7(7), 2068-2072.

Yang, X.Y., Hinnebusch, B.J., Trunkle, T., Bosio, C.M., Suo, Z.Y., Tighe, M., Harmsen, A., Becker, T., Crist, K., Walters, N., Avci, R., Pascual, D.W., 2007. Oral vaccination with *Salmonella* simultaneously expressing *Yersinia pestis* F1 and V antigens protects against Bubonic and Pneumonic plague. J. Immunol. 178(2), 1059-1067.

Yao, X., Walter, J., Burke, S., Stewart, S., Jericho, M.H., Pink, D., Hunter, R., Beveridge, T.J., 2002. Atomic force microscopy and theoretical considerations of surface properties and turgor pressures of bacteria. Colloid Surface B 23(2-3), 213-230.

Zordan, M.D., Grafton, M.M.G., Acharya, G., Reece, L.M., Cooper, C.L., Aronson, A.I., Park, K., Leary, J.F., 2009. Detection of Pathogenic E. coli O157:H7 by a Hybrid Microfluidic SPR and Molecular Imaging Cytometry Device. Cytometry A 75A(2), 155-162.

12

Multiplex TaqMan Real-Time PCR (qPCR) Assay Targeting *prot6E* and *invA* Genes for Fast and Accurate Detection of *Salmonella* Enteritidis

Narjol González-Escalona, Guodong Zhang and Eric W. Brown

Center for Food Safety and Applied Nutrition
Food and Drug Administration, College Park, MD
USA

1. Introduction

Salmonella is an important foodborne pathogen causing significant public health concern, both domestically and internationally (Tirado and Schmidt, 2001; Scallan et al., 2011). According to the latest CDC report *Salmonella* infections affect millions of people every year accounting for 11%, 35% and 28%, of illnesses, hospitalizations and deaths, respectively of the total U.S. foodborne diseases caused by all known foodborne pathogens (Scallan et al., 2011). Among those non-typhoid salmonellosis, S. Enteritidis (SE) has emerged as a major egg-associated pathogen. SE transmission to humans has been linked mainly to consumption of contaminated foods containing undercooked eggs (Rabsch et al., 2000). Fresh shell-eggs can be contaminated easily with SE through cracks in the shell by contact with chicken feces or by transovarian infection (Snoeyenbos et al., 1969). Consequently, the increase of consumption of shell eggs and egg products per capita in the United States to approximately 249 eggs per year (American Egg Board, 2008) may have contributed, in part, to increases in foodborne outbreaks (Altekruse et al., 1997), including a large multistate SE outbreak of SE outbreak associated with eggs in the US in 2010.

Traditional culture methods for SE detection from shell eggs and liquid whole eggs consist of a series of steps including non-selective pre-enrichment, selective enrichment, and selective/differential plating, and finally biochemical and serological confirmation. The traditional microbiological method for SE isolation from liquid eggs is described in detail in Chapter MLG 4.05 "Isolation and Identification of *Salmonella* from Meat, Poultry, Pasteurized Egg and Catfish Products" by the United States Department of Agriculture (USDA) (http://www.fsis.usda.gov/PDF/MLG_4_05.pdf). This method is labor intensive and takes about one weeks to complete the analysis. Consequently, a need exists for the development and validation of faster screening and detection methods for this pathogen in eggs.

The use of PCR or real time PCR (qPCR) for specific pathogen detection in foods has increased in recent years. They are fast and reliable tools for the testing of contaminated foods and had helped in preventing outbreaks. In recent years, numerous methods based on *Salmonella* DNA detection (e.g. *invA* gene) either by conventional or real-time PCR have been developed

(Krascsenicsova et al., 2008; Malorny B et al., 2003; Wolffs et al., 2006). qPCR is faster, is more sensitive than conventional PCR, and provides real-time data avoiding the use of gels (Valasek and Repa, 2005). In particular, the *invA* gene represents a good candidate for *Salmonella* detection as it is present in all pathogenic serovars described to date (Rahn et al., 1992; Boyd EF et al., 1997). The product of this gene is essential for the organism's ability to invade mammalian cells and subsequently cause disease (Galan and Curtiss, III, 1991; Galan JE et al., 1992). In the case of SE in specific, several PCR and isothermal methodologies has also been developed targeting different genes (Seo et al., 2004; Malorny et al., 2007a; O'Regan et al., 2008; Hadjinicolaou et al., 2009). Although isothermal amplification techniques has some advantages over qPCR, such as increased detection limit and lower cost; still has the disadvantage that a single target can be used at a time and lacks internal control for monitoring possible inhibitors of the reaction that might exist in the food matrix analyzed.

In the present study we developed a fast and accurate qPCR assay for the specific detection of SE in eggs. This qPCR contained primers and probes to detect three different targets: the *invA* gene (*Salmonella* genus specific), the *prot6E* gene (SE specific), and the internal amplification control (IAC). A foreign internal amplification control (IAC) was incorporated into the assay with the aim of detecting potential inhibitors present in the matrix analyzed (eggs). *Salmonella* spp. detection in foods is usually achieved after food samples pre-enrichment approaches using overnight incubation (Feder et al., 2001). Consequently, this method described herein is intended as an initial screening of 24 h pre-enrichments for the presence of *Salmonella* in eggs. In turn, this method will dramatically decrease the time and effort required during standard microbiological testing, since only positive pre-enrichment samples will be processed further.

2. Materials and methods

2.1 Bacterial strains and media

Eleven *Salmonella enterica* serovar Enteritidis (SE) strains (CDC 2010K_1543, 13-2, SE12, 18579, 18580, 22689, SE10, SE26, 17905, SE22, and CDC_2010K_1441) (Table 1), were employed in this study for artificial contamination of eggs. Strain CHS44 was employed for determining the detection limit of the real-time PCR (qPCR) assay. These strains are from

Strain	Phage type	Location	Source
SE12	14b	ME	Egg follicle
18579	4	Mexico	poultry
18580	4	Mexico	poultry
22689	8	MD	Chicken breast
SE10	8	ME	Chicken ovary
SE26	13	TX	Chicken viscera
13-2	13	N/A	Chicken
17905	13a	N/A	Chicken
SE22	13a	ME	Poultry environment
CDC_2010K_1441	N/A	N/A	Egg outbreak USA 2010
CDC_2010K_1543	N/A	N/A	Egg outbreak USA 2010

Table 1. Characteristics of *S.* Enteritidis strains used in this study for artificial contamination of eggs.

the FDA, Center for Food Safety and Applied Nutrition (CFSAN), Division of Microbiology's culture collection. Strains were grown overnight in Luria-Bertani (LB) medium at 35°C with shaking (250 rpm). The inclusivity and exclusivity of the qPCR assay for SE was demonstrated with 186 SE (Table 2) and 97 non-SE strains belonging to the FDA's collection (Table 3). Further specificity was demonstrated with 32 non-*Salmonella* species (48 strains) from very closely related genera (Table 4).

S. Enteritidis strains	qPCR multiplex		
	invA qPCR result	*prot6E* qPCR result	IAC qPCR result
54-5431	+	+	+
76-574	+	+	+
19755	+	+	+
22568	+	+	+
53-407	+	+	+
50-5306A	+	+	+
81-2625	+	+	+
78-1757	+	+	+
77-0424	+	+	+
36951	+	+	+
75-2325	+	+	+
60-2506	+	+	+
74-991	+	+	+
77-3493	+	+	+
76-2651	+	+	+
62-1976	+	+	+
77-1427	+	+	+
77-2659	+	+	+
75-199	+	+	+
76-2969	+	+	+
17912	+	+	+
75-1450	+	+	+
78-2938	+	+	+
31952	+	+	+
36388	+	+	+
76-1594	+	+	+
50-3079	+	+	+
50-5646	+	+	+
75-970	+	+	+
576709	+	+	+
639016-6	+	+	+

S. Enteritidis strains	qPCR multiplex		
	invA qPCR result	*prot6E* qPCR result	IAC qPCR result
607307-2	+	+	+
635290-58	+	+	+
640631	+	+	+
60738-9	+	+	+
629163	+	+	+
60-7307-6	+	+	+
622731-39	+	+	+
607308-16	+	+	+
607308-19	+	+	+
8a	+	+	+
98	+	+	+
415	+	+	+
13-1	+	+	+
23b	+	+	+
sz26	+	+	+
416 (pt4)	+	+	+
435 (pt4)	+	+	+
sz6	+	+	+
sz9	+	+	+
418	+	+	+
23a	+	+	+
pt23	+	+	+
421	+	+	+
sz10 (pt8)	+	+	+
436	+	+	+
414 (pt4)	+	+	+
419	+	+	+
CHS14	+	+	+
sz15 (pt8)	+	+	+
420	+	+	+
sz12	+	+	+
sz22	+	+	+
434	+	+	+
CHS44	+	+	+
426	+	+	+
13-3	+	+	+
60481	+	+	+

Multiplex TaqMan Real-Time PCR (qPCR) Assay Targeting prot6E and invA Genes for Fast and Accurate
Detection of Salmonella Enteritidis

251

S. Enteritidis strains	qPCR multiplex		
	invA qPCR result	*prot6E* qPCR result	IAC qPCR result
chs15	+	+	+
chs39	+	+	+
60494	+	+	+
60562	+	+	+
13-2	+	+	+
30663	+	+	+
22689 (pt8)	+	+	+
18570	+	+	+
23711	+	+	+
22705	+	+	+
22599	+	+	+
22706	+	+	+
23698	+	+	+
23703	+	+	+
22581	+	+	+
22600	+	+	+
22606 (pt8)	+	+	+
22574	+	+	+
33944	+	+	+
22690 (pt8)	+	+	+
22619	+	+	+
22601	+	+	+
18580 (pt4)	+	+	+
18572 (pt4)	+	+	+
18671 (pt4)	+	+	+
22532	+	+	+
18577	+	+	+
18512	+	+	+
18575	+	+	+
17924	+	+	+
18511	+	+	+
18578	+	+	+
18579	+	+	+
17927	+	+	+
18510	+	+	+
17929	+	+	+
1793	+	+	+

S. Enteritidis strains	qPCR multiplex		
	invA qPCR result	*prot6E* qPCR result	IAC qPCR result
18568	+	+	+
18567 (pt4)	+	+	+
18518	+	+	+
13183	+	+	+
17912	+	+	+
18088	+	+	+
22621	+	I	+
17914	+	+	+
18514	+	+	+
17923 (pt8)	+	+	+
17921	+	+	+
17905	+	+	+
18509	+	+	+
17918	+	+	+
17919	+	+	+
17917	+	+	+
30661	+	+	+
18569	+	+	+
18081	+	+	+
22568	+	+	+
17930	+	+	+
22701	+	+	+
30658	+	+	+
18574	+	+	+
18516	+	+	+
18573	+	+	+
17931	+	+	+
22510 (pt8)	+	+	+
CDC_2010K_0895	+	+	+
CDC_2010K_0899	+	+	+
CDC_2010K_0956	+	+	+
CDC_2010K_0968	+	+	+
CDC_2010K_1010	+	+	+
CDC_2010K_1018	+	+	+
CDC_2010K_1441	+	+	+
CDC_2010K_1444	+	+	+
CDC_2010K_1445	+	+	+

S. Enteritidis strains	qPCR multiplex		
	invA qPCR result	*prot6E* qPCR result	IAC qPCR result
CDC_2010K_1455	+	+	+
CDC_2010K_1457	+	+	+
CDC_2010K_1543	+	+	+
CDC_2010K_1558	+	+	+
CDC_2010K_1559	+	+	+
CDC_2010K_1565	+	+	+
CDC_2010K_1566	+	+	+
CDC_2010K_1575	+	+	+
CDC_2010K_1580	+	+	+
CDC_2010K_1594	+	+	+
CDC_2010K_1725	+	+	+
CDC_2010K_1729	+	+	+
CDC_2010K_1745	+	+	+
CDC_2010K_1747	+	+	+
CDC_2010K_1791	+	+	+
CDC_2010K_1795	+	+	+
CDC_2010K_1808	+	+	+
CDC_2010K_1810	+	+	+
CDC_2010K_1811	+	+	+
CDC_2010K_1882	+	+	+
CDC_2010K_1884	+	+	+
SE12	+	+	+
SE26	+	+	+
SE22	+	+	+
CDC_07ST000857	+	+	+
CDC_08-0253	+	+	+
CDC_08-0254	+	+	+
02-0062	+	+	+
58-6482	+	-	+
59-365	+	-	+
54-2953	+	-	+
chs54	+	-	+
20036	+	-	+
20035	+	-	+
18845	+	-	+
32393	+	-	+
18685	+	-	+

S. Enteritidis strains	qPCR multiplex		
	invA qPCR result	*prot6E* qPCR result	IAC qPCR result
22558	+	-	+
20034	+	-	+
20037	+	-	+
23710	+	-	+
sz23	+	-	+
sz5	+	-	+
SE-10	+	-	+
Total (186)	186	170	186

Table 2. *Salmonella* strains used for testing SE inclusivity for the *prot6E/invA* multiplex TaqMan qPCR.

Salmonella subspecies and serovars	Strain Numbers	*prot6E* qPCR result	*invA* qPCR result	IAC qPCR result
S. enterica subsp. enterica (I)				
Typhimurium	14	-	+	+
I 4,[5],12:i:-	6	-	+	+
Typhimurium/DT104	4	-	+	+
Newport	1	-	+	+
Heidelberg	1	-	+	+
Typhi	1	-	+	+
4,5,12:b:-	1	-	+	+
Hadar	1	-	+	+
Brandenburg	1	-	+	+
Saphra	1	-	+	+
Rubislaw	1	-	+	+
Michigan	1	-	+	+
Urbana	1	-	+	+
Vietnam	1	-	+	+
Tornow	1	-	+	+
Gera	1	-	+	+
Fresno	1	-	+	+
Brisbane	1	-	+	+
Agona	1	-	+	+
Muenchen	1	-	+	+
Senftenberg	1	-	+	+
Muenster	1	-	+	+

Salmonella subspecies and serovars	Strain Numbers	*prot6E* qPCR result	*invA* qPCR result	IAC qPCR result
Montevideo	1	-	+	+
Johannesburg	1	-	+	+
Javiana	1	-	+	+
Inverness	1	-	+	+
Cubana	1	-	+	+
Cerro	1	-	+	+
Alachua	1	-	+	+
S. enterica subsp. *Salamae* (II)				
II 58:l,z13,z28:z6	1	-	+	+
II 47:d:z39	1	-	+	+
II 48:d:z6	1	-	+	+
II 50:b:z6	1	-	+	+
II 53:lz28:z39	1	-	+	+
II 39:lz28:enx	1	-	+	+
II 13,22:z29:enx	1	-	+	+
II 4,12:b:-	1	-	+	+
II 18:z4,z23:-	1	-	+	+
S. enterica subsp. *arizonae* (IIIa)				
IIIa 41:z4,z23:-	1	-	+	+
IIIa 40:z4,z23:-	1	-	+	+
IIIa 48:g,z51:-	1	-	+	+
IIIa 21:g,z51:-	1	-	+	+
IIIa 51:gz51:-	1	-	+	+
IIIa 62:g,z51:-	1	-	+	+
IIIa 48:z4,z23,z32:-	1	-	+	+
IIIa 48:z4,z23:-	1	-	+	+
S. enterica subsp. *diarizonae* (IIIb)				
IIIb 60:r:e,n,x,z15	1	-	+	+
IIIb 48:i:z	1	-	+	+
IIIb 61:k:1,5,(7)	1	-	+	+
IIIb 61:l,v:1,5,7	1	-	+	+
IIIb 48: z10: e,n,x,z15	1	-	+	+
IIIb 38:z10:z53	1	-	+	+
IIIb 60:r:z	1	-	+	+

Salmonella subspecies and serovars	Strain Numbers	*prot6E* qPCR result	*invA* qPCR result	IAC qPCR result
IIIb 50:i:z	1	-	+	+
S. enterica subsp. *houtenae* (IV)				
IV 50:g,z51:-	1	-	+	+
IV 48:g,z51:-	1	-	+	+
IV 44:z4,z23:-	1	-	+	+
IV 45:g,z51:-	1	-	+	+
IV 16:z4,z32:-	1	-	+	+
IV 11:z4,z23:-	1	-	+	+
IV 6,7:z36:-	1	-	+	+
IV 16:z4,z32:-	1	-	+	+
S. enterica subsp. *Indica* (VI)				
VI 6,14,25:z10:1,(2),7	1	-	+	+
VI 11:b:1,7	1	-	+	+
VI 6,7:z41:1,7	1	-	+	+
VI 11:a:1,5	1	-	+	+
VI 6,14,25:a:e,n,x	1	-	+	+
S. enterica subsp. *houtenae* (VII)				
IV 40:g,z51:-	1	-	+	+
IV 40:z4,z24:-	1	-	+	+
S. bongori (V)				
V 48:i:-	1	-	+	+
V 40:z35:-	1	-	+	+
V 44:z39:-	1	-	+	+
V 60:z41:-	1	-	+	+
V 66:z41:-	1	-	+	+
V 48:z35:-	1	-	+	+
Total	101			

The nomenclatural system used is based on recommendations from the WHO Collaborating Centre for reference and research on *Salmonella*, 9th edition 2007.

Table 3. *Salmonella* strains used for testing SE exclusivity for the *prot6E/invA* multiplex TaqMan qPCR assay.

Organism	No. of strains	*prot6E* qPCR result	*invA* qPCR result	IAC qPCR result
Vibrio parahaemolyticus	4	-	-	+
V. vulnificus	1	-	-	+
Escherichia coli	9[a]	-	-	+
Enterobacter cloacae	1	-	-	+
E. aerogenes (ATCC 13048)	1	-	-	+
Cronobacter sakazakii (former E. sakazakii)	1	-	-	+
Yersinia enterocolitica	1	-	-	+
Y. pseudotuberculosis	1	-	-	+
Hafnia alvei	2	-	-	+
Morganella morganii	1	-	-	+
Edwardsiella tarda	1	-	-	+
Klebsiella pneumoniae	1	-	-	+
Proteus vulgaris	1	-	-	+
Pseudomonas aeruginosa	1	-	-	+
Serratia marcesans	1	-	-	+
Aeromonas hydrophila	1	-	-	+
Citrobacter freundii	1	-	-	+
C. koseri (ATCC 27028)	1	-	-	+
Staphylococcus aureus	1	-	-	+
Streptococcus faecalis	1	-	-	+
Bacillus subtilis	1	-	-	+
B. cereus	1	-	-	+
Listeria monocytogenes	1	-	-	+
L. innocua	1	-	-	+
Shigella sonnei	2	-	-	+
S. flexneri	2	-	-	+
S. boydii	2	-	-	+
S. dysenteriae	2	-	-	+
Achromobacter spp.	1	-	-	+
Providencia stuartii (ATCC 33672)	1	-	-	+
Proteus mirabilis	1	-	-	+
P. hauseri (deposited as P.vulgaris) (ATCC 13315)	1	-	-	+
Total	48			

[a] Five *E. coli* classes (virotypes) that cause diarrheal diseases were included: strain 10009 (enterotoxigenic, ETEC); strains 10010, 10015, 10016, 10017 and 10012 (enteroinvasive, EIEC); strain 10023 (enterohemorrhagic, EHEC); strain 10035 (enteropathogenic, EPEC) and strain ATM395 (enteroaggregative, EAEC).

Table 4. Organisms employed to assess the specificity of the *prot6E/invA* multiplex *TaqMan* qPCR assay for *S.* Enteritidis detection.

2.2 Preparation of SE inocula

Cultures of individual SE strains were prepared by transferring a loopful for three consecutive 24-h intervals to 10 ml of tryptic soy broth (TSB, Difco, Becton Dickinson) at 35 °C. SE cells from an overnight broth culture were centrifuged at 3,000 x g for 15 min at 4 °C. The pellet was washed twice with sterile 0.1% peptone water and re-suspended in sterile 0.1% peptone water. Serial dilutions of the suspension were prepared in sterile 0.1% peptone water to obtain the desired cell populations. The cell number in the inoculum was determined by plating 100µl dilutions (in sterile 0.1% peptone water) on TSA and incubating at 35 °C for 24 h.

2.3 Microbiological assay

All eggs were purchased from local grocery stores in College Park, MD. Analysis of liquid eggs was performed by following USDA procedure with some modifications. Shell eggs were broken by hands aseptically into sterile glass beakers. They were mixed well with a sterile stick by hands for about 2 minutes until it looked uniform. These liquid eggs were inoculated (day 1) at around 5 SE or at > 10^4 cells in 100 g. Each 100 g sample was placed into a 2-liter sterile glass beaker, mixed with 900 ml pre-enrichment broth. Five pre-enrichment broths were used for testing performance of pre-enrichments for SE recovery. They were TSB, TSB plus ferrous sulfate (TSB + Fe), universal pre-enrichment broth (UPB), nutrient broth (NB), and buffered peptone water (BPW). After 24 hr (day 2) pre-enrichment, 1 ml of each pre-enriched sample was transferred to 10 ml of selective enrichment media (Rappaport–Vassiliadis (RV) medium and Tetrathionate broth (TT) (Difco) and incubated for 24 h at 42 °C and 43 °C, respectively. On day 3, tube contents were vortexed for 10 sec, and 10 µl portions of the TT and RV media were streaked on bismuth sulfite (BS) agar, xylose lysine desoxycholate (XLD) agar, and Hektoen enteric (HE) agar and incubated at 35 °C for 24 h. On day 4, the plates were examined for the presence of typical *Salmonella* colonies. Typical colonies were confirmed with *Salmonella* agglutination test kit from BD.

2.4 Design of primers and standards for qPCR

All primers and probes (Table 5) employed in this study were purchased from IDT (Coralville, IA, USA). The targets for qPCR were *invA* gene and *prot6E* gene of SE. Primers and probes for *invA* assay were designed previously (Gonzalez-Escalona et al., 2009). Primers and probes for *prot6E* were designed using Beacon designer v.7 (PREMIER Biosoft, Palo Alto, CA). DNA from strain CHS44 was used to determine the *prot6E/invA* qPCR detection limit. DNA extraction was performed with the DNeasy kit as recommended by the manufacturer (QIAGEN). DNA concentration was determined using Qubit® 2.0 Fluorometer and Qubit™ dsDNA HS Assay Kit following manufacturer's instructions (Invitrogen). The numbers of copies of the qPCR standards were calculated by assuming average molecular masses of 680 Da for 1 nucleotide of double stranded DNA. The calculation was done with the following equation: copies per nanogram = (NL x 10-9)/ (n x mw), where n is the length of the SE strain P125109 complete genome (4,685,848 bp), mw is the molecular weight per nucleotide, and NL is Avogadro constant (6.02 x 1023 molecules per mol).

Multiplex TaqMan Real-Time PCR (qPCR) Assay Targeting prot6E and invA Genes for Fast and Accurate
Detection of Salmonella Enteritidis

259

Target	Name	Sequence (5'-3')[a]	Reference
qPCR primers			
invA	invA_176F	CAACGTTTCCTGCGGTACTGT	(Gonzalez-Escalona et al., 2009)
	invA_291R	CCCGAACGTGGCGATAATT	
prot6E	prot6E-NGE-f	GTAGGTAGCCAGTATAAATC	This study
	prot6E-NGE-r	TCGGTTTCATAATCATTCC	
IAC	IAC-f	CTAACCTTCGTGATGAGCAATCG	(Deer et al., 2010)
	IAC-r	GATCAGCTACGTGAGGTCCTAC	
Probes			
	invA_Tx_208	TX-CTCTTTCGTCTGGCATTATCGATCAGTACCA-BHQ2	(Gonzalez-Escalona et al., 2009)
	prot6E-NGE-FAM	FAM-CACCACAAT/ZEN/ATGCGAATGAACCGT -BHQ3	This study
	IAC-Cy5	Cy5-AGCTAGTCGATGCACTCCAGTCCTCCT-Iowa BlackRQ-Sp	(Deer et al., 2010)

Table 5. Primers and probes employed in this study to detect *prot6E/invA* by qPCR. TX – Texas Red.

2.5 qPCR and data analysis

The qPCR reactions were carried out using the Platinum® Quantitative PCR SuperMix-UDG kit according to the specifications of the manufacturer (Invitrogen). This kit is a ready to use cocktail consisting of a 2X Reaction Mix (Platinum® Taq polymerase, 40 mM Tris-HCl, 100 mM KCl, 6 mM $MgCl_2$, 0.4 mM of each dNTP, 0.8 of dUTP, uracil DNA glycosilase (UDG) and stabilizers). Reactions were scaled down to a final volume of 20 µl. Additional $MgCl_2$ was added to the master mix to a final concentration per tube reaction of 5 mM. Also additional Platinum® Taq polymerase was supplied in order to have 2.5 final units per reaction. Final concentrations of primers in the qPCR mix were 200 nM for *invA* and *prot6E*, and 100 nM for IAC, respectively. Probes were added to a final concentration of 150 nM. qPCR and data analysis was performed on a Mx3005P QPCR System (Agilent Technologies, Inc., Santa Clara, CA) real-time PCR machine. qPCR conditions were as follows: an initial cycle of 2 min at 50°C for UDG incubation, a second cycle of 2 min at 95°C to activate the hot-start Taq polymerase and 35 cycles of denaturation at 95°C for 15 secs, primer annealing and extension at 60°C for 30 secs (the acquisition of dyes Cy5, FAM and Texas Red were performed at the end of this cycle). Two microliters of DNA IAC (10 pg -3,0 * 10^5 copies/2µl) was added to each qPCR reaction.

3. Results

3.1 Evaluation of the *prot6E/invA* multiplex qPCR TaqMan assay

The detection limit of the *prot6E/invA* qPCR was determined using 10-fold dilutions of DNA extracted from *S. enterica* Enteritidis strain CHS44. PCR primers specific for *prot6E* gene (prot6E-NGE-f and prot6E-NGE-r) and *invA* gene (invA_176F and invA_291R) were used (Table 5). Linear calibration curves with a correlation coefficient (R^2) of $\geq$ 0.99 and linear ranges of $\geq$ 5 orders of magnitude for both *prot6E* and *invA* were obtained (Fig. 1A and B). This corresponds to detection limits of about 40 genome copies for both *prot6E* and *invA* genes. The efficiency of the qPCR was $\geq$ 0.99 for both SE targets. The robustness of DNA IAC was observed for all dilutions tested (Fig. 1C). The inclusion of the DNA IAC (internal amplification control) did not affect amplification of either *Salmonella* gene target (Fig. 1C).

3.2 Specificity of the *prot6E/invA* qPCR TaqMan assay

The developed *prot6E/invA* qPCR assay showed 100% (186/186) and 91% (170/186) inclusivity for *invA* and *prot6E* target, respectively, after testing 186 SE strains (Table 2). The strains that rendered a negative result for presence of *prot6E* were: SE-10, 58-6482, 59-365, 54-2953, CHS54, 20036, 20035, 18845, 32393, 18685, 22558, 20034, 20037, 23710, sz23, and sz5. Furthermore, *prot6E/invA* qPCR showed 100% exclusivity, only SE was positive for *prot6E* target, while all *Salmonella* strains tested were positive for *invA* gene (Table 3). Specificity of the new multiplex *prot6E/invA* qPCR assay was examined by testing 48 non-*Salmonella* (Table 4), and was 100% specific for SE. These strains were chosen for specificity testing because many are close phylogenetic kin to the *Salmonellae* and, in several cases, are known to associate with the food supply. False negatives (inhibition of PCR reaction) were also ruled out through the use of a DNA internal amplification control (IAC).

3.3 Performance assessment of different pre-enrichment media for the recovery of SE from eggs using *prot6E/invA* qPCR and USDA microbial culture methods

The usefulness of the qPCR assay developed in this study for detecting SE in eggs was assessed by artificial contamination of eggs with SE. One hundred grams of pooled eggs were artificially contaminated with two different SE strains (CDC-2010K_1543 and 13-2) at high (10^6 CFU/100 g) and low (<10 CFU/100 g) levels (Table 6). We further tested the performance of 5 different pre-enrichment media for SE growth (BPW, TSB, TSB+Fe, NB, and UP). After 24 h, the pre-enrichments were used for detection of SE using both *prot6E/invA* qPCR and USDA *Salmonella* culture method (Chapter MLG 4.05 - "Isolation and Identification of *Salmonella* from Meat, Poultry, Pasteurized Egg and Catfish Products"; http://www.fsis.usda.gov/PDF/MLG_4_05.pdf). Un-inoculated egg samples were used as negative controls. One milliliter of pre-enrichment was boiled and used for qPCR amplification in triplicate. All artificially contaminated egg samples were positive for *Salmonella* using both *prot6E/invA* qPCR and the USDA methodologies (Table 6). We chose to show in the table only lower inoculation levels in order to highlight the sensitivity of this qPCR method. SE levels as low as 5 CFU/100 g were detected after 24 $\pm$ 2 h pre-enrichments. All pre-enrichment media showed fairly similar performances for SE recovery, save for NB which showed less growth after 24h, with SE levels 10-fold lower than other

Multiplex TaqMan Real-Time PCR (qPCR) Assay Targeting prot6E and invA Genes for Fast and Accurate
Detection of Salmonella Enteritidis

261

media (Table 6). Absence of qPCR inhibitors was demonstrated by amplification of the IAC since IAC would not have been amplified had there been PCR inhibitors present in the samples analyzed (Table 6).

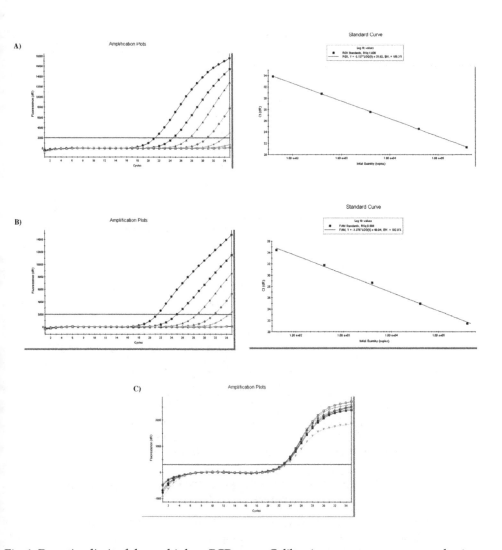

Fig. 1. Detection limit of the multiplex qPCR assay. Calibration curves were generated using 10-fold dilutions of CHS44 DNA (three replicates). A) Average *prot6E* amplification from dilutions 4.0*e10[6] - 40 CHS44 genome copies/reaction tube (FAM channel). B) Average *invA* amplification from dilutions 4.0*e10[6] - 40 CHS44 genome copies/reaction tube (ROX channel). C) Concurrent IAC amplification for each dilution (Cy5 channel). The Cq values were plotted against the nucleic acid target concentration (as copies per reaction for both DNA). The reaction efficiency (E) and R[2] values are also shown.

SE Strain	Pre-enrichment media	Minimum inoculation levels detected by USDA (CFU/100 g)	qPCR		
			prot6E (Cq)	*invA* (Cq)	IAC (Cq)
CDC-2010K_1543	BPW	5	+ (18.71 ± 0.79)	+ (19.98 ± 0.97)	+ (22.96 ± 0.41)
	NB		+ (22.69 ± 0.30)	+ (24.16 ± 0.38)	+ (22.31 ± 0.16)
	UP		+ (21.22 ± 0.55)	+ (20.12 ± 0.88)	+ (24.11 ± 1.30)
	TSB		+ (20.73 ± 0.90)	+ (22.22 ± 1.08)	+ (22.50 ± 0.42)
	TSB + Fe		+ (18.88 ± 0.30)	+ (20.32 ± 0.39)	+ (22.26 ± 0.66)
13-2	BPW	5	+ (17.70 ± 0.64)	+ (19.36 ± 0.72)	+ (22.69± 0.23)
	NB		+ (23.12 ± 1.13)	+ (24.39 ± 1.07)	+ (22.27± 0.32)
	UP		+ (19.63 ± 0.36)	+ (19.29 ± 0.79)	+ (23.83 ± 0.68)
	TSB		+ (20.24 ± 0.88)	+ (21.50 ± 0.62)	+ (22.42± 0.32)
	TSB + Fe		+ (19.43± 0.68)	+ (20.92 ± 0.82)	+ (22.20 ± 0.35)

Cq – Cycle quantification threshold, where the fluorescent is higher than the background.
+ = *Salmonella* positive by the method. In the case of IAC stands for positive signal for IAC.

Table 6. Pre-enrichment medium assessment for *Salmonella* Enteritidis (SE) recovery and detection using *prot6E*/*invA* multiplex qPCR and USDA culture method in pooled eggs artificially contaminated. Cq values are given in parentheses.

3.4 Application of the *prot6E/invA* qPCR assay for *Salmonella* Enteritidis detection in eggs artificially contaminated with different SE strains

After determining the effectiveness of the different pre-enrichment medias, we decided to employ TSB media as pre-enrichment for testing artificially contaminated eggs with 9 additional SE strains (Table 7). One of these strains lacked *prot6E* gene (SE-10). Eighteen pooled egg samples (100 g each) were artificially contaminated with high (~ 10^6 CFU/100 g) and low (<10 CFU/100 g) levels of SE and were analyzed as mentioned previously. After 24h, all artificially contaminated egg pre-enrichments were used for detection of SE using both *prot6E*/*invA* qPCR and USDA *Salmonella* culture method. Un-inoculated egg samples resulted in negative results by both *prot6E*/*invA* qPCR and USDA *Salmonella* culture method. On the other hand, all artificially contaminated egg samples were positive for

Salmonella using both *prot6E/invA* qPCR and the USDA methodologies except for SE-10
strain which was negative for *prot6E* (Table 7). Contrary to what is shown in Table 6, we
showed the results of both high and low inoculation levels Table 7. SE levels as low as 2
CFU/100 g were detected after 24 ± 2 h pre-enrichments. As mentioned previously for the
pre-enrichment media, the absence of qPCR inhibitors was demonstrated by observing no
inhibition of the amplification of the IAC in every sample (Table 7).

SE Strain	Inoculation levels detected by USDA (CFU/100 g)	qPCR (Cq)		
		prot6E	invA	IAC
SE12	-	-	-	+ (23.33 ± 0.38)
	1.15 * 10^4	+ (23.17 ± 1.35)	+ (24.32 ± 0.68)	+ (23.08 ± 0.66)
	3	+ (23.32 ± 0.72)	+ (24.34 ± 0.96)	+ (23.72 ± 1.10)
18579	-	-	-	+ (24.11 ± 0.45)
	1.24 * 10^4	+ (23.56 ± 1.22)	+ (24.45 ± 0.29)	+ (22.91 ± 0.31)
	4	+ (24.47 ± 0.76)	+ (24.72 ± 0.81)	+ (23.57 ± 1.17)
18580	-	-	-	+ (23.45 ± 0.53)
	0.68 * 10^4	+ (22.49 ± 0.59)	+ (24.32 ± 0.64)	+ (22.85 ± 0.43)
	2	+ (22.51 ± 0.72)	+ (23.97 ± 0.60)	+ (22.70 ± 0.14)
22689	-	-	-	+ (23.35 ± 0.37)
	1.35 * 10^4	+ (21.77 ± 0.88)	+ (23.12 ± 0.85)	+ (22.51 ± 0.52)
	4	+ (23.83 ± 0.91)	+ (24.83 ± 0.81)	+ (24.42 ± 1.04)
SE10	-	-	-	+ (23.30 ± 0.36)
	1.32 * 10^4	-	+ (23.20 ± 0.43)	+ (23.01 ± 0.45)
	4	-	+ (24.61 ± 0.75)	+ (22.83 ± 0.11)
SE26	-	-	-	+ (23.04 ± 0.40)
	1.53 * 10^4	+ (21.44 ± 0.28)	+ (22.93 ± 0.43)	+ (22.40 ± 0.37)
	5	+ (23.70 ± 0.91)	+ (24.76 ± 0.69)	+ (23.23 ± 0.48)
17905	-	-	-	+ (23.35 ± 0.44)
	1.52 * 10^4	+ (22.14 ± 0.26)	+ (23.58 ± 0.55)	+ (22.20 ± 0.59)
	5	+ (23.00 ± 0.32)	+ (24.32 ± 0.27)	+ (22.63 ± 0.33)
SE22	-	-	-	+ (23.66 ± 0.37)
	1.39 * 10^4	+ (20.95 ± 0.46)	+ (22.44 ± 0.49)	+ (22.73 ± 0.17)
	4	+ (23.59 ± 0.28)	+ (24.90 ± 0.29)	+ (22.31 ± 1.05)
CDC_2010K_1441	-	-	-	+ (23.40 ± 0.49)
	1.52 * 10^4	+ (22.64 ± 0.24)	+ (23.97 ± 0.28)	+ (23.45 ± 0.90)
	5	+ (22.07 ± 0.80)	+ (23.27 ± 0.79)	+ (22.56 ± 0.02)

Cq – Cycle quantification threshold, where the fluorescent is higher than the background.
+ = *Salmonella* positive by the method. In the case of IAC stands for positive signal for IAC.
- = *Salmonella* negative by the method.

Table 7. *Salmonella* Enteritidis (SE) detection by *prot6E/invA* multiplex qPCR and USDA
culture method in pooled eggs artificially contaminated. Cq values are given in parentheses.

4. Discussion

This study reports the development of a multiplex qPCR TaqMan assay that allowed for the fast and accurate detection of SE cells from eggs. The assay performed comparably to the traditional SE culture methods described in Chapter MLG 4.05 (USDA) for the detection of SE from meat, poultry, pasteurized egg and catfish products. The overall analysis took roughly 24 h, in contrast to the 5 days to 2 weeks that traditional microbiological culture methods often take. It is noteworthy that agreement between the qPCR and the two microbial culture methods was 100% for all artificially spiked samples.

This novel quantitative real-time PCR (qPCR) assay uses specific primers for the detection of *prot6E* and *invA* genes of SE with TaqMan probes. This assay also includes an internal amplification control (IAC) to detect potential PCR inhibitors that may be present in egg samples. It has become increasingly evident that there is a need for internal controls for PCR reaction, to rule out the presence of PCR inhibitors that can cause false negative results for *Salmonella*-positive samples (Hartman et al., 2005; Hoorfar et al., 2004). The inclusion of this internal control did not affect either the amplification or the detection limit of the qPCR assay. The qPCR developed here as opposed to an *invA* single target qPCR method (Feder et al., 2001; Malorny et al., 2004; Malorny B et al., 2003; Malorny et al., 2003; Bohaychuk et al., 2007; Gonzalez-Escalona et al., 2009) is able to detect specifically SE strains by the use of an SE specific marker (*prot6E*). Additionally it is capable to detect other SE that might lack the *prot6E* gene (Malorny et al., 2007a), such as the case for SE-10. The lack of *prot6E* in SE strains has been co-related with the absence of the SE virulence plasmid (~ 55 kb) (Malorny et al., 2007a). Due to the importance of that plasmid in SE virulence (Bakshi et al., 2003), without it SE has a diminish virulence, and such could be the case for SE-10 which was isolated from chicken. Moreover this assay has a further advantage in that it is an open formula assay, whereby no primers, no probes or IAC are patented or proprietary.

Among the most common gene targets used for SE detection by qPCR are: 1) Sdf1, a chromosomal fragment (Agron et al., 2001); 2) *sefA*, encoding for fimbrial antigen SEF14 (Seo et al., 2004); and 3) *prot6E*, encoding for a unique surface fimbriae (Malorny et al., 2007a; Clavijo et al., 2006). Sdf1 is highly specific for SE but is missing in SE phage types (PT) 6A, 9A, 11, 16, 20, and 27 and besides that were only tested on pure cultures (Malorny et al., 2007a). SefA gene is also present in all members of *S. enterica* serogroup D (Gallinarum, Pollorum, Dublin, Rostock, and Typhi, among others) which might lead to false positives results (Seo et al., 2004; Malorny et al., 2007a) and therefore it is not recommended for specific identification of SE. *prot6E* is present in the SE 60 kb virulence plasmid, which is present in most SE (>90%) (Chu et al., 1999; Helmuth and Schroeter, 1994; Clavijo et al., 2006) and therefore was our target of choice for SE specific detection by qPCR.

The detection limit of this qPCR assay was ~ 40 copies of genomic DNA. Usually 1 ml of pre-enrichment is boiled and 2 µl of supernatant is used for qPCR reaction. Thus, a population of approximately $4*10^4$ CFU/ml needs to be reached in the pre-enrichment to render a positive result. Commonly SE levels reach ~ 10^8 CFU/ml in the pre-enriched cultures. Therefore, this assay could be used for identification and/or quantification of SE cells in foods directly after pre-enrichment. It is also important to note, however, that in non-host environments, *Salmonella* persists most likely in a starved and highly stressed state. However, the addition of a requisite pre-enrichment step in culture media substantially increases cell number. Thus, a pre-enrichment culture provides an essential preliminary step in the application of this assay to the reliable detection of SE from eggs.

Rather than performing replicates of several inoculations with the same strain, we opted to spike the eggs in Table 7, with 9 different SE strains. This provided, in our opinion, a more powerful approach than simply repeating the experiment with the same strain multiple times as other investigators usually do. The ultimate goal of the assay is to detect different SE strains. Thus increasing the bio-complexity of testing provided a more thorough and rigorous challenge to the capability of the qPCR method to detect SE, in general. The IAC amplification was not affected in all the samples tested, however a possible failure in samples containing high levels of SE could be expected. That sort of possible failure is not un-expected given the competitive nature of this qPCR reaction, where primers and probes for *Salmonella* two targets are in excess. Thereby favoring SE targets instead of the DNA IAC. Nevertheless, it is important to emphasize that performance of the IAC in the presence of low DNA copy numbers or in the observed absence of SE was robust and reliable for each food sample analyzed, an imperative finding for any *Salmonella* detection qPCR assay (Malorny et al., 2007b).

In addition to being both effortless and reproducible, the use of ready to use mixtures, such as the one used in this study, facilitate performance of the assay. Likewise, conventional PCR methods are incapable of producing products with known identity (*i.e.* DNA sequence), subsequently failing to ensure proper specificity of PCR product(s) (Rahn et al., 1992; Malorny B et al., 2003; Malorny et al., 2003). Additionally, we employed TaqMan probes for our qPCR assay which had several advantages over the use of non-specific (although cheaper) SYBR Green I assays, including greater sensitivity and a probe-based sequence-specific verification of PCR product identity (Wittwer et al., 1997; Fey et al., 2004; Jacobsen and Holben, 2007).

In conclusion, we have developed a method that has the potential to be used as an initial screen for pre-enrichment cultures for SE without precluding the USDA culture method which is deemed necessary to yield a physical isolate that is acceptable to the regulatory process. This assay showed a high selectivity, accuracy and detection capacity. In addition, we believe that this assay will reduce the amount of samples, overall time, and effort expended in the laboratory since only positive samples will be further processed after the initial pre-enrichment step. As an added benefit, this is also a quantitative assay which allows for SE quantification in pre-enrichments or other samples. Last but certainly not least, the inclusion of the IAC makes it useful for rapid diagnosis of SE in foods directly. Moreover, in order to be applied extensively, collaborative studies should be conducted to assess the inter-laboratory reproducibility of this assay.

5. Acknowledgement

This project was supported by the FDA Foods Program Intramural Funds.

6. References

Agron, P.G., Walker, R.L., Kinde, H., Sawyer, S.J., Hayes, D.C., Wollard, J., and Andersen, G.L. (2001). Identification by subtractive hybridization of sequences specific for *Salmonella enterica* serovar enteritidis. Appl. Environ. Microbiol. *67*, 4984-4991.
Altekruse, S.F., Cohen, M.L., and Swerdlow, D.L. (1997). Emerging foodborne diseases. Emerg. Infect. Dis. *3*, 285-293.

Bakshi, C.S., Singh, V.P., Malik, M., Singh, R.K., and Sharma, B. (2003). 55 kb plasmid and virulence-associated genes are positively correlated with *Salmonella* enteritidis pathogenicity in mice and chickens. Vet. Res. Commun. *27*, 425-432.

Bohaychuk, V.M., Gensler, G.E., McFall, M.E., King, R.K., and Renter, D.G. (2007). A real-time PCR assay for the detection of *Salmonella* in a wide variety of food and food-animal matrices. Journal of Food Protection *70*, 1080-1087.

Boyd EF, Li J, Ochman H, and Selander RK (1997). Comparative genetics of the *inv-spa* invasion gene complex of *Salmonella enterica*. J. Bacteriol *179*, 1985-1991.

Chu, C., Hong, S.F., Tsai, C., Lin, W.S., Liu, T.P., and Ou, J.T. (1999). Comparative physical and genetic maps of the virulence plasmids of *Salmonella enterica* serovars typhimurium, enteritidis, choleraesuis, and dublin. Infect. Immun. *67*, 2611-2614.

Clavijo, R.I., Loui, C., Andersen, G.L., Riley, L.W., and Lu, S. (2006). Identification of genes associated with survival of *Salmonella enterica* serovar Enteritidis in chicken egg albumen. Appl. Environ. Microbiol. *72*, 1055-1064.

Deer, D.M., Lampel, K.A., and Gonzalez-Escalona, N. (2010). A versatile internal control for use as DNA in real-time PCR and as RNA in real-time reverse transcription PCR assays. Lett. Appl. Microbiol. *50*, 366-372.

Feder, I., Nietfeld, J.C., Galland, J., Yeary, T., Sargeant, J.M., Oberst, R., and Tamplin, M.L. (2001). Comparison of cultivation and PCR-hybridization for detection of *Salmonella* in porcine fecal and water samples. Journal of Clinical Microbiology *39*, 2477-2484.

Fey, A., Eichler, S., Flavier, S., Christen, R., Hofle, M.G., and Guzman, C.A. (2004). Establishment of a real-time PCR-based approach for accurate quantification of bacterial RNA targets in water, using *Salmonella* as a model organism. Appl. Environ. Microbiol. *70*, 3618-3623.

Galan JE, Pace J, and Hayman MJ (1992). Involvement of the epidermal growth factor receptor in the invasion of cultured mammalian cells by *Salmonella* typhimurium. Nature. *357*, 588-589.

Galan, J.E. and Curtiss, R., III (1991). Distribution of the *invA*, -B, -C, and -D genes of *Salmonella* Typhimurium among other *Salmonella* serovars: *invA* mutants of *Salmonella* Typhi are deficient for entry into mammalian cells. Infect. Immun. *59*, 2901-2908.

Gonzalez-Escalona, N., Hammack, T.S., Russell, M., Jacobson, A.P., De Jesus, A.J., Brown, E.W., and Lampel, K.A. (2009). Detection of live *Salmonella* sp. cells in produce by a TaqMan-based quantitative reverse transcriptase real-time PCR targeting invA mRNA. Appl. Environ. Microbiol. *75*, 3714-3720.

Hadjinicolaou, A.V., Demetriou, V.L., Emmanuel, M.A., Kakoyiannis, C.K., and Kostrikis, L.G. (2009). Molecular beacon-based real-time PCR detection of primary isolates of *Salmonella* Typhimurium and *Salmonella* Enteritidis in environmental and clinical samples. BMC. Microbiol. *9*, 97.

Hartman, L.J., Coyne, S.R., and Norwood, D.A. (2005). Development of a novel internal positive control for Taqman based assays. Mol Cell Probes. *19*, 51-59.

Helmuth, R. and Schroeter, A. (1994). Molecular typing methods for *S*. enteritidis. Int. J. Food Microbiol. *21*, 69-77.

Hoorfar, J., Cook, N., Malorny, B., Wagner, M., De, M.D., Abdulmawjood, A., and Fach, P. (2004). Diagnostic PCR: making internal amplification control mandatory. Lett Appl Microbiol. *38*, 79-80.

Jacobsen, C.S. and Holben, W.E. (2007). Quantification of mRNA in *Salmonella* sp seeded soil and chicken manure using magnetic capture hybridization RT-PCR. Journal of Microbiological Methods *69*, 315-321.

Krascsenicsova, K., Piknova, L., Kaclikova, E., and Kuchta, T. (2008). Detection of *Salmonella enterica* in food using two-step enrichment and real-time polymerase chain reaction. Lett Appl Microbiol. *46*, 483-487.

Malorny B, Hoorfar J, Bunge C, and Helmuth R (2003). Multicenter validation of the analytical accuracy of *Salmonella* PCR: towards an international standard. Appl Environ Microbiol *69*, 290-296.

Malorny, B., Bunge, C., and Helmuth, R. (2007a). A real-time PCR for the detection of *Salmonella* Enteritidis in poultry meat and consumption eggs. Journal of Microbiological Methods *70*, 245-251.

Malorny, B., Hoorfar, J., Hugas, M., Heuvelink, A., Fach, P., Ellerbroek, L., Bunge, C., Dorn, C., and Helmuth, R. (2003). Inter-laboratory diagnostic accuracy of a *Salmonella* specific PCR-based method. Int J Food Microbiol. *89*, 241-249.

Malorny, B., Made, D., Teufel, P., Berghof-Jager, C., Huber, I., Anderson, A., and Helmuth, R. (2007b). Multicenter validation study of two blockcycler- and one capillary-based real-time PCR methods for the detection of *Salmonella* in milk powder. Int. J. Food. Microbiol. *117*, 211-218.

Malorny, B., Paccassoni, E., Fach, P., Bunge, C., Martin, A., and Helmuth, R. (2004). Diagnostic real-time PCR for detection of *Salmonella* in food. Appl Environ Microbiol. *70*, 7046-7052.

O'Regan, E., McCabe, E., Burgess, C., McGuinness, S., Barry, T., Duffy, G., Whyte, P., and Fanning, S. (2008). Development of a real-time multiplex PCR assay for the detection of multiple *Salmonella* serotypes in chicken samples. BMC. Microbiol. *8*, 156.

Rabsch, W., Hargis, B.M., Tsolis, R.M., Kingsley, R.A., Hinz, K.H., Tschape, H., and Baumler, A.J. (2000). Competitive exclusion of *Salmonella* enteritidis by *Salmonella* gallinarum in poultry. Emerg. Infect. Dis. *6*, 443-448.

Rahn, K., De Grandis, S.A., Clarke, R.C., McEwen, S.A., Galan, J.E., Ginocchio, C., Curtiss, R., III, and Gyles, C.L. (1992). Amplification of an *invA* gene sequence of *Salmonella* Typhimurium by polymerase chain reaction as a specific method of detection of *Salmonella*. Mol. Cell Probes. *6*, 271-279.

Scallan, E., Hoekstra, R.M., Angulo, F.J., Tauxe, R.V., Widdowson, M.A., Roy, S.L., Jones, J.L., and Griffin, P.M. (2011). Foodborne illness acquired in the United States--major pathogens. Emerg. Infect. Dis. *17*, 7-15.

Seo, K.H., Valentin-Bon, I.E., Brackett, R.E., and Holt, P.S. (2004). Rapid, specific detection of *Salmonella* Enteritidis in pooled eggs by real-time PCR. J. Food Prot. *67*, 864-869.

Snoeyenbos, G.H., Smyser, C.F., and Van, R.H. (1969). *Salmonella* infections of the ovary and peritoneum of chickens. Avian Dis. *13*, 668-670.

Tirado, C. and Schmidt, K. (2001). WHO surveillance programme for control of foodborne infections and intoxications: preliminary results and trends across greater Europe. World Health Organization. J Infect *43*, 80-84.

Valasek, M.A. and Repa, J.J. (2005). The power of real-time PCR. Adv Physiol Educ. *29*, 151-159.

Wittwer, C.T., Herrmann, M.G., Moss, A.A., and Rasmussen, R.P. (1997). Continuous fluorescence monitoring of rapid cycle DNA amplification. Biotechniques. *22*, 130-138.

Wolffs, P.F., Glencross, K., Thibaudeau, R., and Griffiths, M.W. (2006). Direct quantitation and detection of *Salmonellae* in biological samples without enrichment, using two-step filtration and real-time PCR. Appl Environ Microbiol. *72*, 3896-3900.

13

New Options for Rapid Typing of *Salmonella enterica* Serovars for Outbreak Investigation

Ian L. Ross, Chun Chun Young and Michael W. Heuzenroeder
Public Health Department, Microbiology and Infectious Diseases
SA Pathology (at Women's and Children's Hospital), North Adelaide, South Australia
Australia

1. Introduction

A number of different serovars of *Salmonella enterica* are often implicated in human non-typhoidal outbreaks. Globally, serovars Typhimurium and Enteritidis are often the causative agent of such outbreaks, but other serovars can also be significant (Table 1). Some serovars such as Infantis and Virchow are routinely linked to outbreaks of gastroenteritis. The consumption of raw food products or poor food handling practices and/or storage procedures is often a catalyst for such outbreaks (Behravesh et al., 2010).

Accurate monitoring and tracking of specific *Salmonella* strains is of paramount importance, especially during an outbreak scenario. Large scale outbreaks such as the multistate *S. Saintpaul* outbreak in the U.S.A. in 2008 that implicated peppers as the source of contamination (Behravesh et al., 2010) highlight the need for high resolution testing procedures to enable confident identification of the source(s) of the outbreaks. Incorrect identification of potential sources will delay controlling and limiting the spread and health impact of outbreaks. Typing methods must be in place must not only identify an outbreak strain but also distinguish that strain from closely related but genetically distinct strains of the same serovar. However, many serovars of *Salmonella enterica* can appear clonal, making differentiation of strains difficult. Consequently, a high resolution typing system is required to distinguish individual strains within a serovar.

S. enterica can be subdivided into over 2,500 serotypes and, this, while useful in initial identification does not provide any further information. Classical methodologies such as bacteriophage (phage) typing can provide further subdivision within a serovar. Bacteriophage typing is a widely used phenotypic method for differentiation of clinically significant *Salmonella* serovars including Typhimurium, Enteritidis and Virchow. However, phage typing is a specialist methodology and is often unavailable to laboratories undertaking routine surveillance of *Salmonella*. Furthermore, particular phage types may dominate I a region over a period of time. This potentially makes identification of an outbreak strain difficult. Pulsed-field gel electrophoresis (PFGE) has routinely been employed for subtyping serovars and, where applicable, subtyping phage types of serovars such as Typhimurium, Enteritidis and Virchow. More recently Multiple-Locus Variable-number tandem-repeats (VNTR) Analysis (MLVA) (Lindstedt et al., 2003) and Multiple Amplification of Phage Locus Typing (MAPLT) (Ross & Heuzenroeder, 2005) have been developed as PCR-based methodology for rapid, high resolution subtyping of *Salmonella* serovars.

Serovar	Year(s)	Associated food product	Country
Agona	2008	Processed cereal products	U.S.A.
Anatum	2006	Herbs (basil)	Denmark
Bareilly	2010	Bean sprouts	U.K.
Bovismorbificans	2001	Fast food outlets	Australia
Braenderup	2008	Egg product	Japan
Chester	1999	Cuttlefish chips	Japan
Derby	2006	Restaurant food	Japan
Havana	1998	Alfalfa	U.S.A.
Hvittingfoss	2005	not determined	Australia
Infantis	1999	Poultry	U.S.A.
Kedougou	2008	Infant formula	Spain
Montevideo	2007–2008	Various	Japan
Ohio	2005	Pork	Belgium
Oranienburg	1999	Cuttlefish chips	Japan
Potsdam	2002	Restaurant salad dressing	Australia
Saintpaul	2008	Raw produce	U.S.A.
Saintpaul	2009	Alfalfa	U.S.A.
Schwarzengrund	2006	Dry dog food	U.S.A.
Senftenberg	2007	Herbs (basil)	U.K.
Singapore	2004	Sushi	Australia
Tennessee	2006	Peanut butter	U.S.A.
Virchow	1997-1998	Sun-dried tomatoes/garlic	Australia
Virchow	2004-2009	Various	Switzerland

Table 1. Examples of non-typhoidal or Typhimurium and Enteritidis *S. enterica* serovars implicated with foodborne gastroenteric outbreaks

MLVA targets loci harbouring short tandem repeat sequences, using PCR with the product analysed for fragment length by capillary electrophoresis (Lindstedt et al., 2003) and separates isolates based on the number of tandem repeats in each locus. For *S.* Typhimurium, five loci have been described and a protocol for analysis described by The Institut Pasteur (www.pasteur.fr/recherche/genopole/PF8/mlva/). Loci for other serovars of interest have been described including Enteritidis (Boxrud et al. 2007; Malorney et al., 2008; Ramisse et al., 2004), Typhi (Lui et al., 2003) and Infantis (Ross & Heuzenroeder, 2008) although an agreed MLVA protocol for these serovars is yet to be ratified.

MAPLT is a multiplex PCR-based approach which detects prophage loci located within the *Salmonella* genome (Ross & Heuzenroeder, 2005). The assay is a binary method and is based on the presence or absence of particular loci. Depending on the design of primers for each locus, prophage PCR products can be simply detected by agarose gel electrophoresis, or they may be detected by capillary sequencing in the same manner as MLVA, or by real-time PCR. MAPLT primers have been described for serovars Infantis (Ross & Heuzenroeder, 2008), Typhimurium (Ross et al., 2009) and Enteritidis (Ross & Heuzenroeder, 2009).

While both methods usually provide resolution equivalent to that generated by PFGE, often particular loci within an assay do not provide sufficient allelic diversity for maximum isolate separation. For example, it has been reported that a number of different sized fragments for the plasmid-based MLVA locus STTR-10 in a range of Typhimurium isolates (Lindstedt et al., 2004). Conversely, specific definitive phage types (DTs) of serovar Typhimurium were found to have little or no allelic variation for this locus (Ross & Heuzenroeder, 2005). Routine analysis of human, food and environmental isolates of a range of Typhimurium phage types including untypable and reactive-does not conform (RDNC) isolates suggest that both STTR-9 and STTR-3 display much reduced allelic diversity compared to the other three loci (STTR-5, STTR-6 and STTR-10) (Ross & Heuzenroeder, unpublished data). This is particularly evident within specific definitive types. MAPLT data generated by our laboratory shows a similar phenomenon where some prophage loci are found in >95% of Salmonella isolates or are completely absent. For example, $gtrC_{ST64T}$ is generally found in many S. Typhimurium tested but the $gtrA_{ST64T}$ was rarely detected in the same group of isolates (Ross & Heuzenroeder, 2005).

Salmonella enterica serovar Virchow (S. Virchow) is a relatively less common serovar, showing a prevalence to certain geographic regions. In recent years S. Virchow has ranked among the top 10 serovars among human isolates in countries located in the African, European, Oceania, Latin American and Caribbean regions (Galanis et al., 2006). Australia is one country where S. Virchow has always been endemic, particularly in the Australian state of Queensland and has ranked among the ten most common serovars derived from human source since as early as 1991 (Australian Salmonella Reference Centre [ASRC], 1999-2009). S. Virchow is a ubiquitous organism that can be detected in various food animals and environmental sources such as chickens, pigs, horses and sewage sludge (ASRC, 1999-2009). However poultry and poultry-related products were reported to be the most prevalent reservoir in a number of countries. Over a ten year period to 2009, the majority of S. Virchow isolates received and serotyped by the ASRC were from poultry and eggs. In the United Kingdom S. Virchow has been linked to chickens and chicken-related products (Threlfell et al., 2002; Willocks et al; 1996). S. Virchow is a public health concern as a significant causative agent of food-borne gastroenteric outbreaks and severe extra-intestinal infections. Poultry are the main, but not exclusive reservoir of this serovar (Adak & Threlfall, 2005; Maguire et al., 2000; Semple et al., 1968). While some reported outbreaks in different countries were poultry-associated other food sources implicated in S. Virchow outbreaks included sun-dried tomatoes and processed milk products (Bennett et al., 2003; Taormina et al., 1999; Uresa et al., 1998). Systemic S. Virchow infections in young children have also been reported in Australia and the United Kingdom (Ispahani & Slack, 2000; Messer et al., 1997).

The current international phage typing scheme for S. Virchow was developed in 1987 and comprises 13 typing phages (Chamber et al., 1987). Fifty-seven lysis patterns or phage types have been identified (Torre et al., 1993). Phage types (PTs) 8 and 26 are the most predominant phage types in the UK consisting of 50% of isolates (Torre et al., 1993). Australia and Spain are the other two countries routinely using phage typing routinely. In Spain from 1990-1996 the most frequently isolated S. Virchow phage types were PTs 8, 19 and 31 (Martín et al., 2001), whereas in Australia the same period the most common phage types were PTs 8, 31 and 34 (ASRC, 1999-2009). These results demonstrate the important role of phage typing in the global surveillance of the S. Virchow population. It also indicates that PT8 is a global phage type predominating in endemic countries, whereas PTs 26 and 34 seem to be geographically specific to the UK and Australia respectively (Sullivan et al., 1998). In addition phage typing acts as a long-term epidemiological typing tool revealing any changes in incidence of S. Virchow phage types within a particular source. With respect to S. Virchow in Australia, no significant changes in the incidence of phage types were observed from human source in the decade to 2009 where PT8 was the most prevalent phage type (>50%) (ASRC, 1999-2009). In contrast, there were noticeable changes in the S. Virchow population in chickens and eggs based on the S. Virchow isolates received by ASRC. Even though PT8 was most commonly isolated from chickens and eggs in most years during the same period of time, the proportion of PT8 within these sources has decreased from 81.9% in 2000 to 35.5% in 2009 (ASRC, 1999-2009).

This chapter describes how specific data from MLVA and MAPLT can be combined into a single composite assay, thereby maximizing the resolving power of the assay for closely related isolates. The two classical typing methods, PGFE and phage typing provide a benchmark for determining the efficacy of MLVA and MAPLT, both individually and as a composite methodology. Previously published data for the two most significant non-typhoidal serovars, Typhimurium and Enteritidis have been re-analysed to determine the most variable loci for each protocol and single assays containing these loci have been identified for each serovar. The addition of phage typing data for serovar Enteritidis has also been taken into consideration to determine whether loci selection can be influenced by phage type. MLVA and MAPLT protocols have been developed for serovar Virchow with comparisons with PFGE. A single MLVA/MAPLT hybrid assay for S. Virchow has been developed and described here for the first time.

2. Materials and methods

2.1 S. Virchow strains and culture conditions

A total of 43 epidemiologically unrelated S. Virchow isolates were used for the development of MLVA and MAPLT assays. The isolates were provided by the ASRC, Institute of Medical and Veterinary Science, Adelaide, South Australia. The isolates represented a cross-section of the most commonly identified phage types submitted to the ASRC and were originally isolated throughout Australia between 2005 and 2008. Serotyping and phage typing of all S. Virchow isolates had previously been undertaken by the ASRC. Unless otherwise stated, all isolates were routinely cultured either on XLD agar medium or in bovine heart infusion broth (BHI) (Oxoid) at 37°C.

2.2 MLVA of S. Virchow

MLVA was undertaken utilizing primer sets previously described for Typhi (Liu, et al., 2003), Typhimurium (Lindstedt et al., 2003; Lindstedt et al., 2004) and Enteritidis (Boxrud et al., 2007; Ramisse et al., 2004) Primer sets targeting specific MLVA loci were selected based on their ability to differentiate within particular serovars of Salmonella. The touchdown PCR reaction and thermal cycler conditions were the same as those previously described (Ross & Heuzenroeder, 2005). Confirmation of fragment lengths as determined by genotyping was undertaken by nucleotide sequencing of selected isolates using Big Dye Terminator, version 3-1 (Applied Biosystems, Foster City, Calif.). Both genotyping and nucleotide sequencing were performed on an Applied Biosystems 3700 DNA Analyser. Data were entered into

BIONUMERICS v4.61 software (Applied Maths, Kortrijk, Belgium) as numerical values (fragment lengths in base pairs (bp) and negative PCR results entered as '0'). Dendrograms depicting the genetic similarity of isolates as determined by their MLVA profiles were generated using the categorical multi-state coefficient with zero tolerance and clustering by UPGMA utilising BIONUMERICS v4.61software (Applied Maths).

2.3 MAPLT of S. Virchow

Phages were induced from S. Virchow isolates as previously described (Ross & Heuzenroeder, 2008). Ten microlitres of each phage suspension were spotted onto lawns of epidemiologically distinct S. Virchow indicator isolates, allowed to dry and incubated at 37°C until plaquing could be observed. Phages that generated different lysis profiles (Fig. 1.) were selected for DOP-PCR to detect different phage sequences. DNA was extracted from phage and DOP-PCRs were undertaken as previously described (Ross & Heuzenroeder, 2009). Unique bands (Fig. 2.) were extracted from agarose gels and cloned into the vector PCRs 4-TOPO and transformed into TOPO One Shots TOP10 chemically competent E. coli cells (Invitrogen, Carlsbad, CA) according to the manufacturer's instructions. Amplification of cell lysates using the TOPO primers was followed by sequencing PCR, undertaken with Big Dye Terminator v3-1 (Applied Biosystems, Foster City, CA). Characterization of sequence data was subsequently performed with KODON v3.5 (Applied Maths) and sequences compared with genomic library data for phage identification.

MAPLT analysis was undertaken with the primer combinations derived from prophages ST64B and P22 as published previously (Ross & Heuzenroeder, 2005), as well as loci identified by DOP-PCR from S. Virchow-derived prophages (Table 2). Amplification conditions using touchdown PCR and subsequent analysis were carried out as described previously (Ross & Heuzenroeder, 2005). MAPLT profiles for the S. Virchow isolates were determined based on the presence or absence of PCR product for all loci tested.

S. Virchow V15   S. Virchow V11   S. Virchow V09

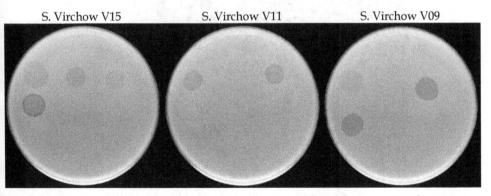

Fig. 1. Detection of different S. Virchow-derived bacteriophages by comparing plaquing patterns on lawns of S. Virchow isolates V15, V11 and V09 (as examples). By detecting differences in these patterns, potentially genetically different phages can then be isolated and identified by DOP-PCR and sequencing. This method results in a range of MAPLT primers that can detect a broad range of phage sequences in S. Virchow.

2.4 PFGE of *S.* Virchow

The protocol for PFGE was based on that of Maslow et al., (1993) as modified by Ross & Heuzenroeder (2005). Agarose-embedded *Salmonella* DNA and the *Staphylococcus aureus* strain NCTC 8325 marker DNA (Tenover et al., 1995) were digested overnight with the restriction endonucleases XbaI and SmaI, respectively (New England BioLabs Beverley, MA). The PFGE running conditions in the BIO-RAD CHEF-DR III System and subsequent comparisons of band profiles were undertaken as described previously (Ross & Heuzenroeder, 2005) using the GELCOMPAR II program (Applied Maths).

? 5 Data analysis

Comparison of the discriminatory power of all typing methods was undertaken using Simpson's index of diversity (Hunter & Gaston, 1988).

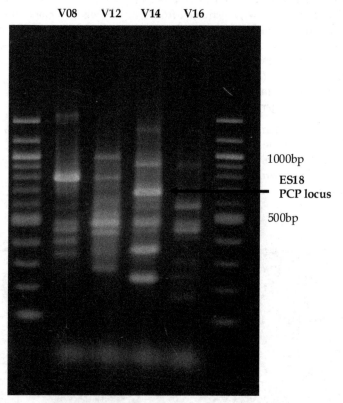

Fig. 2. DOP-PCR amplified phage DNA from *S.* Virchow isolates (V08, V12, V14 and V16). Individual bands were excised, cloned and sequenced to identify phage (see text for details). Phage from *S.* Virchow isolate V08 contained Fels2 sequences, V14 contained sequences from phage ES18 and V16 contained phage sequences from P186. The band containing the ES18 portal capsid protein sequence (PCP) is indicated as an example. No phage sequence was analysed from isolate V12 at time of publication. Molecular weight marker (first and last lanes) is a 100kb ladder.

Phage	Gene or locus	Encoded proteins	Primers	Location in fragment genomes[i]	Sizes (bp)
P22	ninB	ninB protein	ninBF1: AACCTTTGAAATTCGATCTCCAGC ninBR1: CTTCGTCTGACCACTTAACGC	16512 to 16891	380
ES18	gene 9	putative coat protein	PCPF: TGGAACGCACAGCATGATGC PCPR: GGACTGCACCTGAATATTCGG	6368 to 6853	486
Fels2	STM2736	CII protein	Fels2cIIF: TGTATGGAAACGGCAGCCAG Fels2cIIR: GTCACAACATGGCGAAGCTG	2875360 to 2875723	363
Gifsy-1	STM2608	Terminase large subunit	Gifsy1AF: GATCACGCATCCATTATGTTCAC Gifsy1AR: TATTCCCGTACCGCTTACCAC	2756675 to 2757449	775
186	cII	CII protein	P186cIIF: GACATAGCGGGATTAGTCTGC Fels2cIIR: GTCACAACATGGCGAAGCTG	23499 to 23892	394
18b	gene P	Endolysin	P186PF: TCACCGATTACAGCGACCAC P186PR: TGGTGACCAGCTTTTCGAGAC	7877 to 8200	324
ST64B	SB04	Putative portal protein	SB04F: TGTCATACGACACCTATACCG SB04R: TGTTCTGCACCATGTGCAATG	2513 to 3298	786
P7	sit	Putative injection transglyosylase	P7sitF: TGACCTTGATCGCGTACTCAC P7sitR: TAGCCACCAGGAGACATCTG	44668 to 45368	701
V16[ii]	DOP13.7	Possible tail fibre protein	13.7F: CGGTTAGCTCCGTGGTTAAG 13.7R: TAGCCACCAGGAGACATCTG	not described	441

[i] Gene or locus accession numbers as follows:
P22: GeneBank accession no: AF217253
ES18: GenBank accession number AY736146
Fels2: GenBank accession number AE006468 (Prophage sequence of Salmonella Typhimurium strain LT2 from 2844427 to 2879233)
Gifsy-1: GenBank accession number AE006468 (Prophage sequence of Salmonella Typhimurium strain LT2 from 2844427 to 2879233)
186: GenBank accession number U32222.1
ST64B: GenBank accession number AY055382
P7: GenBank accession number AF503408
[ii] Unidentified prophage loci in S. Virchow isolate V16

Table 2. Primers for MAPLT analysis of S. Virchow

3. Results

3.1 Composite data for *S.* Typhimurium

Ten loci comprising seven MAPLT and three MLVA sites were selected for analysis in the development of a combined MAPLT/MLVA protocol; $c1_{ST64B}$ SB06$_{ST64B}$, SB26$_{ST64B}$, SB28$_{ST64B}$, SB46$_{ST64B}$, gene 9_{ST64T}, $gtrC_{ST64T}$, STTR-5, STTR-6 and STTR-10. A dendrogram was generated reflecting analysis by this method (Figure 3). A total of 29 different profiles were generated. As previously observed, *S.* Typhimurium DT126 isolates were distinct from DT108, DT12 and DT12a isolates. The overall Simpson's Index of Diversity (DI) value for all non-DT126 isolates was 0.91, compared with previously published values of 0.83 for MLVA and 0.41 for MAPLT (Ross, et al., 2009). The Simpson's Index of Diversity (DI) value for the DT126 isolates was not calculated as most of these isolates were derived from two outbreaks and therefore would have skewed any statistical analysis due to their clonality.

3.2 Composite data for *S.* Enteritidis

Based on previously published data (Ross & Heuzenroeder, 2009), a combined MAPLT/MLVA was devised based on the most variable loci from each assay. Consequently a universal protocol targeting the following ten loci was devised; SB40$_{ST64B}$, SB21$_{ST64B}$, SB28$_{ST64B}$, SB46$_{ST64B}$, $gtrA_{ST64T}$, $gtrB_{ST64T}$, STTR-3, STTR-5, SE-1 and SE-2. These ten loci can be initially used where no phage typing data is available. Where phage typing data is available, improved separation within a phage type can be achieved. For example, our data shows that, instead of locus SB21$_{ST64B}$, the substitution of the ST64T gene 9 locus at the 5' end (g9:5') (Ross & Heuzenroeder, 2005) improves separation of phage type 26 isolates (Figure 4a) while the composite assay for the phage type 4 isolates indicated that the ten universal loci described above were suitable for this phage type (Figure 4b). The addition of ST64B *immC* gene $c1$ improved separation of the *S.* Enteritidis RDNC isolates and isolates unable to be typed (ut) by phage typing (isolate designations RDNC- and Eut- respectively) (Figure 4c). Simpson's Index figures for the combined MAPLT/MLVA assay and comparisons to the previously published data for individual assays are provided in Table 3.

PT	MAPLT	MLVA	Composite	PFGE
26	0.87 (14)	0.89 (17)	0.99 (21)	0.66 (6)
4	0.83 (10)	0.85 (10)	0.99 (19)	0.48 (4)
ut/RDNC	0.98 (23)	0.96 (20)	0.99 (25)	0.89 (11)

Table 3. Comparative Simpson's Index values for *S.* Enteritidis phage types

Simpson's Index data for separate PFGE, MLVA and MAPLT analyses previously published (Ross and Heuzenroeder, 2009) Figures in brackets are the number of different profiles generated by each assay.

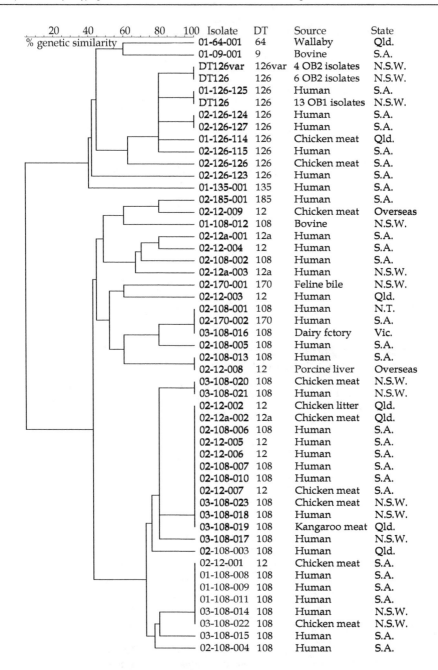

Fig. 3. Dendrogram showing genetic similarity of *S.* Typhimurium isolates. Abbreviations for states are: N.S.W. New South Wales, N.T. Northern Territory, Qld. Queensland S.A. South Australia, Vic. Victoria, W.A. Western Australia.

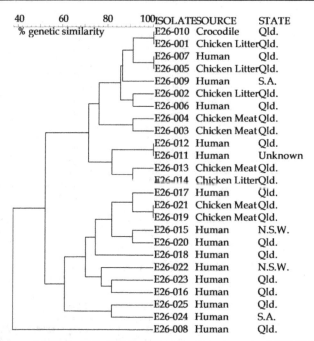

Fig. 4a. Dendrogram of *S. Enteritidis* PT26 analysed with composite MAPLT/MLVA data. No further information available for isolate E26-11

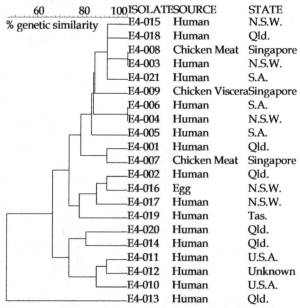

Fig. 4b. Dendrogram of *S. Enteritidis* PT4 analysed with composite MAPLT/MLVA data. All Australian states except where indicated.

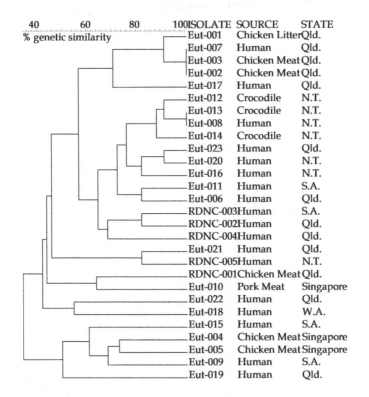

Fig. 4c. Dendogram of untypable and RDNC *S*. Enteritidis isolates analysed with composite MAPLT/MLVA data.

3.3 *S*. Virchow

PFGE analysis of *S*. Virchow divided the 43 isolates into 17 different profiles (Fig. 5). There was no distinct correlation between PFGE profile and phage type. For examples PFGE profiles 1, 3, 9 and 10 were generated from isolates with different phage types. Similarly, isolates of some phage types (17, 19, 31 and 36var1) produced PFGE profiles with 2 to 6 band differences between isolates, indicating that isolates within these phage types could exhibit an extensive genetic diversity. This study included a large proportion of PT8 isolates due to its predominance among all phage types seen in Australia. From the twenty-five PT8 isolates 15 (60%) generated PFGE profile 2. Nearly all PT8 isolates (14 out of 15) had the same MAPLT profile.

MAPLT analysis identified a number of loci derived from various bacteriophages which were useful in distinguishing between *S*. Virchow isolates. Nine MAPLT loci were subsequently chosen for *S*. Virchow differentiation based on the variability of frequency of these loci across the 43 isolates.

Using 15 MLVA primer sets previously described for a range of *S. enterica* serovars, only MLVA locus STTR-5 provided any allelic variation in the 43 *S.* Virchow isolates. The range of fragment sizes for this locus (based on the primer sequences of Lindstedt, et al., 2003) was 217bp (Fig. 6) to 271bp. There was no observed correlation between STTR-5 fragment size and phage type and in particular for PT8 the predominant type.

A composite MAPLT/MLVA dendrogram based on 9 MAPLT loci and the MLVA locus STTR-5 was generated (Fig. 6). This combination significantly improved the separation of the 43 *S.* Virchow isolates both in terms of diversity and number of different profiles generated (Table 4). More importantly, the differentiation of PT8 isolates was improved considerably using the combined method (DI = 0.88) in comparison to PFGE (DI = 0.59).

	MAPLT	MLVA	Composite	PFGE
Number of primers	9	13	10	na
Number of profiles	14	8	23	17
Simpson's DI	0.81	0.79	0.94	0.84

na not applicable

Table 4. Diversity of 43 *S.* Virchow isolates as determined by each method. Composite data based on combined MAPLT and MLVA primers; see Fig. 6 for details.

4. Discussion

The adoption of rapid, high resolution PCR-based typing assays such as MLVA and MAPLT for fine discrimination of closely related isolates of *Salmonella* may provide an alternative to phenotypic assays and current molecular methods such as PFGE. As more data is obtained it is obvious that there are sufficient differences in bacterial genome structure and prophage populations between different serovars of *Salmonella enterica* to necessitate development of such assays on a serovar by serovar basis. While PFGE is not limited by this issue, the development of PCR-based assays for specific serovars of interest is worthwhile due to the likelihood of improved discrimination of isolates and the ease of sharing data between interested laboratories and health authorities.

The combination of separate MAPLT and MLVA data into a single composite assay can provide superior discrimination of isolates than that obtained by either assay alone, as well as by PFGE. In the case of serovar Typhimurium, one of the most significant causative agents of non-typhoidal *Salmonella*-induced gastroenteritis, we have demonstrated that closely related phage types such as DT108 and DT12 can be separated by either PCR-based method, but combining the most variable loci into a single assay provides what may be the optimal separation of isolates. Furthermore, it should be noted that there was no correlation between phage type and clustering by MAPLT and/or MLVA. As mentioned previously the index of diversity for the DT126 isolates was not determined due to the clonality of the outbreak isolates clustering more tightly than would be seen with a group of epidemiologically-unrelated isolates. This however, demonstrates the ability of these PCR-based assays for discriminating outbreak isolates from closely related but epidemiologically distinct strains.

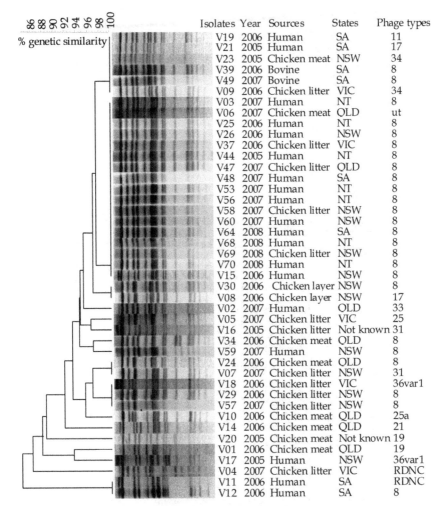

	Isolates	Year	Sources	States	Phage types
	V19	2006	Human	SA	11
	V21	2005	Human	SA	17
	V23	2005	Chicken meat	NSW	34
	V39	2006	Bovine	SA	8
	V49	2007	Bovine	SA	8
	V09	2006	Chicken litter	VIC	34
	V03	2007	Human	NT	8
	V06	2007	Chicken meat	QLD	ut
	V25	2006	Human	NT	8
	V26	2006	Human	NSW	8
	V37	2006	Chicken litter	VIC	8
	V44	2005	Human	NT	8
	V47	2007	Chicken litter	QLD	8
	V48	2007	Human	SA	8
	V53	2007	Human	NT	8
	V56	2007	Human	NT	8
	V58	2007	Chicken litter	NSW	8
	V60	2007	Human	NSW	8
	V64	2008	Human	SA	8
	V68	2008	Human	NT	8
	V69	2008	Chicken litter	NSW	8
	V70	2008	Human	NT	8
	V15	2006	Human	NSW	8
	V30	2006	Chicken layer	NSW	8
	V08	2006	Chicken layer	NSW	17
	V02	2007	Human	QLD	33
	V05	2007	Chicken litter	VIC	25
	V16	2005	Chicken litter	Not known	31
	V34	2006	Chicken meat	QLD	8
	V59	2007	Human	NSW	8
	V24	2006	Chicken meat	QLD	8
	V07	2007	Chicken litter	NSW	31
	V18	2006	Chicken litter	VIC	36var1
	V29	2006	Chicken litter	NSW	8
	V57	2007	Chicken litter	NSW	8
	V10	2006	Chicken meat	QLD	25a
	V14	2006	Chicken meat	QLD	21
	V20	2005	Chicken meat	Not known	19
	V01	2006	Chicken meat	QLD	19
	V17	2005	Human	NSW	36var1
	V04	2007	Chicken litter	VIC	RDNC
	V11	2006	Human	SA	RDNC
	V12	2006	Human	SA	8

Fig. 5. Pulsed-field gel electrophoresis of 43 S. Virchow isloates.

While separate assays may need to be developed for different serovars with unique sets of primers, it is possible that individual loci may provide extra discrimination for particular phage types within a serovar. It has previously been reported that MLVA locus SENTR2 (locus STTR-7 as previously described by Lindstedt et al., 2003) may be useful for improved detection of differences within sample groups of both S. Enteritidis PT4 and PT8 isolates (Malorney et al., 2008). The data for S. Enteritidis presented here further supports this concept. While 10 primers sets formed the basis of a composite MAPLT/MLVA assay for this serovar (as demonstrated for the PT4 isolates), different MAPLT-derived loci proved useful for maximising isolate discrimination (see Fig. 4). This information is more relevant where phage type data is available and pre-selection of primers can be ascertained. However, even in the absence of the phage typing data, the assay may include primers for these extra loci as a matter of course.

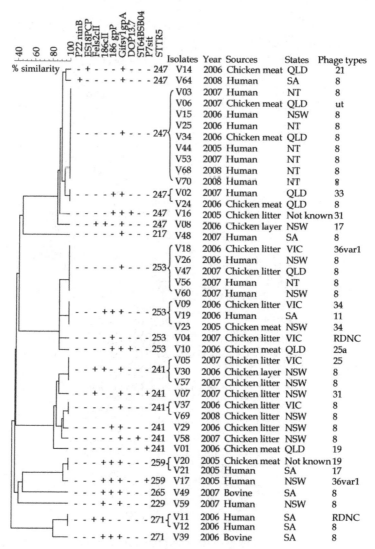

+ MAPLT locus detected by PCR, - MAPLT locus not detected.
Fragment sizes for MLVA locus STTR-5 based on primer locations described by Lindstedt et al., (2003).

Fig. 6. A dendrogram based on composite MAPLT/MLVA data as described in section 3.3. All abbreviations for Australian states as per Fig. 3.

Development of MAPLT and MLVA as well as a composite assay for serovar Virchow has identified the importance of total genomic data being available in genome libraries such as Genbank (www.ncbi.nlm.nih.gov). While a number of suitable MAPLT loci were identified from a range of different prophages isolated from the *S.* Virchow strains with the exception of locus STTR-5, previously described MLVA loci were found to be either homologous in terms of fragment length or not detected by PCR and thus do not provide allelic variation within this serovar. Access to total genomic data on different serovars and strains would facilitate searches for tandem repeat loci that may be unique to that serovar. It is also likely that more than one total genome per serovar may need to be sequenced to enhance the likelihood that most or all MLVA loci present in that serovar are detected. For example, many S. Typhimurium strains do not harbour the plasmid-based STTR-10 locus (Ross & Heuzenroeder, unpublished data). Development of a MLVA assay based solely on the sequence data of one of these isolates may have resulted in one of the most variable MLVA loci being excluded from any devised MLVA protocol. In the case of the development of a MLVA assay for S. Enteritidis, the genomes of two separate isolates of this serovar, LK5 and a phage type 4 isolate (as well as S. Typhimurium LT2) were analysed (Boxrud et al., 2007). Consequently, we conclude that the genomes of suitable S. Virchow isolates may need to be completely sequenced to identify unique tandem repeat loci that provide suitable allelic variation for a MLVA assay. In the interim however, MAPLT loci has provided excellent separation of the S. Virchow isolates while the inclusion of STTR-5 into a composite assay enhanced separation of the isolates, in particular the PT 8 isolates.

The use of PCR-based methodology can be quite useful in outbreak situations where the source of the outbreak must be quickly identified to stop or restrict the spread of the pathogen in the community or environment. Their usefulness is based on the high resolution capabilities, the relatively short time frame required for obtaining data and the simplicity for data sharing. It has been noted that because some MLVA loci of a strain can exhibit subtle mutations in tandem repeat number during the course of an outbreak, some subjective interpretation of data in conjunction with other epidemiological data may be necessary for accurate identification. Boxrud et al., (2007) has suggested that "interpretive criteria that account for genetic variability of MLVA patterns analogous to the Tenover criteria used for PFGE may need to be developed". In Australia, laboratories collaborating in MLVA of *Salmonella* have agreed that minor variations such as one tandem repeat change at two separate loci may not be significant, especially if epidemiological information supports the conclusion. A study on S. Typhimurium DT9 isolates involved in an outbreak in South Australia in 2007 revealed MLVA allelic variability in human-derived isolates that were linked to the outbreak (Ross et al., 2011). Local outbreaks of both DT9 and DT108 during 2011 have also indicated that variability in the three loci STTR-5, STTR-6 and STTR-10 can appear during the course of the outbreak (Ross et al., unpublished data). These variations however, did not prevent the rapid identification of the likely food source of the outbreak. As yet, a comparison with the stability of MAPLT loci in these isolates has yet to be determined.

With the development of PCR-based protocols being undertaken there is a need to ensure consistency of loci identification and nomenclature as well as clear guidelines for data generation. It has been noted that a single locus may be given more than one designation by different laboratories, leading to potential confusion. One example of this has been alluded to previously in this discussion; the naming of the MLVA locus as either STTR-7 or SENTR2 by different laboratories. As the name STTR-7 was documented first we have adopted this description and suggest all subsequent references to this locus be made in accordance with this nomenclature. A different example is where the sequence of a tandem repeat has been published in either direction by two different laboratories. S. Enteritidis MLVA locus SE-2 described by Boxrud et al., (2007), was later described as SENTR6 and published in the reverse direction. In both cases, using different nomenclature for identical loci can generate confusion and unnecessary work for researchers during assay development and/or surveillance programmes.

Standardised guidelines for data generation and interpretation also need to be developed. We have already mentioned previously in the Introduction, guidelines for MLVA of *S.* Typhimurium published by The Institut Pasteur. Even so, there is still a lack of concordance in what constitutes an agreed tandem repeat sequence and whether single nucleotide polymorphisms in flanking tandem repeats disqualify them as being included in a tandem repeat analysis. This laboratory currently reports all *S.* Typhimurium MLVA patterns in terms of total sequence length of the five loci in base pairs in accordance with the primer sequences published by Lindstedt et al., (2003, 2004) and adopted and described in The Institut Pasteur website. This reporting method, lacking tandem repeat numbers, prevents any subsequent misinterpretation of data.

5. Conclusions

Both MAPLT and MLVA offer rapid PCR-based approaches for rapid, high resolution discrimination of phenotypically closely related but epidemiologically distinct *Salmonella* isolates. This level of discrimination is often at least equal to that offered by PFGE. Objective data generated by either PCR method can be easily shared between laboratories and appropriate jurisdictional health authorities for general pathogen surveillance purposes as well as the investigation and control of outbreaks. As either MAPLT or MLVA may be more suited for a particular serovar or, where applicable, phage type, a composite assay comprising multiplex primers from both individual assays targeting the most variable loci in a particular strain can provide the maximum level of isolate separation. This data in the form of universally agreed nomenclature, in combination with epidemiological information, would prove invaluable for detecting sources of outbreaks and thereby restricting their effects.

6. Acknowledgements

The authors would like to thank Dianne Davos, Helen Hocking, and the staff of the Australian Salmonella Reference Centre, Adelaide, for providing phage typed strains for this study. This work was undertaken with the generous assistance of the Rural Industries Research and Development Corporation (Chicken Meat Program) and the National Health and Medical Research Council.

7. References

Adak, B. & E. J. Threlfall, E.J. (2005) Outbreak of drug-resistant *Salmonella* Virchow phage type 8 infection. *CDR Weekly*, Vol.15 pp. 2-3.

Australian Salmonella Reference Centre, Annual Reports 1999-2009. Institute of Medical and Veterinary Science, Adelaide, Australia.

Behravesh, C.B.; Mody, R.K.; Jungk, J.; Gaul, L.; Redd, J.T.; Chen, S.; Cosgrove, S.; Hedican, E.; Sweat, D.; Chávez-Hauser, L.; Snow, S.L.; Hanson, H.; Nguyen, T-A.; Sodha, S.V.; Boore, A.L.; Russo, E.; Mikoleit, M.; Theobald, L.; Gerner-Smidt, P.; Hoekstra, R.M.; Angulo, F.J.; Swerdlow, D.L.; Tauxe, R.V.; Griffen, P.M. & Williams, I.T. (2010) 2008 outbreak of *Salmonella* Saintpaul infections associated with raw produce. The New England Journal of Medicine. Vol.364, pp. 918-927.

Bennett, C. M.; Dalton, C.; Beers-Deeble, M.; Milazzo, A.; Kraa, E.; Davos, D.; Puech, M.; Tan, A. & Heuzenroeder, M.W. (2003) Fresh garlic: A possible vehicle for *Salmonella* Virchow. *Epidemiology and Infection*, Vol.131, pp. 1041-1048.

Boxrud, D.; Pederson-Gulrud, K.; Wotton, J.; Medus, C.; Lyszkowicz, E.; Besser, J. & Bartkus, J.M. (2007) Comparison of multiple-locus variable number tandem repeat analysis, pulsed-field gel electrophoresis, and phage typing for subtype analysis of *Salmonella enterica* seroptype Enteritidis. *Journal of Clinical Microbiology*, Vol.45, pp. 536-543.

Chamber, R. M.; McAdam, P.; de Sa, J.D.H.; Ward, L.R. & Rowe, B. (1987) A phage-typing scheme for *S. virchow*. *FEMS Microbiology Letters*, Vol.40, pp. 155-157.

Galanis, E., Wong, D. M. A.; Patrick, M.E.; Binztein, N.; Cieslik, A.; Chalermchaikit, T.; Aidara-Kane, A.; Ellis, A.; F. J. Angulo, F.J. & Wegener, H.C. (2006) Web-based surveillance and global *Salmonella* distribution, 2000-2002. *Emerging Infectious Diseases*, Vol.12, pp.381-388.

Hunter, P.R. & Gaston, M.A. (1988) Numerical index of the discriminatory power of typing systems: an application of Simpson's index of diversity. *Journal of Clinical Microbiology*, Vol.26, pp. 2465-2466.

Ispahani, P. & Slack, R.C. (2000) Enteric fever and other extraintestinal salmonellosis in University Hospital, Nottingham, UK, between 1980 and 1997. *European Journal of Clinical Microbiology and Infectious Diseases*, Vol.19, pp. 679-687.

Lindstedt, B.-A.; E. Heir, E.; E. Gjernes, E. & Kapperud, G. (2003) DNA fingerprinting of *Salmonella enterica* subsp. *enterica* serovar Typhimurium with emphasis on phage type DT104 based on variable number of tandem repeat loci. *Journal of Clinical Microbiology*, Vol.41, pp. 1469-1479.

Lindstedt, B.-A.; Vardund, T.; Aas, L. & Kapperud, G. (2004) Multiplr-locus variable-number tandem-repeats analysis of *Salmonella enterica* subsp. *enterica* serovar Typhimurium using PCR multiplexing and multicolor capillary electrophoresis. *Journal of Microbiological Methods*, Vol.59, pp. 163-172.

Lui, Y.; Lee, M-A.; Ooi, E-E.; Mavis, Y.; Tan, A-L. & Quek, H-H. (2003) Molecular typing of Salmonella enterica serovar Typhi isolates from various countries in Asia by a multiplex PCR assay on variable-number tandem repeats. *Journal of Clinical Microbiology*, Vol.41, pp. 4388-4394.

Maguire, H.; Pharoah, P.; Walsh, B.; Davison, C.; Barrie, D.; Threlfall, E.J. & Chamber, S. (2000) Hospital outbreak of *Salmonella virchow* possibly associated with a food handler. *Journal of Hospital Infection*, Vol.44, pp. 261-266.

Malorny, B.; Junker, E. & Helmuth, R. (2008) Multi-locus variable -number tandem repeat analysis for outbreak studies of *Salmonella enterica* serotype Enteriditis. *BMC Microbiology*, Vol.8. p. 84.

Martín, M. C.; González-Hevia, M.A.; Alvarez-Riesgo, J.A. & Mendoza, M.C. (2001) *Salmonella* serotype Virchow causing salmonellosis in a Spanish region. Characterization and survey of clones by DNA fingerprinting, phage typing and antimicrobial resistance. *European Journal of Epidemiology*, Vol.17, pp. 31-40.

Maslow, J.N.; Slutsky, A.M. & Arbeit, R.D. (1993) Application of pulsed-field gel electrophoresis to molecular epidemiology, In: *Diagnostic Molecular Microbiology: Principles and Applications*, Pershing, D.H.; Smith, T.F.; Tenover, F.C. & White, T.J. Eds., pp. 563-572, American Society for Microbiology, ISBN-10: 1-55581-056-X, Washington D.C.

Messer, R. D.; Warnock, T.H.; Heazlewood, R.J. & Hanna, J.N. (1997) *Salmonella* meningitis in children in far north Queensland. *Journal of Paediatrics and Child Health*, Vol.33, pp. 535-538.

Ramisse, V.; Houssu, P.; Hernandez, E.; Denoeud, F.; Hilaire, V.; Lisante, O.; Ramisse, F.; Vavallo, J-D. & Vergnaud, G. (2004) Variable number of tandem repeats in Salmonella enterica subsp. enterica for typing purposes. *Journal of Clinical Microbiology*, Vol.42, pp. 5722-5730.

Ross, I. L. & Heuzenroeder M.W. (2005) Discrimination within phenotypically closely related definitive types of *Salmonella enterica* serovar Typhimurium by the multiple amplification of phage locus typing technique. *Journal of Clinical Microbiology*, Vol.43, pp.1604 -1611.

Ross, I. L. & Heuzenroeder, M.W. (2008) A comparison of three molecular typing methods for the discrimination of *Salmonella enterica* serovar Infantis. *FEMS Immunology and Medical Microbiology*, Vol.53, pp. 375-384.

Ross, I. L. & Heuzenroeder, M.W. (2009) A comparison of two PCR-based typing methods with pulsed-field gel electrophoresis in *Salmonella enterica* serovar Enteritidis. *International Journal of Medical Microbiology*, Vol.299, pp. 410-420.

Ross, I.L.; Parkinson, I.H. & Heuzenroeder, M.W. (2009) The use of MAPLT and MLVA analyses of phenotypically closely related isolates of *Salmonella enterica* seovar Typhimurium *International Journal of Medical Microbiology*, Vol.299, pp. 37-41.

Ross, I.L.; Davos, D.E.; Mwanri, L.; Raupach, J. & Heuzenroeder, M.W. (2011) MLVA and phage typing as complementary tools in the epidemiological investigation of *Salmonella enterica* serovar Typhimurium clusters. Current Microbiology, Vol.62, pp. 1034-1038.

Semple, A. B.; Turner, G.C. & Lowry, D.M. (1968) Outbreak of food-poisoning caused by *Salmonella* virchow in spit-roasted chicken. *British Medical Journal*, Vol.4, pp. 801-803.

Sullivan, A. M.; Ward, L.R.; Rowe, B.; Woolcock, J.B. & Cox, J.M. (1998) Phage types of Australian isolates of *Salmonella enterica* subsp. enterica serovar Virchow. *Letters in Applied Microbiology*, Vol.27, pp. 216-218.

Taormina, P. J.; Beuchat, L.R. & Slutsker, L. (1999) Infections associated with eating seed sprouts: An international concern. *Emerging Infectious Diseases*, Vol.5, pp. 626-634.

Tenover, F.C; Arbeit, R.D.; Goering, R.V.; Mickelsen, P.A.; Murray, B.E.; Persing, D.H. & Swaminathan, B. (1995) Interpreting chromosomal DNA restriction patterns produced by pulsed-field gel electrophoresis: Criteria for bacterial strain typing. *Journal of Clinical Microbiology*, Vol.33, pp. 2233-2239.

Threlfall, S. W.; McDowell, J. & Davies, R.H. (2002) Multiple genetic typing of *Salmonella enterica* serotype Typhimurium isolates of different phage types (DT104, U302, DT204b, and DT49) from animals and humans in England, Wales, and Northern Ireland. *Journal of Clinical Microbiology*, Vol.40, pp. 4450-4456.

Torre, E., E.; Threlfall, J.; Hampton, M.D.; Ward, L.R.; Gilbert, I. & B. Rowe, B. (1993) Characterization of *Salmonella virchow* phage types by plasmid profile and IS*200* distribution. *Journal of Applied Bacteriology*, Vol.75, pp. 435-440.

Uresa, M. A.; Rodriguez, A.; Echeita,A. & Cano, R. (1998) Multiple analysis of a foodborne outbreak caused by infant formula contaminated by an atypical *Salmonella virchow* strain. *European Journal of Clinical Microbiology and Infectious Diseases*, Vol.17, pp. 551-555.

Willcox, L.J.; Morgan, D.; Sufi, F.; Ward, L.R. & Patrick, H.E. (1996) *Salmonella* Virchow PT26 infection in England and wales: A case control study investigating an increase in cases during 1994. *Epidemiology and Infection*, Vol.117, pp. 35-41.

Permissions

The contributors of this book come from diverse backgrounds, making this book a truly international effort. This book will bring forth new frontiers with its revolutionizing research information and detailed analysis of the nascent developments around the world.

We would like to thank Yashwant Kumar, for lending his expertise to make the book truly unique. He has played a crucial role in the development of this book. Without his invaluable contribution this book wouldn't have been possible. He has made vital efforts to compile up to date information on the varied aspects of this subject to make this book a valuable addition to the collection of many professionals and students.

This book was conceptualized with the vision of imparting up-to-date information and advanced data in this field. To ensure the same, a matchless editorial board was set up. Every individual on the board went through rigorous rounds of assessment to prove their worth. After which they invested a large part of their time researching and compiling the most relevant data for our readers. Conferences and sessions were held from time to time between the editorial board and the contributing authors to present the data in the most comprehensible form. The editorial team has worked tirelessly to provide valuable and valid information to help people across the globe.

Every chapter published in this book has been scrutinized by our experts. Their significance has been extensively debated. The topics covered herein carry significant findings which will fuel the growth of the discipline. They may even be implemented as practical applications or may be referred to as a beginning point for another development. Chapters in this book were first published by InTech; hereby published with permission under the Creative Commons Attribution License or equivalent.

The editorial board has been involved in producing this book since its inception. They have spent rigorous hours researching and exploring the diverse topics which have resulted in the successful publishing of this book. They have passed on their knowledge of decades through this book. To expedite this challenging task, the publisher supported the team at every step. A small team of assistant editors was also appointed to further simplify the editing procedure and attain best results for the readers.

Our editorial team has been hand-picked from every corner of the world. Their multi-ethnicity adds dynamic inputs to the discussions which result in innovative outcomes. These outcomes are then further discussed with the researchers and contributors who give their valuable feedback and opinion regarding the same. The feedback is then collaborated with the researches and they are edited in a comprehensive manner to aid the understanding of the subject.

Apart from the editorial board, the designing team has also invested a significant amount of their time in understanding the subject and creating the most relevant covers. They scrutinized every image to scout for the most suitable representation of the subject and create an appropriate cover for the book.

The publishing team has been involved in this book since its early stages. They were actively engaged in every process, be it collecting the data, connecting with the contributors or procuring relevant information. The team has been an ardent support to the editorial, designing and production team. Their endless efforts to recruit the best for this project, has resulted in the accomplishment of this book. They are a veteran in the field of academics and their pool of knowledge is as vast as their experience in printing. Their expertise and guidance has proved useful at every step. Their uncompromising quality standards have made this book an exceptional effort. Their encouragement from time to time has been an inspiration for everyone.

The publisher and the editorial board hope that this book will prove to be a valuable piece of knowledge for researchers, students, practitioners and scholars across the globe.

List of Contributors

César Gonzalez Bonilla, Ericka Pompa Mera, Alberto Diaz Quiñonez and Sara Huerta Yepez
División de Laboratorios de Vigilancia e Investigación Epidemiológica, Coordinación de Vigilancia Epidemiológica y Apoyo en Contingencias, Instituto Mexicano del Seguro Social, México

Hiroaki Ohta
CAF Laboratories, Fukuyama, Hiroshima, Japan

Yukiko Toyota-Hanatani
Laboratory of Veterinary Internal Medicine, School of Veterinary Science, Osaka Prefectural University, Izumisano, Osaka, Japan

Marina Štukelj and Zdravko Valenčak
University of Ljubljana, Veterinary faculty, Institute for the Health Care of Pigs, Ljubljana, Slovenia

Vojka Bole-Hribovšek and Jasna Mićunović
University of Ljubljana, Veterinary faculty, Institute for Microbiology and Parasitology, Ljubljana, Slovenia

Anna N. Zagryazhskaya, Svetlana I. Galkina, Zoryana V. Grishina, Galina M. Viryasova and Galina F. Sud'ina
A.N.Belozersky Institute of Physico-Chemical Biology, Moscow State University, Moscow, Russia

Julia M. Romanova
The Gamaleya Research Institute of Epidemiology and Microbiology, Moscow, Russia

Michail I. Lazarenko
National Research Center for Hematology, Moscow, Russia

Dieter Steinhilber
Institute of Pharmaceutical Chemistry, Johann Wolfgang Goethe University Frankfurt, Frankfurt am Main, Germany

Printed in the USA
CPSIA information can be obtained
at www.ICGtesting.com
JSHW011459221024
72173JS00005B/1134